Electrochemical Energy Storage

About the Authors

Slobodan Petrovic is a professor of electrical and renewable energy engineering at the Oregon Institute of Technology, USA. His current areas of research include energy storage devices, solar cell and module reliability, high surface area anodes for lithium-ion batteries, MEMS sensors for smart grids, solar PV technology for developing countries, and electrochemical virus detection.

Peter Kurzweil is a professor of chemistry at the Technical University of Applied Sciences Amberg-Weiden (OTH), Department of Mechanical Engineering and Environmental Engineering, in Germany. His areas of research include chemical analysis, instrumental environmental analysis, and technical electrochemistry.

Jürgen Garche is a senior professor at Ulm University, Institute of Theoretical Chemistry, in Germany. He received his Ph.D. and habilitation (post-doctoral lecturing qualification) in theoretical and applied electrochemistry at Dresden University of Technology. Dr. Garche has worked on batteries and fuel cell research for nearly 50 years.

Electrochemical Energy Storage

Batteries, Fuel Cells, and Hydrogen Technologies

Slobodan Petrovic

Peter Kurzweil

Jürgen Garche

Mc Graw Hill

New York Chicago San Francisco
Athens London Madrid
Mexico City Milan New Delhi
Singapore Sydney Toronto

Library of Congress Cataloging-in-Publication Data

Names: Petrovic, Slobodan (Professor of Engineering), author. | Kurzweil, Peter, author. | Garche, Jürgen, author.
Title: Electrochemical energy storage / Slobodan Petrovic, Peter Kurzweil, Jürgen Garche.
Description: New York : McGraw Hill, [2022] | Includes bibliographical references and index.
Identifiers: LCCN 2021062285 | ISBN 9781260012002 (hardcover ; acid-free paper) | ISBN 9781260012019 (ebook)
Subjects: LCSH: Electric batteries. | Electrochemistry. | Fuel cells.
Classification: LCC TK2896 .P478 2022 | DDC 621.31/242—dc23/eng/20220301
LC record available at https://lccn.loc.gov/2021062285

Electrochemical Energy Storage: Batteries, Fuel Cells, and Hydrogen Technologies

1 2 3 4 5 6 7 8 9 CCD 27 26 25 24 23 22

ISBN 978-1-260-01200-2
MHID 1-260-01200-X

This book is printed on acid-free paper.

Sponsoring Editor
Robin Najar

Editorial Supervisor
Stephen M. Smith

Production Supervisor
Pamela A. Pelton

Acquisitions Coordinator
Elizabeth M. Houde

Project Manager
Radhika Jolly,
KnowledgeWorks Global Ltd.

Copy Editor
Rajesh Mishra,
KnowledgeWorks Global Ltd.

Proofreader
Laura Bowman

Indexer
Michael Ferriera

Art Director, Cover
Jeff Weeks

Composition
KnowledgeWorks Global Ltd.

Contents

Introduction to Electrochemistry

Electrochemistry is a captivating branch of science that deals with interfacial phenomena between chemistry and electricity during processes that produce or consume electrical energy along with chemical changes. Understanding the basics of electrochemistry is essential for studying energy storage systems such as batteries, electrolysis, hydrogen and fuel cells, and electrochemical supercapacitors.

Furthermore, electrochemistry is enormously important for a number of industrial processes, from the production of chlorine and caustic soda in a chlor-alkali electrolysis, the electrowinning of metals, and electroplating, to controlling industrial corrosion and to electrochemical energy storage methods.

Numerous aspects of electrochemistry often appear to be challenging because of the specific components and reaction mechanisms that require advanced understanding of the multidisciplinary concepts. This introduction to electrochemical energy storage attempts to explain the concepts of electrodes, electrolytes, and conductivity.

Historical Progress

The beginnings of the concept of electrochemical energy storage go back to the second part of the 18th century when a number of researchers studied the phenomena arising from the passage of electrical current through different media. Some of these experiments led to the conclusion that chemical changes were permanent upon passage of electricity.

There were very limited choices for creating electrical current in the 18th century, and the best-known device was the Leyden jar, which held a high-voltage electrical charge between two electrical conductors. The device is better described as a capacitor. The Leyden jar was the main instrument a number of researchers from that era were using to conduct early electrochemical experiments. For example, Bekaria used Leyden jars to achieve a discharge over mercury oxide and produce elemental mercury. In 1775, Priestly was able to produce nitric acid by discharging two Leyden jars through air. The first experiment to electrolyze water was demonstrated by Troostwick and Deimann in 1789, and it involved creating sparks through water and generating gaseous products. The subsequent year, Pearson used almost 15,000 discharges from Leyden jars to produce just a few milliliters (0.3–0.5 cubic inch) of gaseous products from water.

The real development of electrochemistry as a branch of science did not take place until the experiments by Galvani in 1791 and Volta in 1792. Galvani first observed that frog legs moved when contacted by two dissimilar metals connected on the other end, and he attributed that to nerve reactions. The following year, Volta explained that such behavior is expected every time that two metals are connected through a medium. Volta, furthermore, constructed in 1799 a first source of electricity or galvanic element by arranging alternately disks of copper and zinc, separated by a cloth soaked in acid—this was the famous Volta pile. It was not too long until Nicholson and Carlisle used this new discovery of the Volta pile to decompose water into hydrogen and oxygen. In 1803, Petrov used a Volta pile comprising 4200 copper and zinc disks to electrolyze water and aqueous salt solutions.

The further development of scientific knowledge saw the works of breakthrough thinkers such as Davy, who discovered the element sodium from wet electrolysis of sodium dioxide, and the discoveries of Faraday, who first defined the laws of electrolysis by connecting the amount of electricity with the quantity of material created in electrolysis. This enabled, for the first time, electrochemical processes to be treated quantitatively and electrolysis to be used for practical processes such as metal electrowinning (e.g., aluminum electrowinning using the Heroult-Hall process), protection, and purification.

In 1839, Grove demonstrated the application of a fuel cell using hydrogen and oxygen, remarkably explaining the complex processes on the electrodes that stand even today. Twenty years later, Plante constructed the first secondary battery based on lead electrodes and sulfuric acid capable of storing electrical energy.

At the beginning of the 19th century, Ritter made the first rechargeable cell battery, comprising layered disks of copper and cardboard soaked in a brine of table salt, and established a connection between galvanism and chemical reactivity; while Daniell made a battery consisting of copper and zinc electrodes in copper sulfate solution, which became the first successful commercial power supply used for telegraphs. Edmund Becquerel invented a solar cell using an electrochemical system, and his father Antoine Becquerel observed the constant current phenomenon when platinum electrodes were immersed in acid.

These early discoveries prompted more questions related to the nature of experiments using electricity in solution, mechanisms of reactions, and potentials forming on the electrodes. Simultaneous with the discoveries of the fundamental thermodynamic principles of the first and second laws of thermodynamics, the depth of knowledge also increased about the nature of energy in galvanic elements. All this led to a milestone work by Walter Nernst in his theory of the electromotive force of galvanic elements and the equation about the cell voltage that stood the test of time and was validated later by the expression derived from thermodynamic values.

The pioneering studies of electrolytic dissociation by Arenhius brought about better understanding of the nature of electrolytic conductivity and further insight into processes in galvanic elements. Subsequent works by the German researchers Debye, Kirchhoff, and Kohlrausch in the 20th century helped explain further the ionic interactions in solution.

The question of electrode polarization in galvanic cells kept intriguing scientists, and Helmholz described around 1870, for the first time, formation of an electrical double layer when a conductor is immersed in an electrolyte solution and a potential difference is established at the two-phase interphase. Then, in 1902, Tafel established empirically the relationship between the current density through an electrode and the voltage difference between the electrode and the bulk electrolyte. It was then that the existence of electrode overpotential was recognized as well. Years later, the Tafel relationship found theoretical justification in the work of Butler and Volmer.

In the 20th century, further advancements deepened the understanding of the electrode processes in galvanic cells through the works of Frumkin on the mechanisms and kinetics of the electrode reactions and of Cottrell, who explained diffusion effects in electrochemical systems. The field of electrochemistry was forever changed by Heyrovsky, Marcus, Yeager, Bard, and Bockris, and most recently by Goodenough, Whittingham, and Yoshino, the Nobel laureates in 2019 for their work on lithium batteries.

Nature of Electrochemical Reactions

Electrochemistry is a distinct branch of science dealing with the interface between chemical reactions and electricity. The reactions that take place with the consumption of electrical energy are called electrolytic (in an electrolyzer), and those that create electricity as a result of chemical reactions are called galvanic or voltaic (in a battery). Both types are the subject of this textbook.

Despite clear similarities, chemical and electrochemical reactions are also very different. In a *chemical reaction*, the reactants must be in direct contact, while electron transfers take place in all directions and over very short distances. Energetic changes in a chemical system are realized through heat variation because the reaction occurs in every part of the system where the reactants come in contact.

In contrast, *electrochemical reactions* take place without direct contact between the reactants, which means that anodic and cathodic reactions are separated in space by the electrolyte. An electrochemical cell comprises two conductive electrodes immersed in an electrolyte (Figure 1).

The two electrodes are connected electrically through the outside circuit, with either an electrical load or a power supply, and through the electrolyte that carries ionic charge. The energetic changes in this case are in the form of electricity drawn from the two electrodes immersed in the electrolyte. As a result, it is necessary for the electrical charge to be transported between the two reaction sites through the electrolyte. Since the electrons cannot simply travel through solution, the medium must be ionized and be capable of carrying electrolytic current (by ions, not by electrons). This enables chemical energy to be transformed into electrical, and at the interface between the solid electrodes

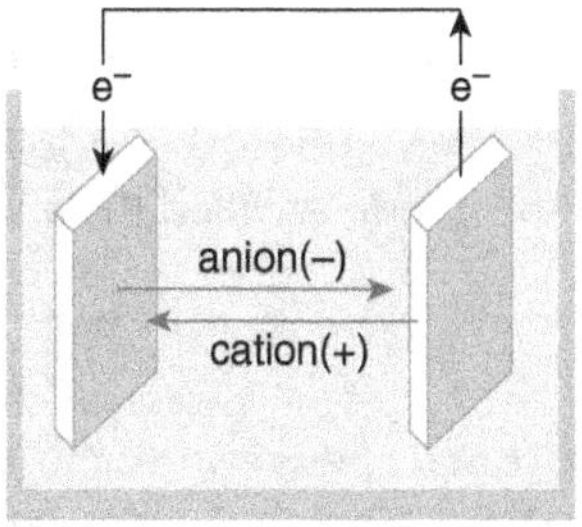

FIGURE 1 Depiction of a simplified electrochemical cell.

and the electrolyte (solid or liquid) a charge transfer reaction takes place. Once the charge enters into the solid electron conductor, that is, the electrode, it is conducted to the outside circuit through which the electrodes are connected and can be used to power a load, in the case of galvanic cells, or can be injected into the electrochemical system by a power supply for an electrolytic process.

Electrochemical systems are truly unique because of the complex interface between the solid electrode and liquid electrolyte and the reactions that occur during the charge transfer at the interface. All this makes electrochemistry a challenging but fascinating field of study.

Anode and Cathode

In an electrochemical reaction, one reactant releases the electrons at the electrode called the *anode*. The electrons are the products of the electrochemical reaction on the anode and they are transmitted to the external circuit. The other electrode, called the *cathode*, accepts those electrons that come through the outside electrical circuit (Figure 1). In this case, electrons are the reactants in the reaction. Electrochemical oxidation takes place on the anode and reduction on the cathode.

In rechargeable or secondary batteries, the location of the anodic or cathodic reaction alternates during charge and discharge. However, the polarity never changes, and the same electrode (anode during discharge) is always negative while the opposite electrode (cathode during discharge) is always positive.

An example of a cathodic reaction, on an electrode called the cathode, is the reduction of oxygen gas in a fuel cell at the positive electrode, where electrons are consumed (Equation 1).

$$O_2 + 4e^- + 4H^+ \rightarrow 2H_2O \tag{1}$$

The negative battery electrode from which electrons leave is called the anode. Oxidation always takes place on the anode and electrons are the reaction products. A common anode reaction is the oxidation of hydrogen at the negative electrode of a fuel cell (Equation 2).

$$2H_2 \rightarrow 4H^+ + 4e^- \tag{2}$$

In an electrochemical reaction, such as in a fuel cell, hydrogen and oxygen never come in direct contact and they are spatially separated, but still the overall reaction demonstrates conversion to a product water, with a release of energy (Equation 3), and it has the same form as a chemical reaction between hydrogen and oxygen.

$$2H_2 + O_2 \rightarrow 2H_2O + \text{Energy} \tag{3}$$

The part of the electrolyte around the cathode is called the catholyte, while the part of the electrolyte around the anode is called the anolyte.

Conductivity

Because there is no direct contact between reactants, electrochemical processes involve transfer of charge between reaction sites on electrodes. The charge transfer occurs in two ways: transfer of electrons between the electrodes through external electrical circuits and transfer of ions through the electrolyte. These represent two different conduction processes, electronic conductions and ionic conductions. The overall principle of preserving neutrality is maintained, so when negatively charged electrons move from one electrode to another initiating a reaction on the electrodes, the opposite charge, that is, positively charged ions (cations), moves in the electrolyte in the same direction (to the negative electrode). Alternatively, the negatively charged ions (anions) move in the opposite direction (to the positive electrode) to ensure that each electrode preserves electroneutrality.

This movement of charge is of critical importance for successful processes in galvanic cells such as batteries or fuel cells, and increasing conductivity is one of the main ways to improve the performance.

In general, there are two forms of conductivity, electronic conductivity and ionic conductivity. It can be said that all substances or materials conduct electricity to some extent, but some conduct electricity with ease and some with great difficulty. Without a clear distinction between the two groups, it is possible to divide all materials into conductors and insulators. For example, copper, which is used as a current collector in lithium batteries, is a good electronic conductor and conducts electricity 10^{22} times better than polypropylene, which is an insulator and is used in lithium batteries as an electronic separator between the two electrodes. Other insulators include glass, rubber, ceramic, porcelain, wood, most polymers, air, and pure water.

Based on the mechanism of conducting charge, conductors are divided into metallic or electronic and electrolytic. There are also gaseous conductors; however, their conductivity is usually very low because the electrons required for conductivity are too dispersed. Metallic or electronic conductors conduct electrical current electronically, which means that the conductivity depends on the flow of electrons and there is no transfer of the material from which the conductor is made. Such conductors are metals, alloys, and carbon. Electronic conductivity of metals is based on their crystalline structure, and loosely bound outermost electrons are free to conduct. But electrons cannot transfer electricity in solution as they quickly react with positive ions from the solution.

Electrolytic conductivity is based on the transfer of electrically charged particles, that is, ions, and is always accompanied by the chemical change that takes place at the site where the current enters the conductor or leaves the conductor. Electrolytic conductors are usually solutions of acids, bases, and salts in polar solvents, as well as molten salts and crystalline ionic substances.

Electronic and electrolytic conductors behave differently as the temperature changes. The resistance, R, increases for *electronic conductors* such as metals (Equation 4), and at near absolute zero some materials become superconductive when their resistance approaches zero. The conductivity, $\sigma \sim 1/R$, is usually expressed in units of S/cm (siemens per cm or m).

$$R(t) = R(t_0) \cdot \left[1 + \frac{\mathrm{d}R}{\mathrm{d}t}(t - t_0)\right] \quad (4)$$

The second type of conductors is *electrolytic conductors*, and they can be described as belonging to two groups. The first and most important group is liquid electrolytic conductors that include ionic solutions made of a polar solvent and one or more dissolved substances. The solvent is usually water, but other polar solvents, including organic solvents such as alcohols, formamide, acetonitrile, diethyl ether, tetrahydrofurane, and others, are also used. These solvents can be used in combination with salts such as NaCl, $LiBF_4$, $HClO_4$, and others.

The mechanism of forming electrolytes that conduct ionic charge has the essential step of dissolving the salts or breaking down the attractive forces (Coulomb forces) in the crystal lattice. Large amounts of energy are needed to break down the crystal lattice, but the attractive forces become weaker when the substance is mixed with a solvent, for example, water, and the process of solvation creates freely moving ions (e.g., Na^+ and Cl), which carry the charge in solution (Equation 5).

$$NaCl_{(s)} \rightarrow Na^{+}_{(aq)} + Cl^{-}_{(aq)} \tag{5}$$

Furthermore, it is easier to dissolve salts with larger ionic radii (r), because the Coulomb force (F) is inversely proportional to the square of the radius ($F \sim r^{-2}$). For example, salts with large anions (e.g., PF_6^-) are used in combination with organic electrolytes used in Li batteries.

In contrast to electronic conductivity, the resistance of electrolytic conductors decreases with increasing temperature. In all electrochemical processes, both those that use electrical energy to accomplish chemical change (electrolytic) and those that generate electrical energy through a chemical process (galvanic), it is of critical importance to have the least electrical resistivity. High conductivity, that is, low resistivity, means that voltage is lower in electrolytic processes and higher in galvanic processes. In both cases, this is called voltage loss or overpotential. Based on this, in electrochemical energy storage devices, such as batteries, fuel cells, or electrochemical capacitors, it is always desirable for the electrolyte to have high specific ionic conductivity, that is, low electrolyte resistance. In some cases, low specific ionic conductivity of the electrolyte can be compensated for by short electrode distances and large electrodes.

The electronic conductivity of an electrolyte can cause an electrical short or leakage current. The boundary between electronic and electrolytic conductivity is not very sharp, but electrons, which are the charge carriers in electronic conductors (electrodes), cannot pass through the solid-liquid interface (electrolytic double layer) and conduct in solution (electrolyte). At the interface, electrons rapidly react with positive ions from the solution. Similarly, ions, which conduct charge in solution, cannot conduct electricity in solid electronic conductors. The problems arise when some electrolytes display small electronic conductivity and allow "leakage" current between the electrodes, which lowers the efficiency of electrochemical processes.

In an electronic conductor such as metal, the electrons are already capable of conducting and they reside in the interstitial spaces between the atoms, as shown in Figure 2.

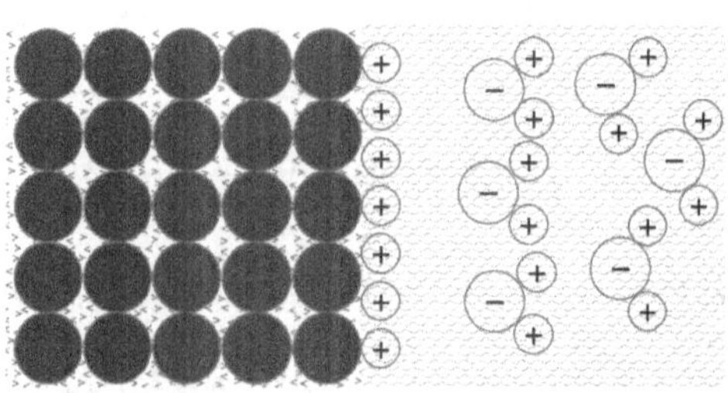

FIGURE 2 Metal-solution interface illustration. Electrons (<) are shown between metal atoms (●) of the electrode; cations (⊕) and anions (⊖) are in the electrolyte.

This figure also depicts the formation of the so-called *electrolytic double layer* in the solution around the electrode, which is responsible for specific behavior of electrodes in solution. The formation of a double layer when a metal conductor is inserted in the electrolyte is initiated by the ionization of the metal, which results in positively charged ions on the electrode surface and, in response, orientation of mostly negatively charged ions.

An electrolyte can also be made without a solvent such as water. The process involves melting a salt and dissociating it into ions. A molten carbonate fuel cell, for example, uses a molten salt composed of lithium and potassium carbonates at 650°C. When potassium carbonate melts in the absence of water, it dissociates into ions capable of conducting ionic charge (Equation 6).

$$K_2CO_3 \rightarrow 2K^+ + CO_3^{2-} \tag{6}$$

Another type of fuel cell, the solid oxide fuel cell, utilizes ZrO_2, a solid ceramic substance, as the electrolyte. These solid electrolytes contain ions, for example, O^{2-} ions, capable of moving through the structure and conducting ionic charge under the influence of an electric field at temperatures above 800°C.

Finally, there are numerous examples of electrolytes based on polymers, from industrial electrochemical processes and water electrolysis, to use in fuel cells and batteries. These polymers are modified to enable conduction of ions such as H^+, Na^+, or Li^+.

A complete *electrochemical cell* is made when two electronic conductors called electrodes (e.g., metals, carbon materials, or semiconductors) are immersed in an electrolyte, that is, electrolytic conductor, and an electrical field is established between the electrodes (from an external power supply or from a spontaneous process) that causes movement of ions and transport of charge.

PART 1

Batteries

CHAPTER 1

Battery Performance Factors

Introduction

The first battery, also called an accumulator, based on a copper anode and copper oxide (CuO) cathode, was invented in 1801 by Johann Wilhelm Ritter (1776–1810). Soon after, other systems were proposed but were not practically realized for some time. The breakthrough came with the development of the lead-acid battery about 50 years later by Gaston Planté (1834–1889). At the beginning of the 20th century, E. W. Jungner (1869–1924) developed the nickel-cadmium battery and T. A. Edison (1847–1931) developed the nickel-iron battery. Ever since, different battery systems have been developed, but only some of these systems are still in production (Figure 3).

Among the systems shown in Figure 3, this book covers the most important ones, lead-acid batteries, nickel-based systems, and the lithium-ion batteries (LIBs). In addition, precommercial and post-Li-battery systems will be evaluated.

The colloquial language in the battery field considers a cell as the smallest unit of an electrochemical power source, while a battery is an interconnection of two or more cells (nonrechargeable or chargeable). The term *battery* was adopted from the artillery, meaning several guns (see New York's Battery Park), and has nothing to do with the working principle of an electrochemical power source.

During the discharge in a battery, the negative electrode is the anode and oxidation takes place on that electrode, while the positive electrode is the cathode and reduction takes place on that electrode. During the charge process, the negative electrode becomes the cathode and the positive electrode the anode because the redox reactions are reversed. However, in colloquial language, the negative electrode is always considered an anode and the positive electrode the cathode, for both charge and discharge reactions.

Capacity

Capacity is the electrical charge that can be drawn from the cell or a battery. Symbols C or Q are commonly used. For easier understanding and comparison, the capacity is related to the mass; volume and capacity are expressed, for example, in kg or per L. The gravimetric capacity density is usually called specific capacity in Ah/kg, while the volumetric capacity density is called capacity density.

The importance of these specific capacity parameters depends on the application. For applications with restricted volumes, for example, consumer applications, the volumetric

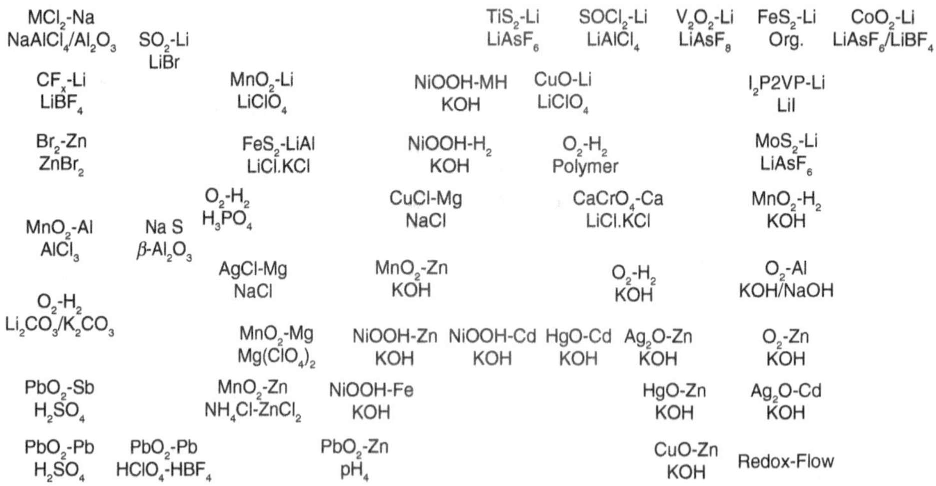

FIGURE 3 Overview of developed battery systems. (The arrangement of system symbols in shape of letters has no meaning.)

capacity density is more important. For mobile applications, the battery weight strongly influences the car range, and gravimetric capacity density, or specific capacity, is more important. For smaller cars, the battery volume can also be important. For stationary applications, the volume values determine the footprint of the battery energy storage system (BESS) and are more important than the mass values.

The theoretical capacity of an electrode or a cell is calculated from Faraday's first law of electrolysis assuming 100% utilization of the active masses. Some theoretical specific capacities are given in Table 1.

Battery System	Theoretical Specific Capacity, Ah/kg
Rechargeable lead-acid	83
Ni-Cd	180
Ni-H_2	289
Ni-MH	209 (depending on the metal hydrid)
Na-S	377
Na/$NiCl_2$	305
$LiNiO_2$/graphite	158

(Data from *Handbook of Batteries*, ed. by D. Linden, 2nd ed., McGraw-Hill.)

TABLE 1 Theoretical Specific Capacity of Selected Rechargeable Battery Systems

The theoretical values serve as a starting point for the design of new systems. More important are the practical values, which are considerably lower than the theoretical ones, as a consequence of an incomplete use of the active masses and because of inactive cell components. The practical capacity Q is derived as an integral of the discharge current (I) over time, from a fully charged to a fully discharged state, where the cutoff discharge voltage U_{cod} is reached. If a constant discharge current is used, the capacity is proportional to the discharge time t_D (Equation 7).

$$Q = \int_0^{t_D} I(t)\,dt = I \cdot t_D \text{ [for constant current]} \tag{7}$$

The voltage for a constant-current discharge shows typical behavior with three regions: the initial voltage decline region due to charge transfer reaction, the linear portion as a result of resistive losses, and sharp decline at high current because of the mass transfer limitations (Figure 4).

Nominal or rated capacity is the capacity of a battery under defined discharge conditions of current, cutoff voltage, and temperature. This nominal capacity is given by the manufacturer for their products and is based on empirical data. The nominal capacity rarely has the same value as actual measured capacity of a cell because it is strongly dependent on the temperature and discharge current. As with any other chemical reaction, the increased temperature leads to increased capacity due to higher reaction rate.

The following basic equation describes relation between the actual cell voltage during discharge, U, thermodynamic (no loss) cell voltage, U_0, resistance, R_Ω, due to slow ion transport and electrical resistance, and a term commonly called polarization, η, which describes voltage loss due to slow charge transfer reactions at the interface between solid electrode and liquid electrolyte.

$$U = U_0 - I \cdot R_\Omega - \eta \tag{8}$$

As shown in Equation 8 and in Figure 5, with increasing temperature, the ohmic resistance (R_Ω) and the polarization on electrodes (overpotential η) are reduced (upper curve in the figure), and the cut-off voltage is higher or reached later, t_{dch1}.

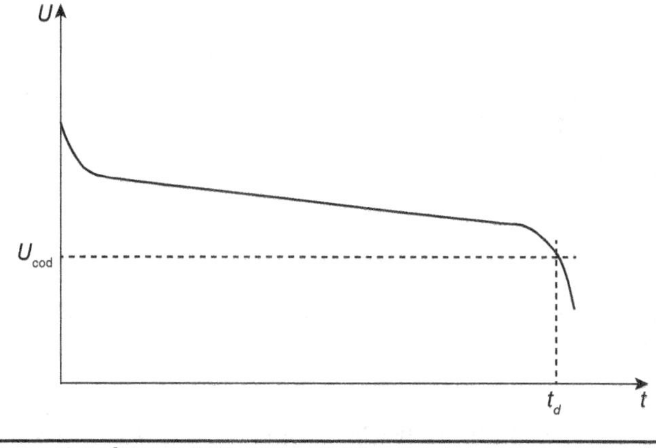

FIGURE 4 Voltage vs. time for constant current discharge.

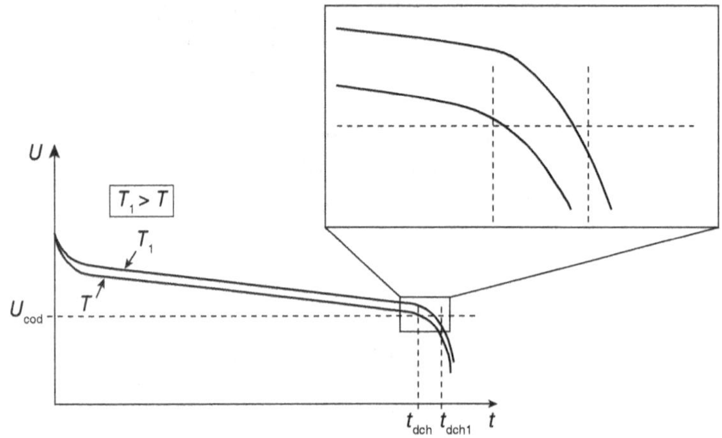

FIGURE 5 Voltage for a hypothetical battery vs. time for a constant current discharge at two temperatures ($T_1 > T$). t_{dsch} is the time when end-of-charge (EOC) voltage is reached.

The temperature dependency of the capacity is shown for different cells, such as Ni-MH, lead-acid, Ni-Cd, and Li-ion, and for different discharge rates (Figure 6).

The capacity of an AA cell sharply increases with the initial temperature increase to room temperature, followed by a plateau and slight decrease (Figure 7).

With increasing discharge current (I), voltage (U) drops as a result of larger resistive losses (IR_Ω) and higher electrode polarization, η (Figure 8).

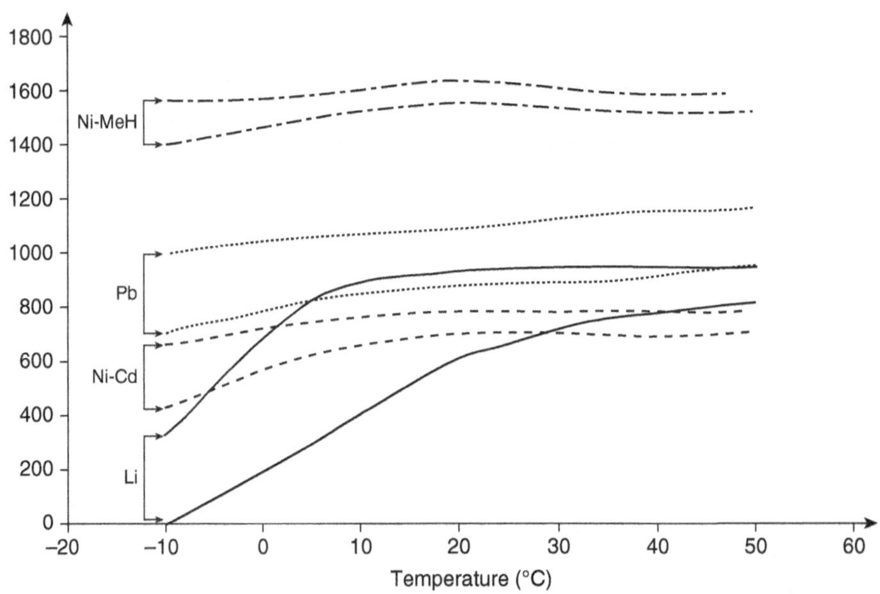

FIGURE 6 Capacity of various batteries as a function of temperature. (*Courtesy of R. Groiß, BASYTEC.*)

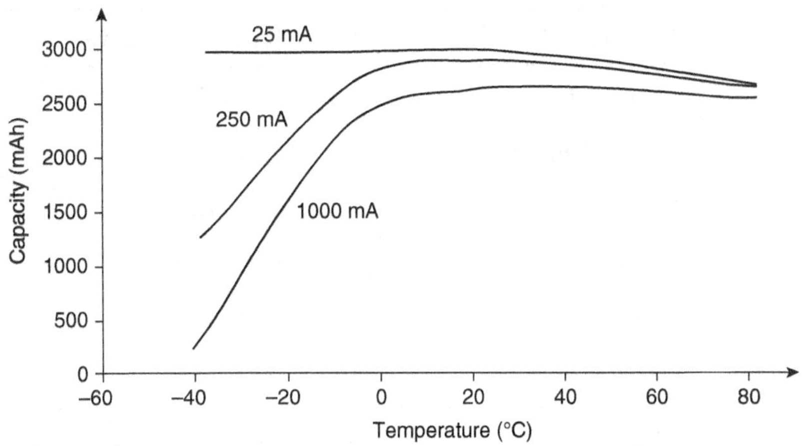

FIGURE 7 Capacity of an AA cell (ENERGIZER L91, Li/FeS_2, primary cell) as a function of temperature at three constant discharge currents (U_{cod} = 0.8 V). (*Courtesy of ENERGIZER.*)

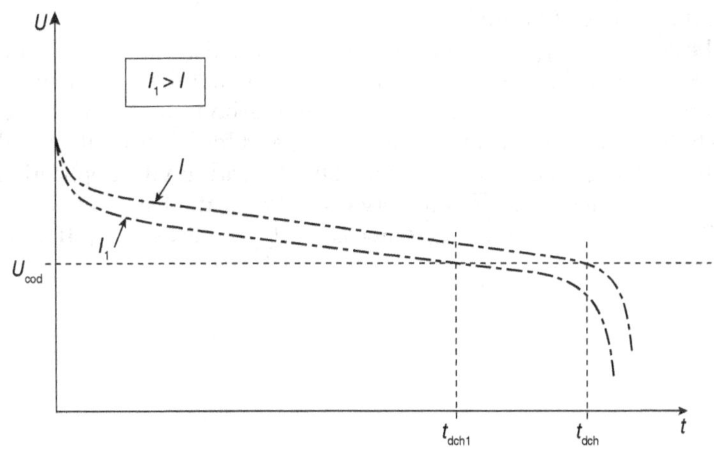

FIGURE 8 Battery voltage variation with time for two discharge currents ($I_1 > I$).

The discharge current effect on the capacity is given by the Peukert equation. Based on a known capacity (Q_n), at a certain discharge current I_n, the capacity (Q_{n1}) at the discharge current I_{n1} can be calculated (Equation 9).

$$Q_{n1} = Q_n \left(\frac{I_n}{I_{n1}} \right)^{p-1} \tag{9}$$

The calculated capacity, Q_{n1}, is mainly dependent on the Peukert coefficient (p), which is related to the internal resistance of the cell. The internal resistance not only varies by battery type but is also based on design features for the same battery chemistry. The values given in Table 2 are approximated values, which are furthermore dependent on the temperature and the age of the battery.

Battery	Peukert Coefficient
Lead-acid flooded	1.2–1.4
Lead-acid VRLAB	1.15–1.25
Ni-Cd	1.10–1.20
Ni-MH	1.05–1.15
Li-ion	1.05–1.1
Ideal battery	1.0

Table 2 Peukert Coefficient (p) for Different Battery Systems and Types for Room Temperature

Battery capacity declines with increasing discharge current and is dependent on the Peukert coefficient (Figure 9).

The capacity is typically specified for a certain discharge temperature and discharge current or C-rate. The discharge rate, xC, means that the cell is completely discharged in $1/x$ hours. For example, a 2C-rate refers to 1/2-hour discharge; for a cell with capacity of 1 Ah, the discharge current would be 2 A, and for a 5-Ah cell it would be 10 A.

The discharge current or C-rate usually depends on the application (Table 3).

The percentage of the full battery capacity is called state-of-charge (SOC—100% = full; 0% = empty), and its inverse value is the depth-of-discharge (DOD—100% = empty; 0% = full).

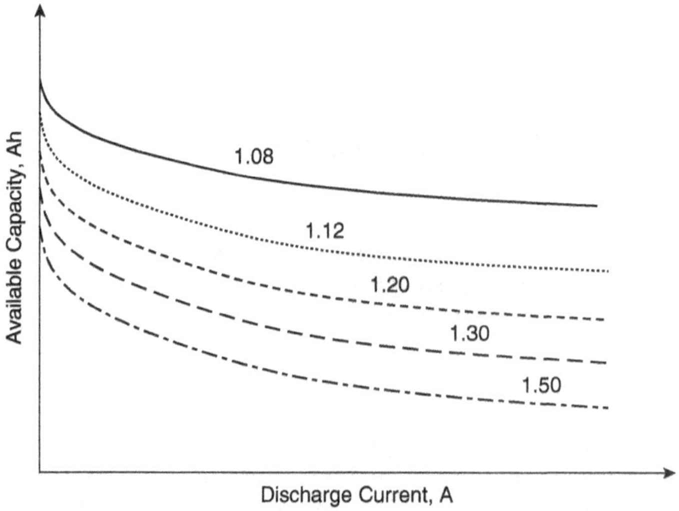

Figure 9 Capacity of a hypothetical battery as a function of discharge current for different Peukert coefficients.

Battery Application	Typical C-rate
Battery EVs	3 h discharge, C/3 capacity
Hybrid EVs	1 h discharge; 1C capacity
Stationary application	10 h or 5 h discharge; C/10 or C/5 capacity
Portable application	1 h discharge; 1C capacity
Car starter battery	20 h discharge; C/20 capacity

TABLE 3 Typical Time of Discharge and C-rates for Selected Battery Applications

For most practical cells, the anode capacity is in the first approximation equal to the cathode capacity. In some cells, one electrode is oversized to mitigate safety problems. In that case, the smaller electrode determines the cell capacity.

Energy

The energy content of a battery system (in watt-hours, Wh) determines its operating time. The energy is expressed per mass or volume of an electrode or a cell: specific energy or gravimetric energy density (Wh kg^{-1}) and energy density or volumetric energy density (Wh L^{-1}). The practical specific energy and energy density can be easily calculated by dividing the empirically obtained energy of a battery by the weight or volume of the cell.

The energy, W, is also the product of cell voltage, U, and the cell capacity, Q (Equation 10). Symbol W for energy is not to be confused with the unit for power, watt, W.

$$W = \int U(Q)\, dQ = \bar{U} \cdot Q \tag{10}$$

The theoretical energy values are calculated from the theoretical capacity values and the equilibrium voltages. These values are significantly lower in practice because of the cell voltage reductions by electrochemical polarization and ohmic resistances during current flow, incomplete utilization of active materials, and increase in the cell mass by electrolytes, separators, current collectors, containers, connectors, and other components (Table 3a).

The practical specific energy of most batteries is about 20–25% of the theoretical. Additional system components that contribute to weight increase and further specific energy reduction include cell connectors, container, cooling system, safety devices, and a battery management system. For most batteries, the weight of auxiliary components is 60–80%, which means that active masses amount to 20–40%. The practical energy density is also affected by the cell design and can be significantly different for cells of the same type.

System	Theoretical Specific Energy (Wh kg^{-1})	Practical Specific Energy (Wh kg^{-1})	Percentage of Theoretical Specific Energy (%)
Lead-acid	167	33	20
Ni-Cd	240	45	19
Ni-MH	300	79	26
Ni-Zn	320	80	25
ZnBr	435	90	21
Li-ion	719	110	15
$NaNiCl_2$	719	110	15
NaS	795	90	11

Table 3a Theoretical and Practical Specific Energy for Selected Battery Systems

The specific energy for a given cell also depends on the temperature and the discharge current because of their effect on the active mass utilization and the discharge voltage. Consequently, the specific energy and energy density values should be always specified for a certain temperature and discharge C-rate.

Power

The power (P) is equal to the rate at which the energy is used and is calculated from the current and the voltage of the cell (Equation 11).

$$P = I \cdot U \tag{11}$$

The power depends mainly on the battery chemistry and the design of the cell, which together determine the total internal resistance of the cell. One of the design factors affecting the internal resistance is the electrode thickness. The thinner electrodes have lower resistance and higher power, but less energy. While energy and power do not directly affect one another, the design of a battery usually favors one of these performance attributes. In most cases, an increase in power leads to a reduction in energy. This is caused by the lower utilization of the active masses, lower voltage due to limited mass transport, and an inhomogeneous current distribution. In a few applications such as regenerative braking for electrical vehicles (EVs), very high currents are generated, meaning that a battery must have high current capability (Table 4).

Lower temperature and SOC of a battery affect its power as a result of slower reactions and increase in current density.

The indirect connection between energy and power is shown in the so-called Ragone diagram (Figure 10).

Efficiency

Battery electrical efficiency (η_{EE}) is the ratio between electrical energy obtained in the discharge reaction, W_d, and electrical energy used in the charge reaction, W_{ch} (Equation 12). Electrical efficiency is always <1 because of the various losses that take place.

$$\eta_{EE} = W_d / W_{ch} \tag{12}$$

Parameter	High Power	High Energy
Anode thickness	30 μm	150 μm
Cathode thickness	40 μm	200 μm
Separator thickness	20 μm	30 μm
Weight	330 g	1050 g
Dimension	170 × 160 × 8 mm	250 × 210 × 11 mm
Capacity	6 Ah	40 Ah
Voltage	3.5 V	3.5 V
Energy	21 Wh	140 Wh
Max. current	20C–120 A	5C–200 A
Spec. energy	64 Wh kg^{-1}	133 Wh kg^{-1}
Spec. power	1250 W kg^{-1}	660 W kg^{-1}

Table 4 Comparison of Battery Operational Parameters for Different Electrode Thickness of Li-Ion batteries

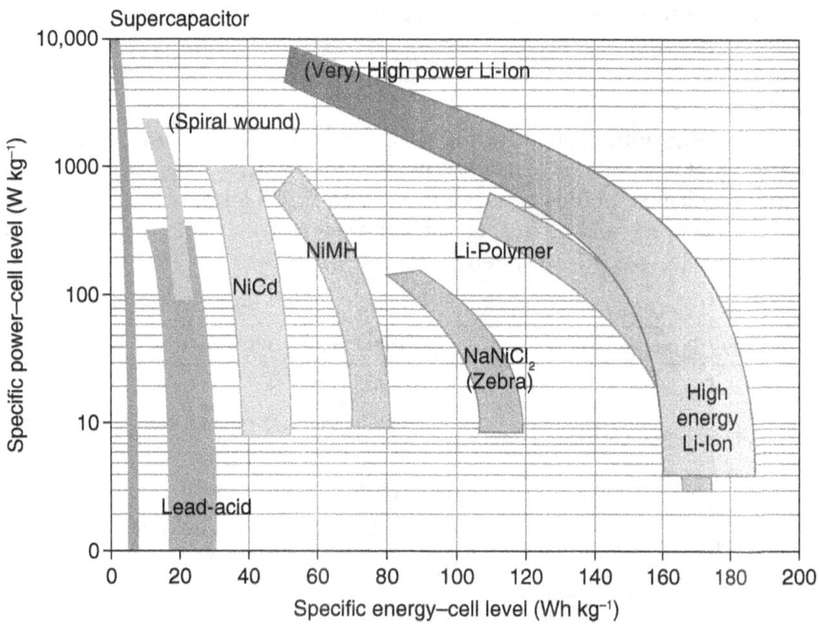

Figure 10 Ragone diagram of average values for different battery systems on a cell level. (Adapted from R. S. Scharner, P. Lamp, E. Hockgeiger, *AABC 2010*, Orlando (USA), May 17–21, 2010, courtesy BMW.)

The η_{EE} is related only to the battery and excludes the charger losses and those of the whole storage system. The electrical efficiency can also be defined as the product of the voltage efficiency, η_U, and the Coulomb efficiency, η_C (Equation 13).

$$\eta_{EE} = \eta_U \cdot \eta_C \tag{13}$$

The voltage efficiency is the ratio of the average discharge voltage, $\overline{U_d}$, and average charge voltage, $\overline{U_{ch}}$ (Equation 14).

$$\eta_U = \overline{U_d} / \overline{U_{ch}} \tag{14}$$

Because the discharge and charge voltages depend on the temperature and the current density, the voltage efficiency depends also on the temperature and the current density.

The differences between the discharge and charge voltages are caused by the internal resistance of the battery.

The Coulomb efficiency is the ratio of the average discharge current to average charge current (Equation 15).

$$\eta_c = \overline{I_d} / \overline{I_{ch}} \tag{15}$$

The Coulomb efficiency depends on the extent of side reactions mostly during the charge process and in particular for higher SOC and higher charge voltage. These side reactions are, for example, electrolyte decomposition and corrosion of electrode components. Normally, self-discharge processes are not taken into account for determination of η_C and η_{EE}.

Temperature and charge current also affect the Coulomb efficiency.

Self-Discharge

Self-discharge refers to the tendency of every battery to lose some of its capacity in the rest period and without any external current. This capacity loss is primarily caused by thermodynamic instabilities of active masses and cell components. The self-discharge rate is usually expressed as percentage of capacity loss per month or year. The self-discharge rates for selected battery systems are given in Table 5.

Battery Type	Self-Discharge Rate
Li-ion	<2% per month
Ni-MH (low self-discharge type)	~6% in the first month, decreasing to below 1% after a few months
Ni-Cd	~15–20% per month
Lead-acid batteries	~2–5% per month initially and up to 15–25% per month for aged batteries
Li-Mn_2O_4 (nonrechargeable)	~1–2% per year

(Reproduced from H. Wenzl, Self-Discharge, in *Encyclopedia of Electrochemical Power Sources*, ed. by J. Garche et al., Elsevier, Amsterdam, 2009, Vol. 1, p. 407.)

TABLE 5 Typical Self-Discharge Rate Values at Room Temperature

The reactions leading to self-discharge are temperature dependent. To normalize capacity losses during the rest period, a self-discharge coefficient (c_s) is used (Equation 16):

$$c_s = \frac{1}{t}\frac{\Delta Q}{Q_0} \tag{16}$$

where Q_0 is the starting capacity, ΔQ is capacity loss, and t is rest period.

This definition of self-discharge does not consider processes that reduce the discharge voltage, for example, formation of passivating layers. In this case, a self-discharge coefficient related to the energy loss per month or year would be more appropriate.

Lifetime

The lifetime or end-of-life of a battery is affected by several factors, each corresponding to a function that it has to fulfill. Failure to fulfill any one of these functions means that the product has reached the end of its useful life. The criteria for end-of-life include capacity, energy, self-discharge, charge acceptance, internal resistivity, and occasionally others, but usually the end-of-life is reached when the capacity drops to less than 80% of the rated capacity. Sometimes the increase in the internal resistance by a factor of 1.3–2.0 is also used as the end-of-life indicator.

There are three stages of the aging processes analogous to considerations for electronic devices: early lifetime, useful life, and wear-out. The early life failures are caused by production issues such as short circuits between the poles, or electrical interrupts between cells of a battery, while the useful life and wear-out stages can be considerably affected by the application and conditions of use, such as temperature, vibrations, and electrical usage profile.

Cycle Life

The cycle lifetime or cycle life, given in numbers of cycles, is one of the most important properties of a battery, and it is used to define the lifetime of a battery in applications where a high number of charge and discharge cycles is important.

The requirements and standards for cycle life testing differ for each application. For EV batteries, there are standards available, for example, ISO 12405-2, IEC 62660-1, under which the battery is cycled at a constant current regime (charge: $I = C/3$, discharge: $I = C/3$, 60 min rest). But this constant current load profile does not match the real car driving load and because of that there are also standards that cover the average driving characteristics of cars, during different stages of driving such as acceleration, cruising, deceleration, and stop. The most common standardized driving cycles are those targeting passenger vehicles in urban driving conditions, such as the U.S. Federal Test Procedure, FTP-75 (Figure 11a), the New European Driving Cycle (NEDC) Europe, or the Japanese JC08. Based on these regimes, the power demand of an EV battery was derived with a maximum power value in W kg^{-1} or A kg^{-1}, adjusted for mass of the vehicle. The adopted standard load profiles in the United States include the FTP-75 driving cycle, the Federal Urban Driving Schedule (FUDS), and a simplified FUDS version called the Dynamic Stress Test (DST) Schedule (Figure 11b).

Other standards have been developed for hybrid EV batteries, which consider higher power pulses characteristic for discharge acceleration and charge braking

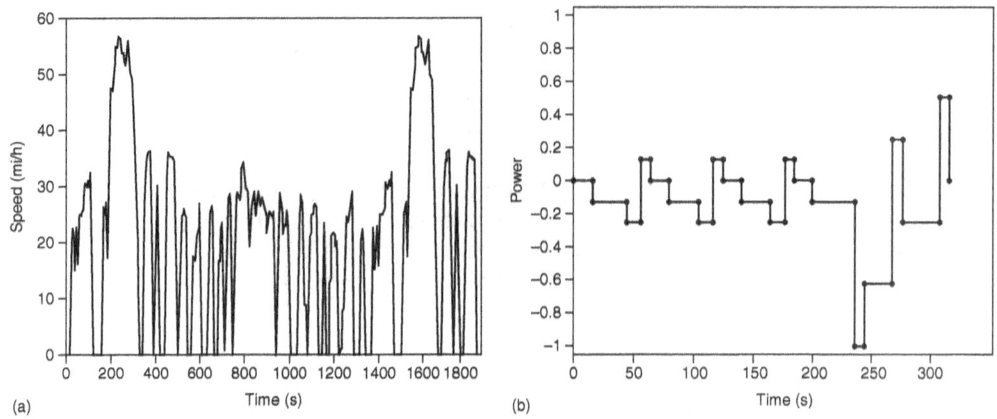

FIGURE 11 (a) United States FTP-75 driving cycle profile. (b) United States Dynamic Stress Test (DST) Schedule. (Reproduced from E. Cabrera Castillio, in *Advances in Battery Technologies for Electric Vehicles*, ed. by B. Scrosati, J. Garche, W. Tillmetz, Woodhead Publishing-Elsevier, Amsterdam, 2015, pp. 469–494.)

["Hybrid pulse power characterization" (HPPC) test, INL/EXT-07-12536 Rev. 2, DOE/ID-11069, or "Electrically propelled road vehicles—Test specification for lithium-ion traction battery packs and systems—Part 1: High-power applications," ISO 12405-1].

The main causes for cycle aging are mechanical strain, changes in the microstructure of the active materials, and current collector corrosion. The deeper the discharge, the larger the mechanical strain and therefore the shorter the cycle life. The connection between the depth of discharge (DOD) and cycle life is not linear, but exponential due to only small changes in the active mass structure at low discharge, which largely retains the original structure. Besides the DOD, the cycle life is affected by the charge/discharge currents and temperature. This is especially the case for battery chemistries where the discharge/charge reaction is a solution-precipitation reaction, as in lead-acid batteries. In such reactions, the whole active mass structure is changed. If the discharge/charge reaction is an intercalation reaction as in LIBs, the active mass structure retains, to a large extent, the structure of the host material.

Calendar Lifetime

As previously explained, batteries reveal a capacity loss even without current load, a phenomenon called self-discharge. The reactions leading to self-discharge in thermodynamically unstable active masses and cell components result in formations of passive components or component degradation. When the capacity loss exceeds 20% of the rated capacity, it can be considered that the battery reached its calendar lifetime. In most cases (but not all), this capacity loss is reversible, and the battery regains the capacity back after charging. In some cases, after prolonged storage and self-discharge, charging is suppressed by the formation of passive structures (layers, crystals). In some cases the calendar lifetime is related to the time during storage without recharging and it is called shelf-life.

The calendar life is strongly dependent on the SOC and storage temperature through reactions related to the Arrhenius law.

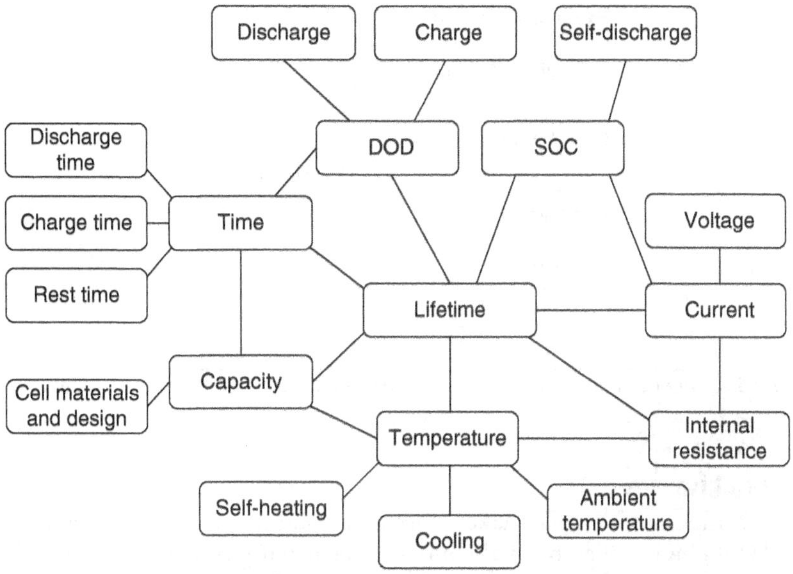

FIGURE 12 Most important lifetime stress factors for Li-ion batteries and their interactions. (Reproduced from M. A. Danzer, V. Liebau, F. Maglia, in *Advances in Battery Technologies for Electric Vehicles*, ed. by B. Scrosati, J. Garche, W. Tillmetz, Woodhead Publishing-Elsevier, Amsterdam, 2015, pp. 388.)

Accelerated Lifetime Testing

The lifetime demand for EV batteries is about 15 years, and it is obvious that direct lifetime tests would not be practical. Accelerated tests can reduce the testing time and provide a prediction of a battery life.

The central premise of an accelerated test methodology is that the principal aging parameter does not change by the accelerated conditions, which could be higher temperature, current, or SOC. The interactions between the stress factors have to be considered as well. It is understood, for example, that a current increase leads also to a cell temperature increase, as a result of the JOULE heat evolution, and battery heating in turn leads to faster aging. It should be equally obvious that advanced understanding of the aging reaction mechanisms, acceleration factors and their interactions, is critical for the successful design of accelerated tests. The most important lifetime stress factors and their interactions are shown in Figure 12, for Li-ion batteries.

Safety

The safety of a battery depends on its energy content and the propensity for an uncontrolled release of this energy outside of the normal battery operation during which electrical energy is produced. A fully charged battery comprises one component of chemical energy convertible into electricity and another one convertible into thermal energy, under unwanted, but favorable conditions and outside of normal battery operation. These conditions lead to several types of hazards.

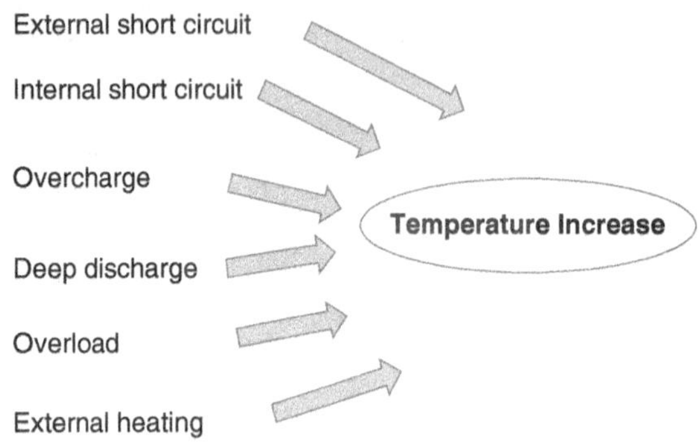

FIGURE 13 Operating parameters contributing to temperature increase in a battery.

Thermal Hazard

A cell temperature increase takes place as a result of one or more factors (Figure 13).

While higher than normal temperatures usually contribute to short lifetime, they do not directly affect the safety. However, at some critical temperature, exothermal chemical reactions become accelerated, leading to a further temperature increase, fire, or explosions in the worst case. When the rate of heat evolution exceeds the rate of removal or dissipation, a process called thermal runaway begins.

Kinetic Hazard

Heat evolution and temperature increase in a battery often lead to decomposition reactions and formation of gaseous products, followed by increase in gas pressure. If the heat is not dissipated, the pressure increase can lead to explosion, blast waves, and material ejection from a cell.

Chemical Hazard

Chemical hazard during a safety breach event is related to release of substances harmful to health and/or the environment.

Electrical Hazard

The effect of current passing through a human body depends on the current intensity, voltage, contact area, contact pressure, and body resistance. Human body resistance varies with contact area, contact pressure, thickness of skin, and presence of moisture. The perception threshold of current is at 0.2–1 mA; 10–15 mA current causes muscle reaction, 15–25 mA can produce diaphragm spasm, and 60–100 mA a ventricular fibrillation or a heart rhythm problem. Currents through the body of about 1 A cause a heart stop. Such high currents hardly occur if the battery voltage is less than 60 V, but some EV and especially BESS batteries could generate much higher voltages.

As the size of a battery or the number of cells increase, more energy is available that can be converted to heat, and the safety risk increases. The propagation of heat and fire can also progress from one cell to the next.

Battery fire can be extinguished with the help of an extinguishing agent, which lowers the cell temperature and the oxygen concentration. Water, with high molar heat of evaporation (~41 kJ/mol), is the common extinguisher agent and together with steam reduces oxygen concentration at the origin of a fire. Fires from Li-battery with metallic Li anode, Na, Mg, and Al batteries, as well as sodium high-temperature batteries, may not be extinguished with water but with metal fire extinguishing agents. If there is no danger of fire propagation to the surroundings, the battery fire can also be left to burn out.

The extent of risk due to battery fire and explosion can be experimentally determined. The safety events in the field are various, and it is difficult to reproduce these events in a standardized test. Nevertheless, tests have been developed, which include mechanical tests (e.g., crash, drop, nail penetration), electrical tests (e.g., short circuit, overcharge, overdischarge), thermal tests (e.g., thermal stability, thermal shock, fire), and environmental tests (e.g., vibration, mechanical shock, underwater).

Battery Cost

Capital expenditure, CAPEX, is the sum of the investment costs necessary to build a facility for a cell, a battery, or a battery system manufacturing. The costs for a battery system are determined by the CAPEX due to the battery, the balance of plant (BOP), and the power conditioning system (PCS). The costs for auxiliary systems are currently estimated at \$250/kWh, with the goal of reaching \$120 kWh^{-1}.

The costs to build the cells include nature of material and quantity necessary for the intended design, specific energy, and production costs. The battery prices are determined by the cost, profit, depreciation, recycling costs, and warranty expenditures and are typically 1.4–1.6 times the costs.

Operational expenditure (OPEX) or operation costs for batteries are maintenance and electricity costs for charging. Maintenance costs are incurred for charge balancing and controlling capacity, and in a case of some aqueous systems, for refilling the electrolyte. Electricity costs clearly depend on the energy efficiency.

Total expenditure (TOTEX) is the sum of CAPEX and OPEX at a given time. Normally, TOTEX is related to the whole lifetime of a battery.

Throughput costs are calculated by dividing the TOTEX by the battery energy in kWh, delivered over the lifetime of a battery. The throughput costs for different battery chemistries (i.e., battery types) for the battery and system level are given in Table 6. The values in the table have been derived using the following methods:

Maximal DOD: at DOD values <100%, the system has to have more capacity than the usable capacity, increasing CAPEX.

Energy efficiency: energy losses during charge and discharge have to be compensated by charge energy from the grid, increasing OPEX. Grid energy costs: 0.20 EUR/kWh.

Energy in kWh over lifetime: based on 300 cycles/a.

The throughput kWh system costs are calculated from the battery throughput kWh costs plus the BOP and PCS costs; BOP&PCS: \$240/kWh.

Depreciation and OPEX for BOP&PCS have not been used in this calculation.

Parameter	Lead Acid	Li-Ion	Na-HT (High T)	Redox-Flow
DOD_{max}/%	70	90	90	90
Energy efficiency/%	80	92	90	75
CAPEX/EUR kWh^{-1}	250	200*	600	800
Lifetime/years	6	15	10	15
kWh throughput costs single battery/EUR kWh^{-1}	0.24	0.07	0.24	0.25
kWh throughput cost for system/EUR kWh^{-1}	0.48	0.16	0.36	0.34

*Realized for EV batteries.

Table 6 Cost-Related Parameters for Selected Battery Systems

No system, at the present, meets the critical value for BESS of \$0.11 kWh^{-1}. But in the future, a LIB with DOD up to 90%, η_{EE} of 95%, CAPEX of \$115 kWh^{-1}, and lifetime of 20 years could be reached, resulting in kWh throughput costs for the battery of \$0.034 kWh^{-1} and \$0.08 kWh^{-1} for BESS.

Lifecycle Analysis

The lifecycle analysis is an inclusive process of analyzing all battery components and factors during the cradle-to-gate sequence, that is, from raw material acquisition through production, use, end-of-life treatment, recycling, and final disposal. The analysis includes the cumulative energy demand (CED) and global warming potential (GWP). Other parameters such as respiratory inorganics substances, ozone depletion agents, acidification chemicals, ionizing radiation, and marine and terrestrial eutrophication are also included in the consideration. The CED and GWP values for cradle-to-gate stage process, excluding end-of-life factors, are given in Figure 14. These values are normalized to 1 MWh, delivered over the lifetime of a battery. The data indicates that the lowest energy and least environmental problems are caused by LIB and redox-flow batteries.

Battery Selection

Operational parameters used for battery selection are shown in Table 7. These are average values from different battery systems, under typical usage conditions. A proper approach would use specific data for each battery that normally comes from the manufacturers. The values of these parameters are dependent on load profiles, temperature, SOC, and other factors characteristic of a specific application. The operating temperature, for example, has strong influence on the energy and lifetime of a battery. If the temperature is not in its operating range, then cooling or heating of a battery is necessary.

Depending on the application, different parameters may have different importance when selecting a battery. For example, the most important parameters for electromobility,

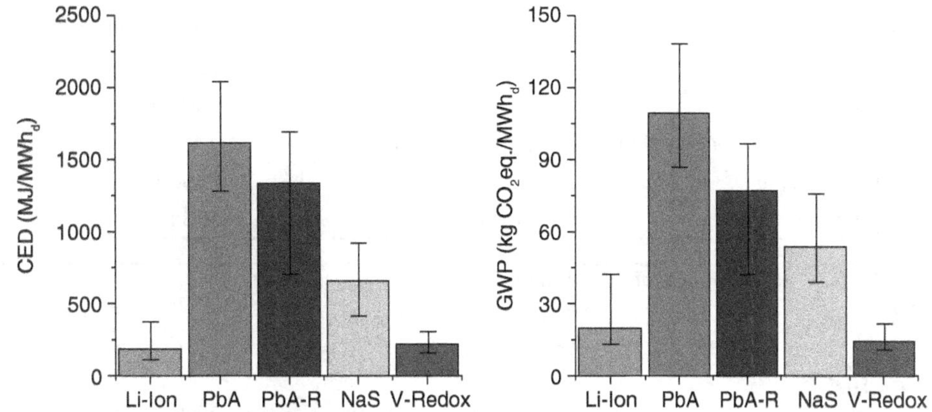

FIGURE 14 Average cradle-to-gate values per delivered MWh for CED and GWP for different battery chemistries (PbA-R: battery with 30% primary lead and 70% recycled lead). (Reproduced from M. Hiremath, K. Derendorf, Th. Vogt, Comparative life cycle assessment of battery storage systems for stationary applications, *Environ. Sci. Technol.* 49 [2015] 4825–4833.)

Battery Type	Specific Energy, Wh kg^{-1}	Specific Power, W kg^{-1}	Cycle Life, Number of Cycles	Efficiency, %	Self-Discharge, % Per Month	Cost, \$ kWh^{-1}
Lead-acid	35–40	200	800	80	3 = 5	SLI—75 Industrial—250
Ni-Cd	40–50	400	1500	70	20	Consumer—200 Industrial—4500
Ni-MH	70–90	350	1000	70	30	Consumer—200 Industrial—600
Ni-Zn	75	500	400	70	20	350
Redox flow	50*	25	3000	75	Low**	800
HT-Na	90–110	150	2000	90	Only thermal	600
Li-ion	100–150	400	3000	93	2–3	200

*Depends on tank volume.
**Only stack electrolyte could be self-discharged.

TABLE 7 Average Operational Parameters for the Most Important Battery Systems

that is, electrical vehicles, are specific energy and power; for stationary applications, CAPEX, cycle life, efficiency, and self-discharge; and for portable applications, energy density is the most important.

Cost is more important for larger batteries. Portable applications in the Wh range are less cost sensitive than stationary applications in the MWh range. Battery voltage is also important for the battery selection.

Parameter	First-Generation EVs	Second-Generation EVs
Specific battery energy	100 Wh kg^{-1}	150 Wh kg^{-1}
Battery weight	300 kg	500 kg
Specific EV consumption	0.1 Wh km^{-1} kg^{-1}	0.1 Wh km^{-1} kg^{-1}
Car weight	1500 kg	1650 kg
Range	200 km	455 km

Table 8 Driving Range of EVs of the First and Second Generation

EV Application

The most important considerations for electrical vehicles include safety, speed, range, and cost. The speed and acceleration of a vehicle depend on the electrical motor revolutions per minute (rpm). This determines the battery voltage, which has to match the top speed of a vehicle. The driving range of an electrical vehicle with one charge is the most commonly used comparator (Equation 17):

$$d = W_m \cdot m \cdot \frac{1}{E} \cdot \frac{1}{m_t} \tag{17}$$

where d is distance a car drives in km, W_m is the specific energy (Wh kg^{-1}), m is the battery mass, E is the specific consumption (Wh km^{-1} kg^{-1}), and m_t is mass of car (kg).

If the specific energy of a battery, the specific electric consumption of the car, and the car total weight (curb weight including battery and payload) are known, the driving range can be calculated. It is clear that a battery should be as light as possible, so it doesn't significantly contribute to the car weight, and have the capacity as high as possible. The driving range is shown along with battery and car parameters for first- and second-generation electrical vehicles in Table 8.

The energy requirement for the battery depends on the drive resistance, that is, the rolling resistance, which depends on the wheel quality, air resistance, and speed of the vehicle. Figure 15 shows one-charge driving distances for different driving conditions.

Stationary Applications

The most important parameters for stationary applications are safety, throughput kWh costs (determined by investment cost, CAPEX), lifetime, electrical efficiency, maximum allowable DOD, energy and power density, and self-discharge. The best battery candidate systems for stationary applications include LIB, LAB, Na-HT, and redox-flow systems.

Charge and Discharge in Batteries

Discharge

The discharge of a cell or a battery is a spontaneous reaction producing electricity when a battery is connected to a load. The reaction proceeds until a predetermined end-of-discharge voltage is reached, at which time the battery control system disconnects the load.

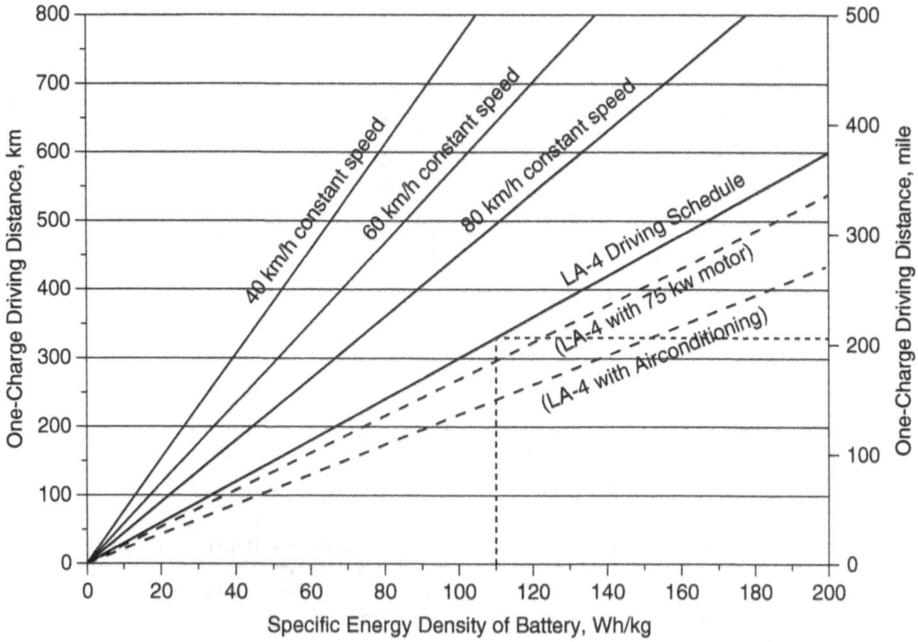

Figure 15 One-charge driving distance of EVs (car total weight 1500 kg, battery weight 300 kg) vs. specific battery energy in dependence on driving conditions. LA: Los Angeles driving cycles.

Because the discharge voltage depends on the discharge current and temperature, the end-of-discharge voltage responds to the discharge conditions, for example, normal or heavy duty (high current, deep temperature) (Table 9).

The different modes of discharge are constant current, constant resistance, and constant power (Figure 16), and the selection of the mode depends on the requirements of the application.

Because the discharge voltage depends on the discharge current and temperature, the end-of-discharge voltage adjusts to discharge conditions.

Cell Reversal

In some cases, one of the single cells in a battery of cells has a lower capacity than the other cells. Reasons for that could be an assembly of the battery with cells of different

Voltage	LMO	LFP	LAB	Ni-Cd/Ni-MH
Nominal/V	3.6	3.2	2.0	1.2
End-of-Discharge				
Normal discharge/V	3.0–3.3	2.7	1.75	1.0
Heavy-duty discharge/V	2.7	2.45	1.4	0.3

Table 9 End-of-Discharge Voltages for Different Battery Chemistries and Loads (Normal and Heavy Duty: High Current, Deep Temperature)

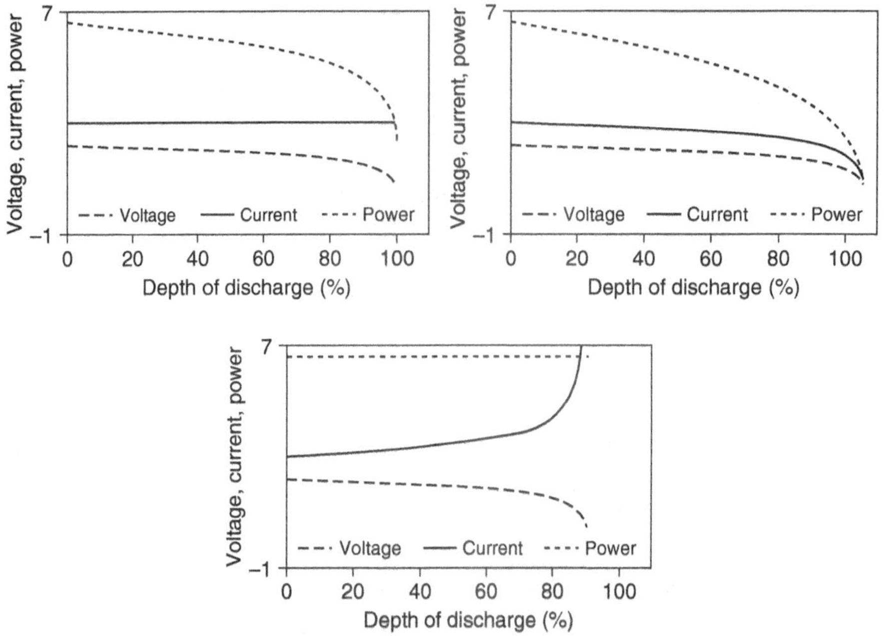

FIGURE 16 Schematics of the constant current discharge (top-left), constant resistance discharge (top-right), and constant power discharge (bottom). (Reproduced from D. U. Sauer, Batteries: Charge–Discharge Curves, in *Encyclopedia of Electrochemical Power Sources*, ed. by J. Garche et al., Elsevier, Amsterdam, 2009, Vol. 1, p. 443.)

capacities (mismatching) and divergence of the cell capacity during the service life. During a discharge of a battery with mismatched cells, the current is the same through all cells, and the cell with the lower capacity may be forced into a cell reversal. To prevent cell reversal, batteries comprise cells with a maximal capacity deviation of 2.5%. But even a small deviation may increase during the battery cycle life. The cell reversal effect can also occur if the cells in a battery are installed with wrong polarity.

The cell reversal almost always leads to irreversible chemical reactions, resulting in permanent damage to the cell. These reactions logically result in negative electrode material reduction and positive electrode material oxidation beyond their natural states. In addition, the electrolyte can be decomposed in these reversed reactions.

Charge

From a practical point of view, the battery charge process should be the shortest possible, but complex battery reactions and material interactions control the rate of charge and optimal charging times. The optimized charge time helps to prolong battery lifetime and charge efficiency. As previously discussed, shorter lifetime is related to faster aging processes usually triggered at higher temperatures, which in turn are caused by heat evolution during the charge process. Lower charge efficiency and parasitic side reactions are also the consequence of higher charge currents.

The charge current (I_{ch}) should correspond to conversion of the active mass (I_{ch}^{am}), but in practice there are also side reactions (e.g., electrolyte decomposition, corrosion) and some of the current (I_{ch}^{sr}) is used in those reactions (Equation 18).

$$I_{ch} = I_{ch}^{am} + I_{ch}^{sr} \tag{18}$$

If the charge and side reactions have similar electrochemical potentials, then both reactions take place. This situation typically occurs at the EOC and for higher SOC values, when the charge voltage increases and reaches the potential region as the side reaction (U_{ch}^{sr}).

If side reactions take place along the main charge reaction, the charged capacity (Q_{ch}) is divided between active mass (Q_{ch}^{am}) component and the side reactions (Q_{ch}^{sr}) component. The relation of Q_{ch}^{am} to Q_{ch} at the EOC (100% charged AM) determines the charge efficiency (η_{ch}), which is the reciprocal of the charge factor (F_{ch}).

$$\eta_{ch} = \frac{1}{F_{ch}} = \frac{Q_{ch}^{am}}{Q_{ch}^{am} + Q_{ch}^{sr}} \tag{19}$$

In the case of lead-acid batteries, the charge factor is about 1.3, while for lithium-ion batteries, the factor is much lower.

Overcharge

If a battery is fully charged, meaning that active mass is completely converted, then the charge current goes only toward the side reactions, $I_{ch} = I_{ch}^{sr}$, which leads in most cases to the decomposition of electrolyte and heat generation. In batteries with aqueous electrolytes, the electrolyte is decomposed into hydrogen and oxygen. This is normally not a big problem in open systems, and loss of water can be controlled by refilling. In closed, valve-regulated systems, the gases internally recombine back into water.

In batteries with organic electrolytes, a recombination of gaseous products back to original electrolyte or refilling is not possible, so electrolyte decomposition results in pressure increase and heat. The heat increase, in turn, could lead to a thermal runaway, as previously discussed in the safety section.

At the EOC, when SOC is 100%, all active mass is converted into a charged product; so 100% SOC is physically determined. But in some cases, 100% SOC is not physically determined but by convention based on experience about an optimum of capacity and service life. For example, for lithium-ion batteries with layered cathode materials, Li^+ removal from the cathode active mass is still possible even after 100% SOC (by convention) is reached. More Li^+ removal from the cathode material by increasing the EOC voltage is physically possible and leads to higher capacity, but at the expense of shorter lifetime and heightened safety risk.

An example of the cycle life and capacity trends as a function of the EOC voltage for lithium batteries is shown in Figure 17. For the reasons previously explained and connected with the stability of the active materials at the beginning of lifetime, the curve for the cycle life is exponential, while the capacity increase is predominantly linear with increasing EOC voltage because more active material is being converted. However, one of the biggest safety concerns with lithium-ion batteries arises as a result of too high voltage and parasitic side reactions that occur in the overcharge phase.

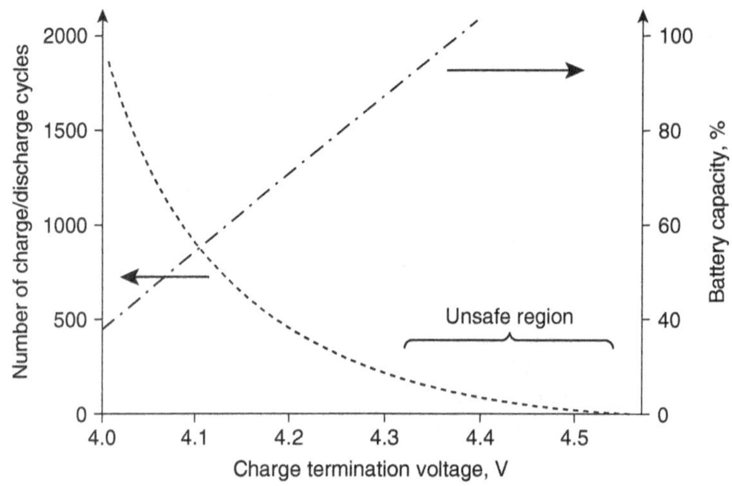

FIGURE 17 Battery cycle life and capacity vs. EOC voltage for a lithium-ion batteries with layered cathode materials.

Design of Cells and Batteries

Active Mass

The active masses are involved in the redox reactions and have the properties of electrochemically highly positive or highly negative electrode material, which results in high cell voltage. The active mass electrochemical potentials should not change too much during charge and discharge processes in the cell. The essential condition for this is small-size active mass particles and a resultant large surface area (measured by BET); which in turn means that the diffusion lengths for both electrons and ions are short. The mean diffusion time of ions in the active mass (τ), which determines the reaction rate, is related to diffusion coefficient (D), and the diffusion length (L) (Equation 20).

$$\tau = \frac{L^2}{2D} \qquad (20)$$

As expected, higher capacities are observed for smaller particle size. A more complex situation arises when the particle size is further reduced down to a nanometer size. The reduced particle size affects not only kinetical parameters but also creates higher free surface energy condition and larger potential. For example, a 62 mV higher potential was measured for 25 nm rutile TiO_2 compared to 2 μm particle size material. The larger particle BET surface means increased contact area to electrolyte and other particles such as current collector. It should be noted, however, that a smaller particle size heightens certain disadvantages such as low thermodynamic stability and strong side reactions.

If the active mass or its discharge products are nonconductive, then electronic conducting additives are incorporated into or coated on the active mass. The active masses should also be thermally stable under the battery operating conditions and chemically stable in contact with the electrolyte, additives, binders, current collector, and other components.

Other considerations for the active masses include availability of resources, low prices, low overall production energy, low CO_2 footprint, and low toxicity.

Electrodes

Electrodes support the redox reactions by creating best connection of the active masses with the electrolyte and current collector. Electrodes must have sufficient porosity to ensure electrolyte wetting and contact with the active mass, but also optimal compression and contact between particles of the active mass to ensure good electronic conductivity. The electronic conductivity can be improved by addition of electronic conducting additives (e.g., carbon), and binders can improve contact between the particles and mechanical stability of the electrode. A representative design of an electrode is shown in the cross-section in Figure 18.

Most electrodes are constructed by coating active mass onto a current collector, for example metal grids, thin metal foil, or metal screen. The resulting porous layer of active mass, typically 50–80 μm for Li batteries, is in good contact with the metal and immersed or soaked with electrolyte. The ensuing points and surfaces where three phases meet, solid current collector, active mass, and the electrolyte, are essential for the optimal battery function.

The electrodes designed for high energy should be able to encompass large quantities of active material in a thick form factor and still maintain mechanical stability, as well as good porosity for functional ionic pathways. With most electrode designs, the large thickness results in longer ionic transportation pathways and consequently in lower specific power. Secondary effects in thicker electrode structures include lower mechanical stability and reduced lifetime.

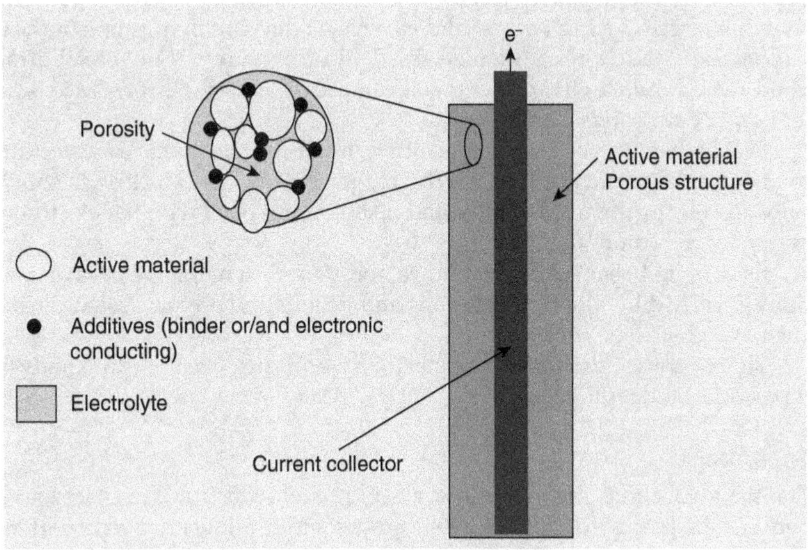

Figure 18 Cross-section of a representative electrode design.

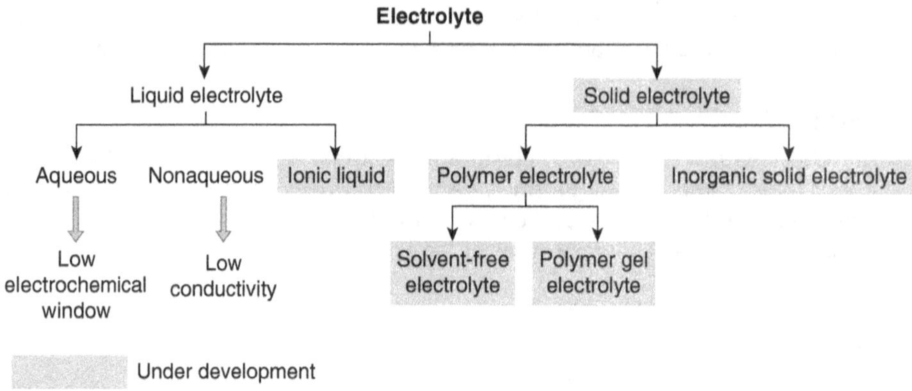

Figure 19 Battery electrolyte classification.

Electrolyte

The electrolyte is an essential component in a battery, responsible for ionic transport of charge between the electrodes in a cell. The transport is carried out by cations (positive ions, e.g., H^+, Li^+) and anions (negative ions, e.g., OH^-, PF_6^-, SO_4^{2-}). The fundamental property of an electrolyte is the ability to contain and conduct ions, which are formed mostly by dissociation of salts in solvents, for example, water or an inorganic solvent. Besides liquid electrolytes, there are also solid materials such as polymers or ceramics modified to conduct ions. Electrolyte classification is illustrated in Figure 19.

The most commonly used electrolytes in batteries are water-based and have small electrochemistry windows where no oxidation or reduction takes place due to water decomposition reaction. Nonaqueous electrolytes have a much wider electrochemical window, but typically suffer from low ionic conductivity. Some new developments, such as ionic liquids, have large electrochemical windows and a good ionic conductivity but come at a high cost. Solid electrolyte development is one of the critical innovations expected to revolutionize the field of batteries. A novel solid material, with good ionic conductivity, could greatly improve battery performance, safety, lifetime, and other parameters.

Besides being a good ionic conductor, the electrolyte must be an electronic insulator, to avoid leakage current and self-discharge of a cell. The electrolyte should be chemically and electrochemically inert, and it should not react with the electrodes, separator, and other cell components.

Electrolyte behavior often follows non-thermodynamic pathways, creating conditions governed by kinetic limitations and enabling, therefore, use of unstable systems such as lead-acid battery.

The electrolyte should also be inexpensive, nontoxic, environmentally friendly, and, if possible, biodegradable.

Separator

The main functions of a separator are to prevent electronic contact and short circuit between the positive and negative electrodes and to allow ionic transport inside the cell. The separator must also be resistant to physical penetration by active mass particles, which in some battery systems propagate from one electrode to another.

Separator	Membrane	Nonwoven
Polymer	Mostly polyolefin	Polyolefin, polyimide, etc.
Structure	Sheet with holes	Fibers around free space
Pore volume, %	~40	50–80
Maximum pore size, µm	<25	80–300
Pore structure	Hole	Labyrinth
Thickness, µm	<25	80–300
Air permeability at 200 Pa, $Lm^{-2}s^{-1}$	<<1	50–500

(Reproduced from J. Cook, P. Kritzer, Alkaline Battery Separators, in *Encyclopedia of Electrochemical Power Sources*, ed. by J. Garche et al., Elsevier, Amsterdam, 2009, Vol. 4, p. 424.)

Table 10 Parameters of Membrane and Nonwoven Separators

The separator is typically a microporous material, with high ionic conductivity that enables a high-power performance. It should also be thin, have low tortuosity (meaning straight through pores), and exhibit high ab/adsorption and retention of electrolyte (reservoir effect). Good initial and permanent wettability with electrolyte is also required and can be accomplished by the selection of separator material, special surface treatment, or by use of wetting agents.

Most separator materials are mechanically stable, with small pores, and have high puncture resistance. They are also highly manufacturable. The separator must also be highly inert toward the electrode materials and the electrolyte and has to be thermally stable to achieve long lifetime.

Some separator parameters must simultaneously satisfy contradictory requirements. For example, large thickness is needed for better electrical isolation between electrodes and better prevention of particle penetration, but thinner separators with lesser tortuosity provide better ionic conductivity.

The materials for separators are membrane- and nonwoven-type. A membrane can be envisioned as a sheet with holes, while a nonwoven separator has fiber-type structure with void volume. Membrane-type separators have porosities of about 40% (e.g., Li-ion battery), while nonwoven separators have up to 80% pore volume (e.g., alkaline batteries). The differences between both separator types are shown in Table 10.

No additional separator is needed in cells with solid electrolyte.

Types of Cells

Button Cells

Button cells are constructed by placing a separator between the two electrodes and inserting such assembly into a small container. A contact spring is placed on the top of the electrode stack and a gasket assembly is added, before filling the cell with electrolyte and sealing the cell by mechanically collapsing the sides of the cylinder around the lid (Figure 20). Thin-form factor button cells are called coin cells.

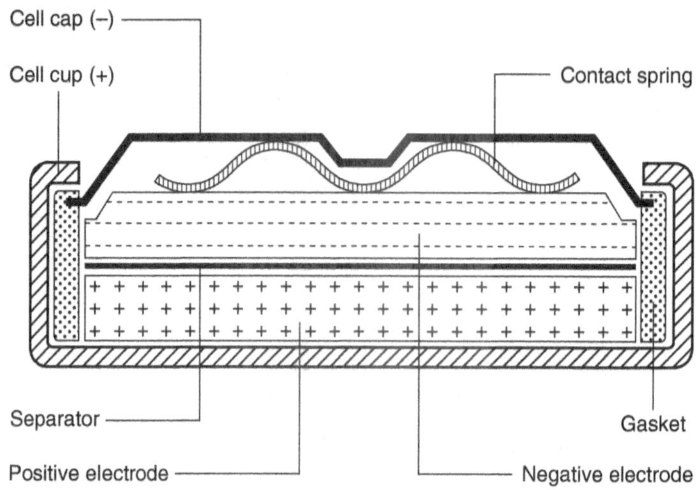

FIGURE 20 Cross-section of a button cell (Ni-MH). (Reproduced from M. Fetcenko, J. Koch, M. Zelinsky, chap. 6, Nickel–Metal Hydride and Nickel–Zinc Batteries for Hybrid Electric Vehicles and Battery Electric Vehicles, in *Advances in Battery Technologies for Electric Vehicles*, ed. by B. Scrosati, J. Garche, W. Tillmetz, Woodhead, 2015, p. 117.)

Cylindrical Cells

The cylindrical cells are assembled by rolling two electrodes divided by a separator and inserting such assembly into a cylindrical container (Figure 21). The cell is then filled with electrolyte and sealed by attaching the top assembly (a lid, a resealable safety vent,

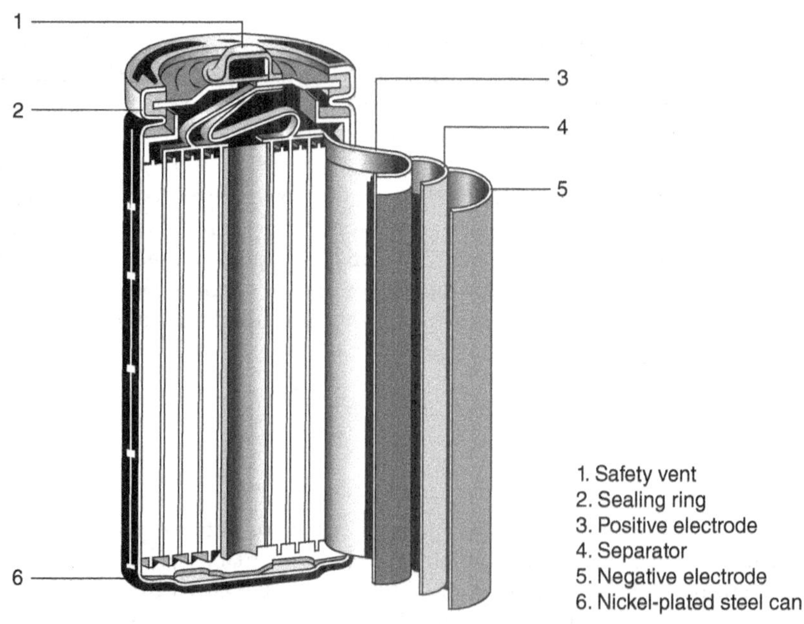

FIGURE 21 Sketch of a sealed cylindric Ni-Cd cell in expanded view.

a terminal cap, and a plastic gasket) onto the can. The most common cylinder material is metal, which also serves as the negative terminal, while the lid is the positive terminal. These are mechanically strong cells because metal cylinders are able to withstand internal gas pressures of more than 13 bar (200 psi).

Cylindrical design is used for cells with lower capacity and lower power as this type of packaging does not provide efficient heat removal and thus is not suitable for higher current rates. These are relatively low-cost cells due to mature manufacturing and high productivity packaging processes.

Prismatic Cells

Prismatic configuration consists of several layers of planar electrodes interspaced by separator sheets. The contact tabs are attached to the anodes on one side of the rectangle assembly and to the cathodes on the other, before welding all tabs to cell terminals. The assembly is then inserted into a prismatic container and sealed by welding the top (consisting of a resealable safety vent, terminals, and plastic gaskets) to the container (Figure 22).

The container material for prismatic cells is metal or plastic and the selection depends on the application. Plastic cells provide a safer environment for high-voltage batteries because of the voltage isolation and they are also less expensive, while metal containers are preferred for cells where high internal gas pressure could arise.

Pouch Cells

The newest type of packaging for modern batteries is the so-called pouch cell (Figure 23). It is similar in form to a prismatic package, but the container material is a thin (70–160 μm) and flexible multilayer foil (nylon, Al, and cast poly propylene). The foil is sometimes referred to as a laminated film and the cell is called a "laminated cell." The pouch cell design was first used for Li-ion cells with polymer electrolytes and is sometimes called lithium polymer cell or "LiPo."

The assembly proceeds in the similar manner as for prismatic cells. The planar electrodes are first cut into desired size and contact tabs are welded using a resistance or arc welding. The separator on a roll is then sandwiched between pairs of electrodes and wrapped around for the next pair of electrodes. The stack is then inserted into a pouch, which is open on one side. Since this type of packaging is used mostly for lithium-ion batteries, the next steps are performed inside a "dry box," in a controlled atmosphere; the electrolyte is first added, and the pouch is then sealed by closing the top, around the contact tabs.

The pouch material is light and flexible and provides good thermal conductivity. Because it is very thin, there are some concerns with humidity diffusion into the cell. Low mechanical stability allows swelling of the cell. The packing efficiency of pouch cells reaches 90–95%—the highest among all packaging types.

Battery Design

Batteries are connections of at least two cells in series or in parallel. To achieve higher operating voltages, several cells are connected in series; higher power cells are connected in parallel. The connected numbers (x) of cells are abbreviated as xS in series connection, xP in parallel connection, and xSxP in series and parallel connections. For example, a notebook battery (14.4 V; 5 Ah) consists of eight Li-ion cells (3.6 V; 2.5 Ah)

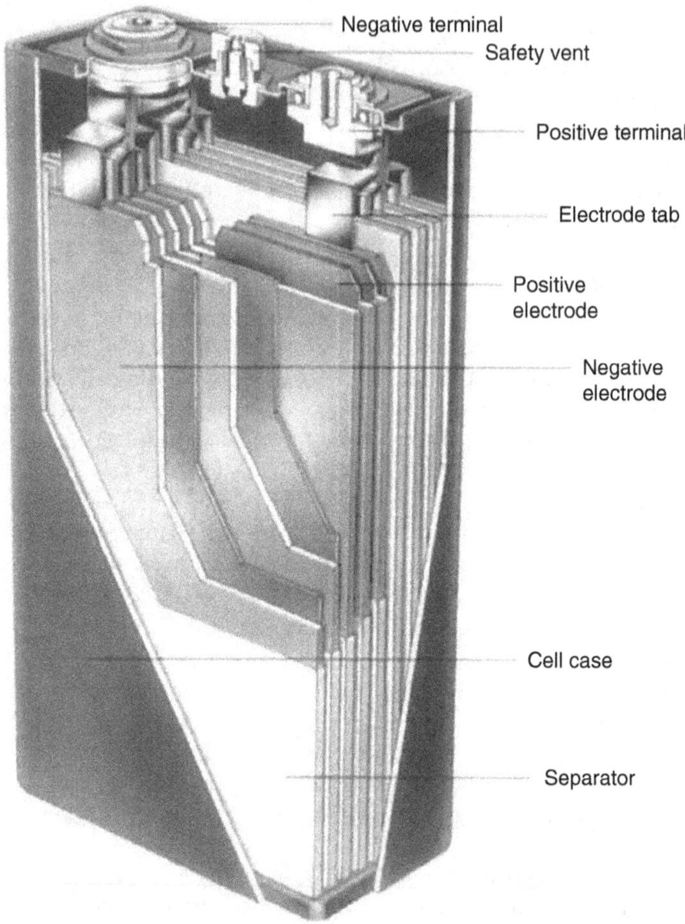

Figure 22 Prismatic Ni-MH cell. (Reproduced from M. Fetcenko, J. Koch, M. Zelinsky, chap. 6, Nickel–Metal Hydride and Nickel–Zinc Batteries for Hybrid Electric Vehicles and Battery Electric Vehicles, in *Advances in Battery Technologies for Electric Vehicles*, ed. by B. Scrosati, J. Garche, W. Tillmetz, Woodhead, 2015, p. 103.)

with four cells connected in series and two of these four-cell series strings connected in parallel, so that the battery configuration is 4S2P.

An example of a battery comprising single cells is shown in Figure 24.

When individual cells are connected in series or in parallel, it is essential that all the cells are of the same type and same capacity, otherwise an imbalance could occur. In case of a defective cell with lower capacity, that cell would be discharged first, lowering the total battery voltage on the discharge. The other cells in a string would then drive current through a faulty cell and force it into cell reversal, which almost always results in irreversible cell damage and a safety risk.

It is recommended that all cells in a battery have capacities that differ less than 2.5%, to avoid mismatching problems. But, even for cells in a battery that start with a similar capacity, it is likely that their capacities will diverge over time and cause imbalance.

Figure 23 A pouch cell.

Figure 24 480-piece, 2-V VRLA gel cells in two strings used in the Padre Cocha hybrid RAPS system. (Reproduced from D. Rand, P. Moseley, chap. 13, Energy Storage with Lead-Acid Batteries, in *Electrochemical Energy Storage for Renewable Sources and Grid Balancing*, ed. by P. Moseley, J. Garche, Elsevier, 2015, pp. 201.)

It is often necessary to mitigate the problem by charge-balancing routines and leveling the SOC of all cells.

If one cell in a serial connection has high resistance, then the whole string of cells is likely to fail as the power would be dissipated to overcome the resistance in a bad cell. To avoid this problem, a solid-state switch is normally used to bypass the high-resistance cell. This method obviously leads to lower overall string voltage, but this is usually tolerated by the application and is generally not a problem. A replacement of one or more aged or low-capacity cells is ordinarily not considered because a new replacement cell would have higher capacity and lower resistance than the other cells in a string, causing another mismatch. In most cases, rather than trying to replace a failing cell, the whole string is replaced with new cells.

In parallel connections, the voltage is the resultant voltage of all cells. A cell mismatch for parallel connection causes capacity reduction, but it is not a catastrophic mismatch condition. However, it could be a serious safety hazard in case of a short circuit because the energy from all good cells is dissipated in the lower performance cell to overcome the resistance, which increases the risk of a thermal runaway (Li-ion cells).

Modules comprise several prismatic cells connected by an intercell connector through the cell wall. They normally contain a common plastic container, a single vent assembly, and two electrical terminals, for example, lead-acid starter batteries for cars.

A battery pack contains two or more modules, electrical interconnects, interfaces for cooling, cell electronics including voltage circuit protection, and communication (Figure 25). For a battery pack of 60 V DC or higher, a manual shut-off function (service disconnect) may be included. All components are typically placed in an impact resistance case.

The cells can be serially connected in a monopolar or bipolar fashion. As seen in Figure 26, a positive electrode from the first cell is connected to a negative electrode of the next cell using a conductor.

A bipolar cell connection comprises positive and negative electrodes from two adjacent cells sharing a conductive divider (Figure 26). On one side of this electronic divider, a positive active material is deposited and on the other side the negative. The bipolar

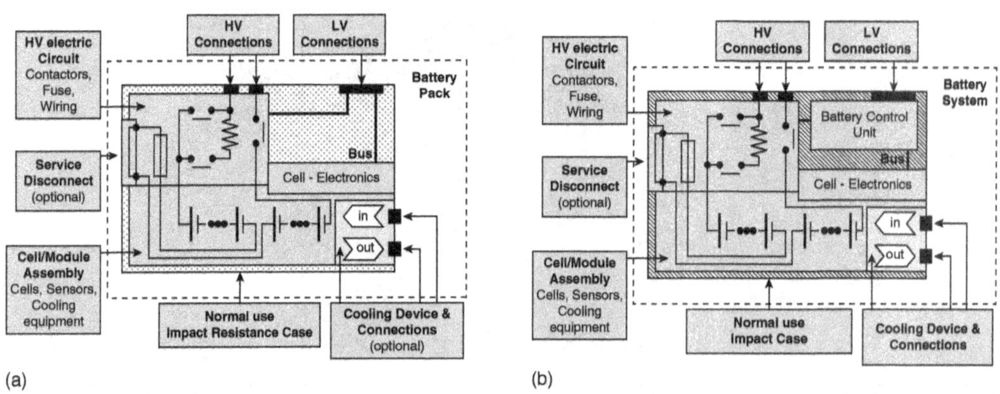

FIGURE 25 Typical configurations of (a) a battery pack and (b) a battery system. (Reproduced from ISO 12405-2:2012, Electrically propelled road vehicles—Test specification for lithium-ion traction battery packs and systems—Part 2: High-energy applications.)

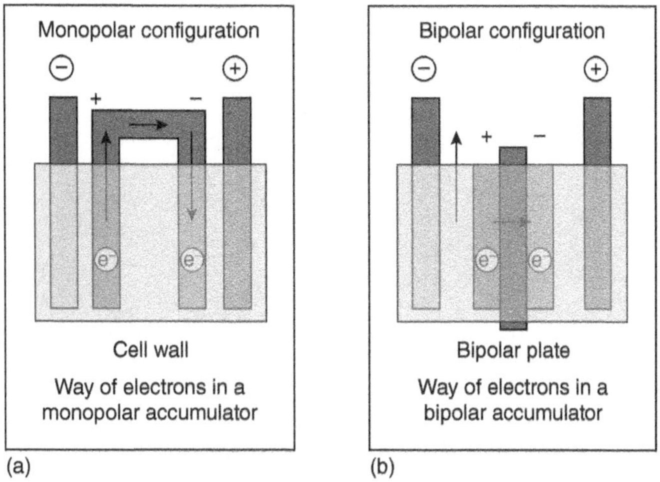

FIGURE 26 Schematics of (a) monopolar and (b) bipolar cell connections.

connection enables a short conductive path connecting the two neighboring cells, very uniform current distribution across the electrode plates, and in turn high power.

In stacks with bipolar electrodes, the main configuration is serial connection for obvious reasons, so the capacity of a stack cannot be increased compared with single cell (unless strings are connected in parallel). The material for the electronic divider of a bipolar electrode must have good conductivity and corrosion resistance, and be pinhole-free to prevent ionic communication between the cells.

The bipolar concept has been successfully used for fuel cell and redox flow battery electrodes, where gaseous or liquid reactants are supplied from outside of the cell and differences in capacity can be compensated by adjusting the flows. But that is not the case for battery cells with bipolar electrodes that contain fixed amounts of solid active masses, and equalization of capacity differences has to be done by overcharge, which is not always the best solution and could lead to electrolyte decomposition.

Finally, thermal balance is critically important when designing batteries, modules, and systems, as individual cells can develop differences in the SOC and cause the failure of a complete system. In some cases, an active thermal management subsystem is necessary to maintain homogeneous temperature distribution for all battery cells.

CHAPTER 2

Lead-Acid Batteries

Introduction

In 1859, over 50 years after Johann Wilhelm Ritter first proposed an accumulator (i.e., battery) based on a Cu anode and a CuO cathode, a breakthrough invention of a lead-acid battery (LAB) started the new age in energy storage, which at the present time has an enormous importance. The working principle of the LAB was invented in 1854 by Wilhelm Joseph Sinsteden (1803–1891), and the first working LAB was built in 1859 by Gaston Planté (1834–1889). One of Planté's designs included cylindrical and planar cells where the anodes and cathodes were separated by rubber strips. The second design already at that time proposed a 12-V battery built from cylindric cells (Figure 27).

The active masses of the negative and positive electrodes were electrochemically prepared on lead plates, as it is done even today for large surface area plates, called Planté plates. In 1881, C. A. Fauré (1840–1909) covered lead plates with porous active mass. In the same year, E. Volckmar invented grid plates, the foundation for the starting-lighting-ignition batteries (SLI). In 1888, G. Philippart produced new types of electrodes, well known today as iron-clad electrodes. The history of LABs reveals that at the end of the 19th century, most of the technological developments used today were already established. These developments were naturally optimized and improved over the years: in 1957, Otto Jache introduced the first lead-acid gel battery, and the first absorbed glass mat (AGM) battery was patented by Gates Rubber Corporation in 1972. From the market point of view covering all application segments, the LAB is still the most popular and over 50% of the overall battery revenue comes from LAB systems.

Principle of Operation

The following reactions take place on LAB electrodes during the discharge process:

Negative electrode:

$$\mathrm{Pb} + \mathrm{HSO_4^-} \rightarrow \mathrm{PbSO_4} + \mathrm{H^+} + 2\mathrm{e^-} \qquad E^\circ = -0.355\ \mathrm{V} \tag{21}$$

Positive electrode:

$$\mathrm{PbO_2} + \mathrm{HSO_4^-} + 3\mathrm{H^+} + 2\mathrm{e^-} \rightarrow \mathrm{PbSO_4} + 2\mathrm{H_2O} \qquad E^\circ = 1.686\ \mathrm{V} \tag{22}$$

Cell:

$$\mathrm{PbO_2} + \mathrm{Pb} + 2\mathrm{H_2SO_4} \rightarrow 2\mathrm{PbSO_4} + 2\mathrm{H_2O} \qquad E^\circ_{\mathrm{cell}} = 2.041\ \mathrm{V} \tag{23}$$

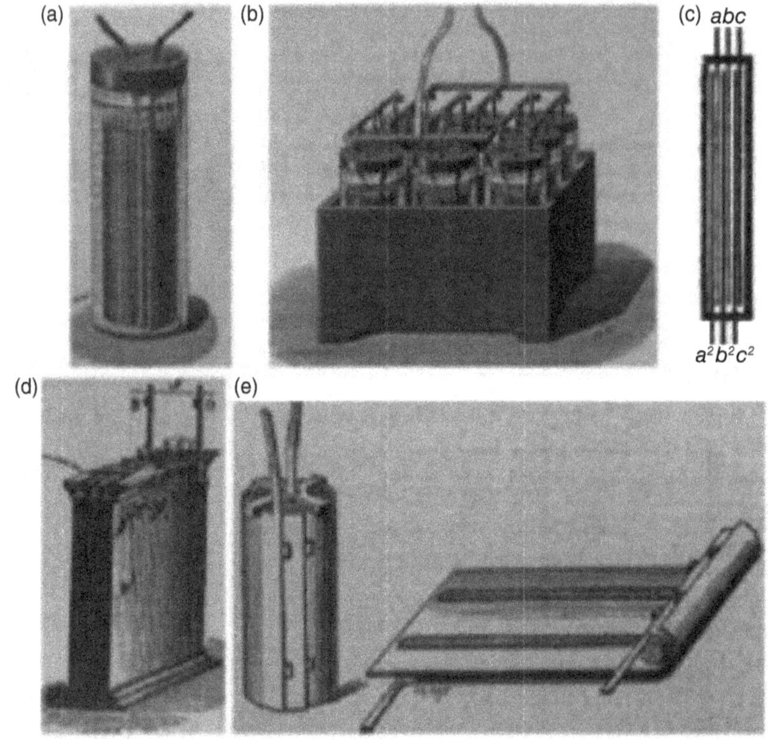

FIGURE 27 Original drawings of Planté's lead-acid systems: (a, e) cylindric cells, (c, d) planar cells, and (b) 12-V battery.

The reactions are reversed on the charge. The immediate observation from the reaction equations above reveals that the product of reaction on both electrodes during discharge is lead sulfate ($PbSO_4$), the fact observed as early as 1882 by Gladstone and Tribe.

The reaction mechanism of all charge and discharge reactions involves processes that proceed in the electrolyte. During the discharge, Pb^{2+}-ions formed in the electrode reactions react with SO_4^{2-}ions from the electrolyte to produce $PbSO_4$, which precipitates as crystal at the electrodes. This solution-precipitation process produces, in every cycle, a change in morphology of the electrodes, causing mechanical stress and consequently leads to low cycle life. We will see in the later chapters that lithium-ion battery electrodes rely on a different mechanism and do not undergo significant structural changes during cycling.

The equilibrium cell voltage for LAB at 25°C is given by Equation 24, which together with reaction equations above shows that the electrode potential, E and cell voltage, $U_0 = \Delta E$, depend on the concentration of H_2SO_4 electrolyte.

$$\Delta E = \Delta E^0 + \frac{RT}{2F}\log\frac{c(HSO_4^-)^2\ c(H^+)^2}{c(H_2O)^2} = 2.041\ V + 0.059\ \log\frac{c(H_2SO_4)}{c(H_2O)} \tag{24}$$

In the LAB technology, the concentration of the electrolyte is usually expressed using its density ($\rho_{H_2SO_4}$) and the expression for the open-circuit cell voltage (U_o) of a LAB becomes:

$$U_0 = \left[0.84 + \frac{\varrho(H_2SO_4)}{g\ cm^{-3}}\right] V \tag{25}$$

Because the sulfuric acid is consumed during the discharge (its concentration changes), the cell voltage obviously depends on the state-of-charge (SOC) and the SOC of an open battery can be determined by measuring the density of the acid, for example, using a hygrometer. The sulfuric acid concentration for a fully charged battery is about 1.28 g cm^{-3} (~38%).

$$U^0 \propto SOC \propto \rho_{H_2SO_4} \tag{26}$$

The equilibrium voltage of H_2O is 1.23 V, which is below the equilibrium voltage of the LAB system (2.04 V). Therefore, the LAB is thermodynamically unstable and it is because of the overvoltage that the gas evolution reactions occur at much higher voltages than the theoretical thermodynamic values (Equations 27 and 28).

The charging process at LAB electrodes proceeds, at first, by conversion of the active masses. Near the end of the charge process, the side reactions take place as well and two competing processes run in parallel. Once the main reactions of the active mass conversion have completed, only the gas evolution reactions are observed.

Negative electrode: $$2H^+ + 2e^- \rightarrow H_2 \qquad E° = 0\ V \tag{27}$$

Positive electrode: $$H_2O \rightarrow \frac{1}{2}O_2 + 2H^+ + 2e^- \qquad E° = 1.23\ V \tag{28}$$

Gas evolution also takes place during open-circuit periods without load because of the thermodynamic instabilities. The rates of these self-discharge reactions are very low.

Besides the gas evolution reaction, there are other side reactions on electrodes. The most consequential side reaction is the corrosion of Pb grid or current collector on the positive electrode, which takes place during charge and also in the open-circuit period (Equation 29).

$$Pb + 2H_2O \rightarrow PbO_2 + 4H^+ + 4e^- \tag{29}$$

The above-discussed reactions are summarized in Figure 28.

The vertical dashed lines on each side of the graph imply constant voltage charge and discharge conditions and demonstrate the relation to side reactions of gas evolution and corrosion. The shaded area in the graph represents the voltage region where water decomposition should occur, but the actual hydrogen and oxygen evolution is shown to take place outside of this voltage window due to overvoltage limitations. This phenomenon enables the LAB despite thermodynamic contradiction. Also shown in Figure 28 on the right side is the positive electrode Pb-grid corrosion region.

The gas generating side reactions naturally lead to pressure increase, preventing sealing of a cell. However, a method was developed whereby oxygen gas generated at the positive electrode near the end of the charge and in the overcharge phase is reduced

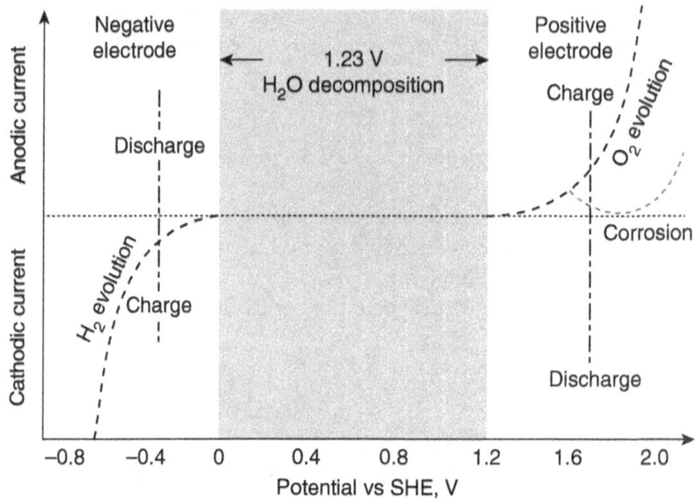

FIGURE 28 Conceptual representation of LAB current vs. potential showing voltage regions where main and side reactions take place.

(i.e., recombined) at the negative electrode. The cathodic reactions during charge at the negative electrode are shown in Equations 29a, 29b, and 30.

$$PbSO_4 + H^+ + 2e^- \rightarrow Pb + HSO_4^- \qquad E^\circ = -0.355\text{ V} \tag{29a}$$

$$2H^+ + 2e^- \rightarrow H_2 \qquad E^\circ = 0\text{ V} \tag{29b}$$

$$\frac{1}{2}O_2 + 2H^+ + 2e^- \rightarrow H_2O \qquad E^\circ = 1.23\text{ V} \tag{30}$$

During overcharge phase in LABs, oxygen is generated at the positive electrode; it then diffuses through the interspaces between the electrodes and the electrolytes and is reduced at the negative electrode. Oxygen reduction has the most positive potential of the three reactions shown above and will be a dominant reaction as long as oxygen is available.

Oxygen formation at the positive electrode and removal or reduction at the negative electrode is called the oxygen cycle and is the foundation of sealed LABs. The reaction is diffusion limited, and to achieve high reaction rates of oxygen reduction on the negative electrode, it is critical to design a battery that will enable fast transport of oxygen through the electrolyte. The current related to oxygen reduction reaction is proportional to oxygen diffusion coefficient in electrolyte (D), surface area for reaction (S), solubility constant (K_s), the partial pressure of oxygen (p), and the thickness of the diffusion layer on the electrode surface (δ).

$$I(O_2) = 4F\frac{D(O_2)}{\delta}K_S\, p(O_2)\, S \tag{31}$$

The reaction area and the diffusion layer thickness are strongly influenced by the nature of the gas pores in the electrolyte layer. It should be noted that sealed LABs are designed with no free liquid electrolyte, which instead is held in the pores of a

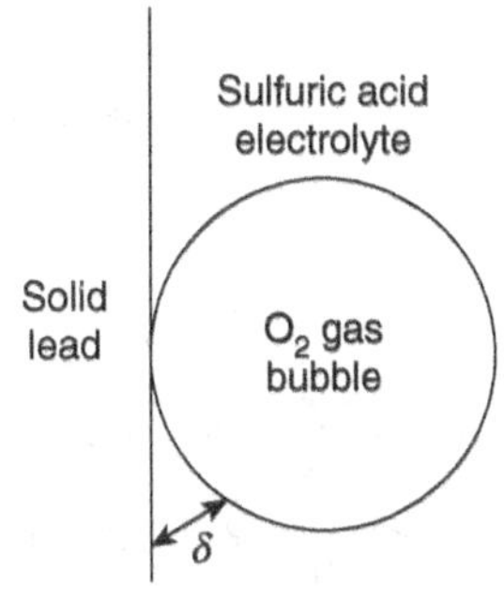

FIGURE 29 Illustration of a three-phase boundary between oxygen gas, sulfuric acid electrolyte, and solid lead electrode.

suitable substrate. The reaction area or so-called three-phase boundary between the solid lead electrode, liquid electrolyte, and oxygen gas is essential for the fast reaction of oxygen removal, and it depends on the nature of the interface, the gas pore area, and the amount of electrolyte in the pores. The desired three-phase boundary (Figure 29) exists at the interface between the solid electrode, the liquid electrolyte, and a layer of gas bubbles. The perfect thickness of the gas diffusion layer is about one third of the radius of an oxygen bubble for the case where the electrolyte is adsorbed in a porous matrix.

It should be obvious that the oxygen reduction reaction must be faster than the rate of oxygen generation, $v_{O_2\text{-gen}} \leq v_{O_2\text{-red}}$ to remove all oxygen generated on the positive electrode and prevent pressure increase in the cell. A gas pressure increase can take place at higher charge currents because the rate of gas evolution is proportional to charge current, $v_{O_2\text{-gen}} \sim I_{ch}$. To prevent mechanical damage to cell housing in case of pressure increase, sealed lead-acid cells are equipped with a pressure relief valve built into the cell lid. These types of cells are called valve-regulated lead-acid battery (VRLAB).

Lead-Acid Battery Components

Negative Electrode

The negative electrode comprises active mass and the current collector. The negative mass is porous lead, which is manufactured from a mixture of lead and lead oxide powder, the "leady oxide." The "leady oxide" is then mixed with H_2SO_4, H_2O, and additives to a consistency of a paste and is then coated onto the current collector or grid. The paste is next cured at a temperature of 40–55°C to produce a sulfate, before electrochemical conversion to porous lead active mass is achieved in a so-called formation charge. The BET surface (Brunauer, Emmett, and Teller surface determination method, based on the gas adsorption data) of the negative active mass is typically 0.5–1.0 m^2g^{-1} and the porosity is 50–60%.

The discharge product of the porous lead active mass is $PbSO_4$, which is an electrical isolator and can passivate the electrode surface preventing further reaction. To avoid dense $PbSO_4$ passivation, a layer called an expander is added to the negative paste. The expander typically comprises an organic component (e.g., lignosulfonate), $BaSO_4$, and

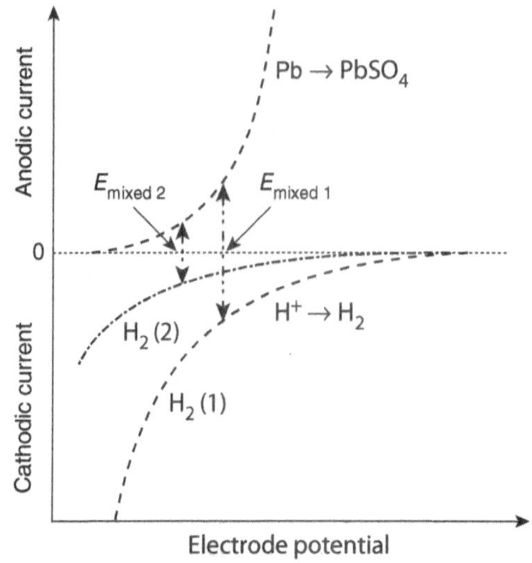

FIGURE 30 Schematic of the mixed potential formation at the negative LAB electrode for two different hydrogen development rates. (Adapted from *Valve-Regulated Lead-Acid Batteries*, ed. by D. A. J. Rand, P. T. Moseley, J. Garche, C. D. Parker, Elsevier, 2004, p. 129.)

carbon, which are adsorbed on the lead surface to enable formation of porous $PbSO_4$, instead of a continuous passivating layer.

The negative lead electrode immersed in sulfuric acid electrolyte assumes a mixed potential between the $Pb/PbSO_4$ and the H_2/H^+ electrode (Figure 30).

The graph in Figure 30 shows two different cathodic hydrogen development rates and one anodic lead oxidation rate. The open-circuit potential of the negative electrode forms at the point where anodic and cathodic currents are equal in magnitude. As a result, the electrode potential is more positive in case of higher hydrogen development rate (lower hydrogen overpotential). The mixed negative electrode potential results in a self-discharge reaction, consumption of active mass, and hydrogen generation.

$$Pb + H_2SO_4 \rightarrow H_2 + PbSO_4 \tag{32}$$

The lower is the hydrogen overpotential, the greater are the rates of the hydrogen evolution and self-discharge reaction. Under most circumstances and electrode designs, the hydrogen evolution rate is low because of the high overpotential. But some metallic impurities on the lead electrode can lower the overpotential and cause increase in the rates of both the hydrogen evolution and self-discharge. The impurities come from the production process, oxidation of Pb grid that contains antimony (Sb), contamination in electrolyte, or additions of unclean water (for flooded-type cells). Any of these sources of impurities or a combination can result in a significant self-discharge, typically encountered in case of older LABs. Elevated temperature and higher than normal sulfuric acid concentration also contribute to increase in self-discharge.

The hydrogen reaction overpotential and the above factors affect in the similar manner the hydrogen evolution in the overcharge phase by encouraging hydrogen evolution at lower overpotential and higher charge current.

Negative electrode of an LAB is the limiting factor for the battery lifetime due to its sulfation. In recent years, carbon additives have been used to reduce sulfation effect.

A special type of negative electrode is used in the so-called UltraBattery, comprising carbon and lead. This type of battery is a hybrid between a supercapacitor (with negative carbon electrode) and conventional LAB (with negative lead electrode). The hybrid electrode configuration enables high energy from lead, while carbon reduces sulfation.

Positive Electrode

The positive electrode comprises active mass and the current collector. The positive active mass is porous lead dioxide, manufactured in a similar way to the negative active mass except for different additives. Analogous and parallel to formation of Pb on the negative electrode, the sulfates on the positive electrode are converted to PbO_2 in the electrochemical formation charge (Equation 33). Additives to "leady oxide" often include Pb_3O_4, which shortens the formation time due to chemical reaction with sulfuric acid to form PbO_2.

$$Pb_3O_4 + 2H_2SO_4 \rightarrow PbO_2 + 2PbSO_4 + 2H_2O \qquad (33)$$

The BET surface of the positive active mass is found to be 3–6 m^2g^{-1} and the porosity is 50–60%. The discharge product in the positive electrode reaction is $PbSO_4$, which is difficult to recharge in case it forms large crystals. Nucleation additives such as $CaSO_4$ or Na_2SO_4 have been used to reduce formation of large crystals. Other additives such as C, Ti, Bi, and conducting oxides have been used to improve conductivity, and porous materials have been used to enhance electrolyte transportation through the active mass by forming local reservoirs.

The potential of the positive electrode in sulfuric acid electrolyte is a mixed potential between the $PbSO_4/PbO_2$ and the H_2O/O_2 electrode (Figure 31).

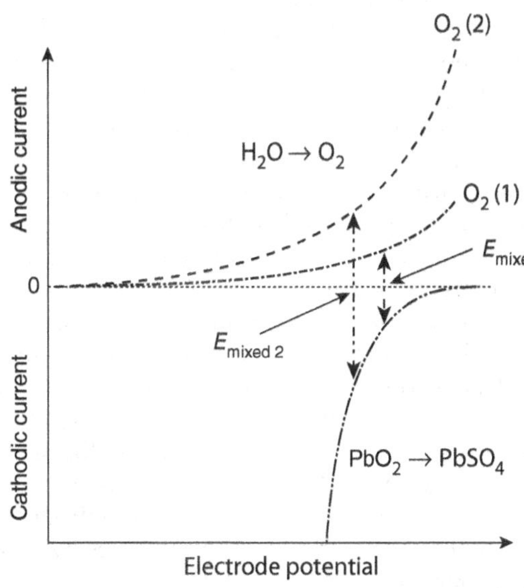

FIGURE 31 Schematic of the mixed potential formation at the positive LAB electrode for two different oxygen development rates ($O_2\uparrow$).

The formation of a mixed electrode potential at the positive electrode affects self-discharge, consumption of lead dioxide active mass, and formation of oxygen.

$$PbO_2 + H_2SO_4 => PbSO_4 + \frac{1}{2}O_2 + H_2O \quad (34)$$

This self-discharge rate is greater in case of lower oxygen oxidation overvoltage. The oxygen evolution rate at the lead dioxide electrode is usually quite low because of the high oxygen overpotential (curve 1 in Figure 31), but impurities can decrease the oxygen overpotential and lead to higher oxygen generation rate.

The oxygen evolution is also observed in the overcharge phase—oxygen evolution is encouraged at lower overpotential and higher charge current.

If the sulfation of the negative electrode is controlled, then the positive electrode will become the lifetime limiting factor for LABs because of the shedding of the active mass caused by loss of particle contact and corrosion of the positive grid.

Current Collector

Material

The active masses in charged states are PbO_2 for the positive electrode and Pb for the negative electrode. On both electrodes, the active mass in the discharged state is $PbSO_4$. These active masses are fixed into the current collector, which provides the electronic conductivity path between the active materials and the cell terminal and supports mechanical stability for the porous active masses during production and assembly of the cells. The critical requirement for the current collector material is stability in the electrolyte at high potential of the positive electrode ($E° = 2.041$ V).

Metallic lead is the most commonly used material, and it is obviously different from the active masses. Pure lead has high corrosion stability but comparatively low mechanical stability. To increase its mechanical stability, lead is alloyed with other metals, primarily antimony, used up to a maximum of 12%. Besides better mechanical stability, antimony also made grid casting easier. As the antimony content increases, the corrosion susceptibility increases as well and the hydrogen overvoltage at the negative electrode decreases over the lifetime of a battery.

Figure 32 shows current versus voltage, for a constant current charge. For any point on the current axis, a voltage of 2.4 V indicates a start of an undesired side reaction of gas evolution. The voltage of 2.4 V, where hydrogen is generated, is reached at lower current for a new battery.

The reason for reduction of hydrogen generation overpotential is the leakage of antimony from the alloy with lead as a result of corrosion and plating (or deposition) of Sb on the electrode, and subsequent changes in the electrochemical characteristics and overvoltage. An obvious solution is to reduce the Sb content to 1–3% (low antimony alloys) and introduce other alloys such as PbCaSn, with 0.05–0.07% Ca and up to 3% Sn, as well as alloys with Ag and Bi. One of the successful solutions proved to be a Pb grid alloyed with Ca, which produced a reduction in hydrogen generation rate. This formula became the choice for maintenance-free batteries (which do not require refilling of water during the battery life).

Pure lead grids can be made thinner because they have high corrosion resistance. This technology that requires advanced mechanical processing is referred

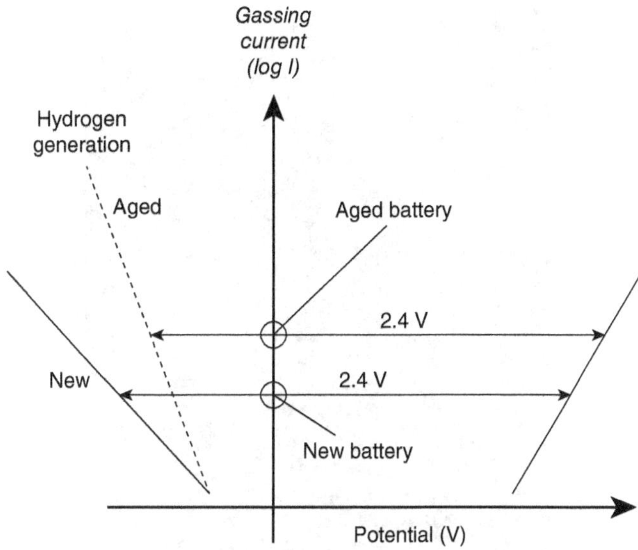

Figure 32 Current-voltage graph for battery charging, showing hydrogen overpotential regions for new and aged battery with PbSb current collector. (Reproduced from D. U. Sauer, Lifetime Determining Processes, in *Encyclopedia of Electrochemical Power Sources*, ed. by J. Garche et al., Elsevier, Amsterdam, 2009, Vol 4, p. 805.)

to as Thin Plate Pure Lead (TPPL) and is especially used in stationary applications that require a long battery life.

There are alternative construction methods for current collectors, for example, grids with plastic cores, copper, or aluminum; titanium grids; current collectors based on conductive and nonconductive fibers; carbon foam; and carbon honeycomb grids. The material choice depends on the electrode because of different electrochemical behavior, in particular the voltage stability windows. For example, carbon materials and copper rapidly corrode (i.e., are oxidized) in the electrode potential range of the positive electrode, while titanium corrodes if it is used as a negative grid. Some successful commercial batteries employ Pb-plated Cu grids for the negative electrode or surface-modified titanium electrode for the positive.

Electrode Design

The current collector design determines the electrode design, and the following types are considered: Planté, flat plate, and tubular plate.

Planté plate design consists of a lead plate which has thin vertical grooves and horizontal cross-ribs to increase the surface area. The active mass (PbO_2) is formed by an electrochemical process from the lead plate and, in same subsequent fabrication methods, by the charge-discharge formation cycles (Figure 33).

The Planté plate is characteristic for long lifetimes and high reliability, and it is used in stationary batteries such as backup battery in power plants. The thickness of the PbO_2 layer is about 0.2 mm, providing average specific capacity.

The flat-plate current collector design has higher specific energy and is used for transportation applications. The plates consist of a current collector in form of a grid in which the active mass is incorporated (Figure 34).

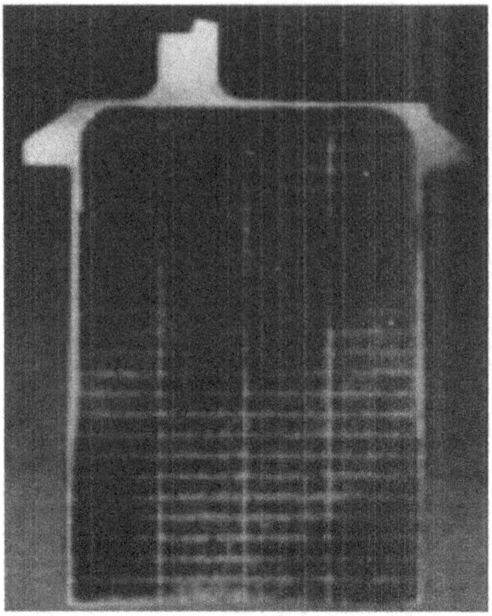

Figure 33 Image of a Planté plate. (Reproduced from R. Wagner, chap. 8, Positive Active-Materials for Lead-Acid Battery Plates, in *Lead-Acid Batteries for Future Automobiles*, ed. by J. Garche, E. Karden, P. T. Moseley, D. A. J. Rand, Elsevier, 2017.)

In some other cases, grids could also be spirally wound, ensuring higher mechanical stability of the electrode (Figure 35), while pure lead can also be used without any modifications, despite lower mechanical stability, but for its high corrosion-resistance properties.

The grids are normally produced by a casting process, in which a liquid material is poured into a mold, which contains a hollow cavity of the desired shape, and then allowed to solidify. The process can be controlled either by the gravity or under pressure for larger-size grids.

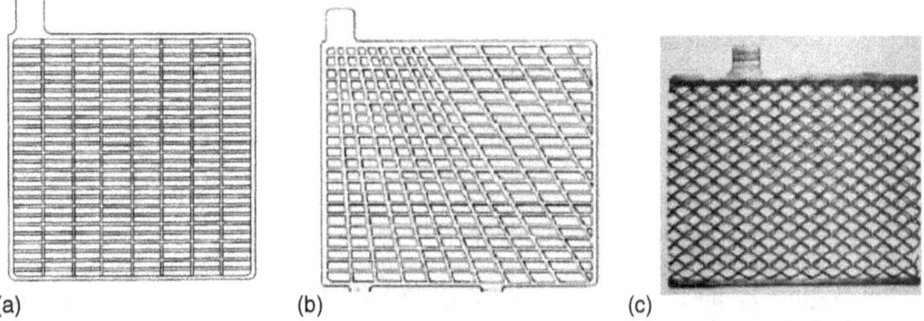

(a) (b) (c)

Figure 34 Illustration of conventional grid design (a), grid with diagonal wires directed to the terminal (b) (Reproduced from D. Pavlov, *Lead-Acid Batteries Science and Technology*, Elsevier, 2011), and expanded grid design (c). (Reproduced from *Lead-Acid Batteries for Future Automobiles*, ed. by J. Garche, E. Karden, P. T. Moseley, D. A. J. Rand, Elsevier, 2017.)

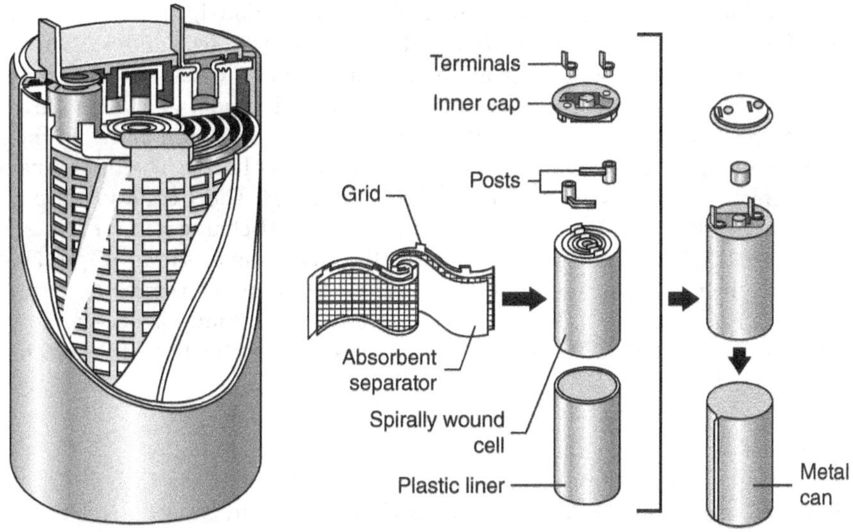

Figure 35 Illustration of the construction of a spirally wound electrode and cell. (Reproduced from D. A. J. Rand, P. T. Moseley, Lead–Acid Batteries—Overview, in *Encyclopedia of Electrochemical Power Sources*, ed. by J. Garche et al., Elsevier, Amsterdam, 2009, Vol 4, p. 550.)

Other continuous technologies are the punched grid production and the drum-casting production process (CONCAST).

For the protection of active masses, tubular plates are made from woven or nonwoven cylindrical (i.e., tubular) envelopes around a vertical rod. This is an improved design compared to flat plates where the active mass is prone to shedding and softening over lifetime (Figure 36).

The tubular design is used mostly as a positive electrode because active mass shedding has been observed at the positive electrode.

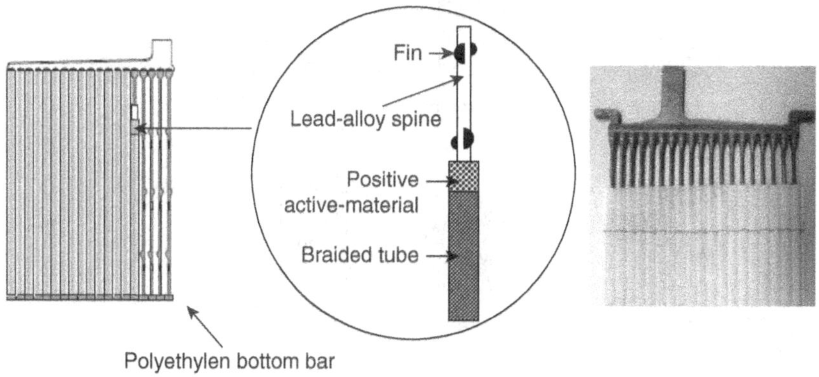

Figure 36 Illustration of a multitube electrode. (Reproduced from D. A. J. Rand, P. T. Moseley, Lead–Acid Batteries—Overview, in *Encyclopedia of Electrochemical Power Sources*, ed. by J. Garche et al., Elsevier, Amsterdam, 2009, Vol 4, p. 550.)

The four main electrode designs include conventional, tubular, expended metal, and fiber designs. It is intuitive that grid mesh size, pore size, and pitch are important parameters influencing the utilization of the active mass, distance to a current collector for electronic conduction, and formation of effective boundary between the active mass and electrolyte. The grid mesh size affects the average electron pathway for electronic conduction from the active mass to the grid current collector. The minimization of the electronic resistance, which is accomplished through short conductive pathways, is even more critical during later stages of the battery discharge and for higher depth-of-discharge (DOD). In those situations, the active mass converts to $PbSO_4$ and becomes an insulator, increasing the resistivity of the electrode. Grids with small mesh sizes contribute to overall higher conductivity of an electrode and minimize the resistive voltage drop even at higher DOD. This, in turn, enables better utilization of the electrode active mass (Figure 37).

Electrolyte

Contrary to other battery systems, the H_2SO_4 electrolyte of the LAB is a component of the active masses, and it takes part in the charge/discharge reactions. Not surprisingly, the cell voltage depends on the acid concentration and is affected by its concentration changes through the charge and discharge processes, as well as the electrolyte volume of the cell. It could be shown theoretically, based on the Faraday's law of electrolysis, that in the context of LABs 1 Ah of electricity corresponds to 3.7 g of H_2SO_4.

As a result of these complex interactions in the cell as part of the electrochemical reactions, the H_2SO_4 electrolyte concentration is optimized in relation to the cell voltage, the ionic conductivity, and the electrochemical stability (corrosion, passivation) of the active and inactive cell components. In the LAB terminology, the acid concentration is usually given as the relative density (i.e., ratio of the density of the H_2SO_4 solution in g/cm^3 to the density of water at the same temperature). It should also be apparent and not surprising that different types of batteries, with different performance expectations, electrode sizes, and serving different application, would have different requirements for the acid concentration (Figure 38).

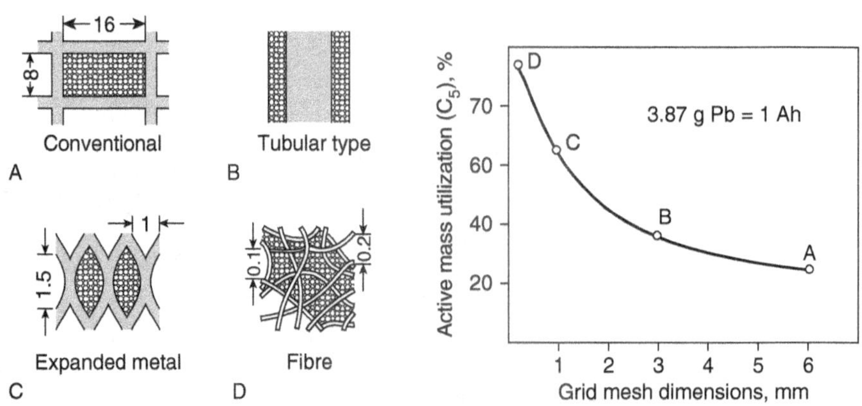

Figure 37 Schematics of the four different grid types (left) and graph of active mass utilization vs. grid mesh dimensions (right). (Reproduced from P. Faber, *Power Sources*, ed. by D. H. Collins, Vol 4, Oriel Press, Newcastel upon Tyne, 1974, pp. 525–538.)

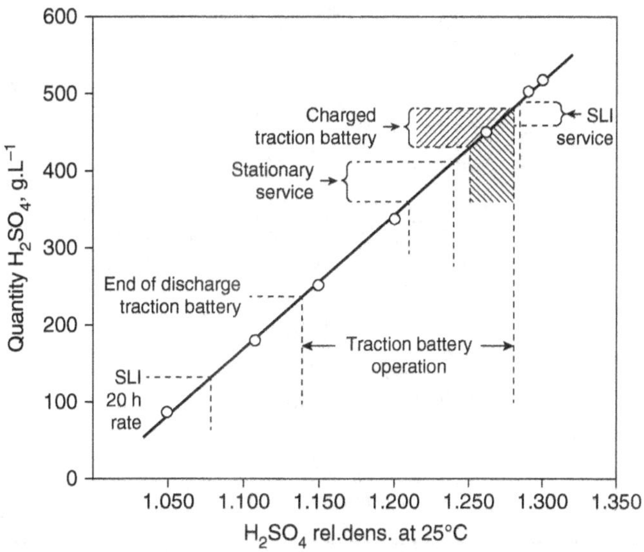

FIGURE 38 Sulfuric acid concentration ranges for different battery applications. (Reproduced from M. Barak, *Electrochemical Power Sources*, Peter Peregrinus Ltd., England, 1980, p. 159.)

The electrolyte must be of high purity to maintain the hydrogen and oxygen overvoltages high, and only distilled water should be used to refill water in flooded batteries. As previously explained, lowering hydrogen and oxygen evolution overvoltages encourage gas evolution and self-discharge.

During the charge process, pure H_2SO_4 is generated, and because of its high density it stratifies vertically in the cell and a concentration gradient is formed with high concentration at the bottom and lower concentration at the top of the battery container. Because the electrolyte concentration affects the electrode potential, a vertical potential gradient is also formed as a consequence of the electrolyte stratification. During the subsequent open-circuit phase, an equalization process ensues with the formation of $PbSO_4$ crystals (sulfatation) at the bottom where the acid concentration is higher (Figure 39). This can be envisioned as a current-less discharge process.

The acid stratification process during charge effectively causes the discharge in the part of the electrode at the bottom of the container. The equalization takes place in the subsequent rest period in the open-circuit voltage regime, when the acid concentration becomes uniform again from top to bottom, while the electrode is more charged at the top than at the bottom. One method to prevent stratification involves deliberate overcharge to produce gases, which would stir and mix the electrolyte.

The issue of acid stratification has been effectively resolved for AGM and gel VRLA (Valve-Regulated Lead-Acid) batteries because these batteries do not have free liquid electrolyte and the stratification of the acid does not take place (Figure 40). There is only a minimal stratification and no additional overcharge to produce gas bubbles is necessary for electrolyte mixing, eliminating one factor that contributes to grid corrosion of AGM and gel battery electrodes.

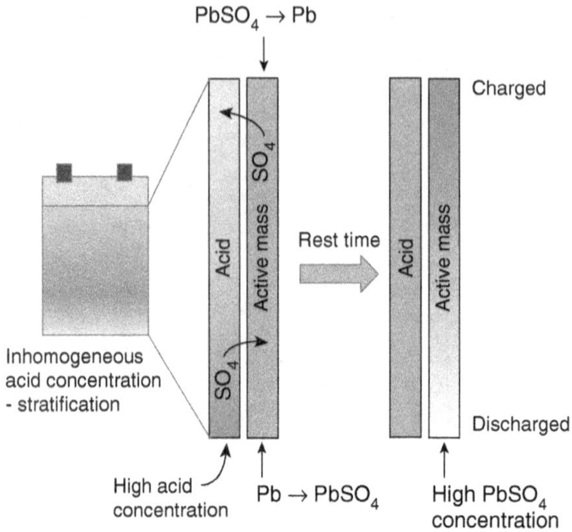

FIGURE 39 Schematic representation of the effects of stratification processes in LAB electrodes. (Adapted from J. Albers, E. Meissner, S. Shirazi, *J. Power Sources* 196, 2011, 3993–4002.)

Separator

The main functions of a separator in LABs are to prevent contact between the two electrodes, to conduct the electrolyte ions, to provide sufficient amount of electrolyte for charge/discharge reactions, to exert enough pressure on the active masses to reduce their shedding as a result of volume change during cycling ($PbO_2 \Leftrightarrow PbSO_4$, $Pb \Leftrightarrow PbSO_4$), and to prevent gas evolution.

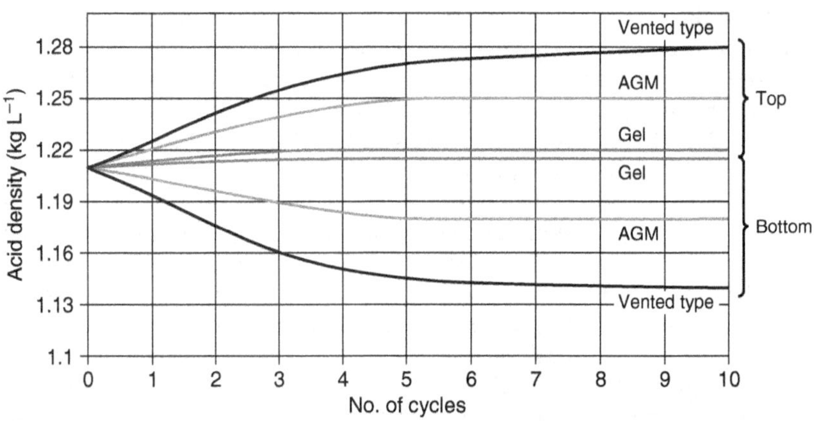

FIGURE 40 Electrolyte density vs. cycle number for different LAB types. (Reproduced from F. Kramm, H. Niepraschk, Valve-Regulated Batteries: Gel, in *Encyclopaedia of Electrochemical Power Sources*, ed. by J. Garche et al., Elsevier, Amsterdam, 2009, Vol 4, p. 727.)

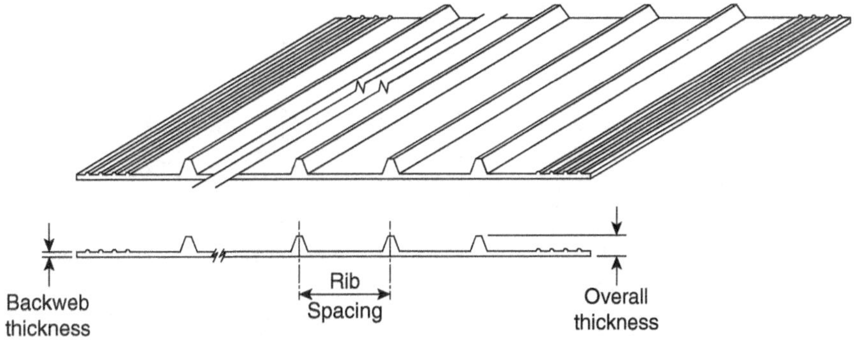

FIGURE 41 Separator and its cross-section for LABs. (Reproduced from W. Boehnstedt, J. K. Whear, Separators, in *Encyclopedia of Electrochemical Power Sources*, ed. by J. Garche et al., Elsevier, Amsterdam, 2009, Vol 4, p. 620.)

A separator is typically a polymer such as PVC (polyvinyl chloride) or PE (polyethylene) and is constructed with a so-called backweb, which prevents electrical shorts and ribs protruding in the central part of the structure, which exert a pressure on the active masses (Figure 41).

The backweb thickness is about 200 μm for PVC separators and 400 μm for PE separators; and it has different morphology on two sides or toward two electrodes because of different electrolyte volume requirements. On the positive electrode, the electrolyte volume must be about 1.6 times larger than at the negative electrode. While reactions in both active masses consume HSO^-_4 in the discharge, it is evident from the basic reaction equations that two H_2O molecules are formed at the positive electrode and the electrolyte becomes more diluted. Because of that, larger volume of electrolyte has to be accommodated by the separator and its ribs in the vicinity of the negative electrode.

Mechanical properties of a separator such as puncture resistance, tensile strength, elongation, and dimensional stability are critically important to ensure stable battery performance from cycle to cycle and prolong battery lifetime. Especially important is the puncture resistance because of some electrode designs that involve expanded metal grids. In general, separator properties of interest for a successful design are type of material, thickness, porosity, pore size, and pore shape.

As part of the overall ionic conduction path, a separator increases the ohmic resistance of the electrolyte by reducing the ion passageway area and its length because ions have to transmit charge through the pores and not freely in the electrolyte (Figure 42).

To minimize the resistive losses in the separator, a high porosity and a low tortuosity factor of a separator are preferred. On the other hand, such a design would increase the risk of short circuit between the electrodes due to particles shedding from the active masses and the dendrite growth. Dendrites are metallic particle that form in a branch-like shape, propagating from one electrode to another. As a result of the contradictory requirements, separator design usually reflects compromised features and moderate porosity and tortuosity, with pores of <1 μm diameter.

After the first separators made of wood, the material for separators evolved to polymers such as phenol-formaldehyde resin, polyvinyl chloride (PVC), and glass fiber leaf. The higher demands on a separator were created by the introduction of expanded-grid metal electrodes, which enable lower battery weight and material cost savings but

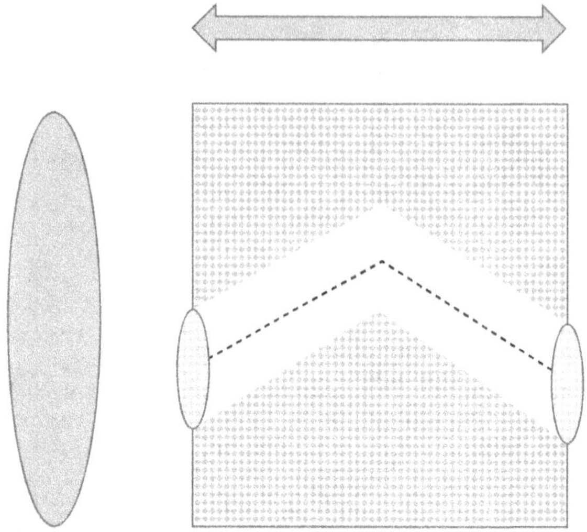

Figure 42 Simplified scheme of the separator pore demonstrating passageway area and length (patterned small area of the pore opening), compared to normal area and distance for ion transport in the electrolyte (darker-shaded surface area and length).

contribute to higher active mass shedding and higher risk of short circuit by particles called the burrs. The size of burrs of only several microns could result in their growth through the separator from positive to negative electrode.

Polyethylene (PE) has high microporosity and a combination of properties best suited for reducing the risk of electrical shorts by the metal burrs, while providing sufficient electrolyte pathways. Polyethylene separators are also flexible and could be easily mechanically joined to form a three-sided envelope or pouch, which can enclose the electrodes and prevent wandering of the shedding particles by collecting them at the bottom.

In AGM VRLA batteries, the separator is made of glass fiber. For gel VRLA batteries, which do not contain free liquid electrolyte, separators are still necessary and cross-linked phenolic formaldehyde resin or microporous PVC separators are used. Both types offer a relatively high porosity and a low ionic resistance.

A separator should exhibit good electrical, mechanical, and chemical stability in the course of battery lifetime, and it should be completely inert in the electrolyte, that is, no chemical species should be emitted that can degrade battery performance.

Lead-Acid Battery Types

Flooded Lead-Acid Batteries

As previously shown, LABs exhibit gas generation in the charge phase and during the open-circuit phase as well. This type of a cell cannot be sealed because gas from water decomposition has to leave the cell. There is an excess of electrolyte in the cell to compensate for water loss, and consequently this type of cell is called a flooded cell. The electrolyte level should always be maintained above the electrodes to avoid their damage. Sometimes, this type is also called a vented battery because the cell container is not sealed.

Valve-Regulated Lead-Acid Batteries (VRLABs)

A particular complex reaction known as oxygen-cycle offers a possibility to realize a sealed lead-acid cell by recombining oxygen produced on the positive electrode in the overcharge phase. The two essential requirements for effective oxygen recombination reaction are fast oxygen gas transport from the positive electrode through the electrolyte and to the negative electrode, and fast chemical reaction of oxygen at the electrode. In most cases, the overall rate of recombination reaction is low and is controlled by the low oxygen solubility in the electrolyte. Methods to improve oxygen transport rate through the electrolyte include additives to form a thixotropic gel, which over time starts "flowing" under the effects of shaking or agitation. The additives typically used include about 6% of high surface (~600 $m^2 g^1$) SiO_2 or Al_2O_3 powder. An alternative is an absorbent-glass-mat (AGM), which is partially filled with electrolyte.

In both cases (SiO_2, AGM), the electrolyte is absorbed on the surface of the SiO_2 powder or the glass fiber and immobilized while forming gas channels for the oxygen transport (Figure 43).

The oxygen transport rate in the electrolyte and the overall reduction rate depend on the gas channel area and the degree of channel or pore saturation with electrolyte. It should be intuitive that the amount of electrolyte in the separator pores and in the active mass has to be optimal. Too much electrolyte means that the pores would be flooded, slowing oxygen transport, while not enough electrolyte means that insufficient active mass is in contact with the electrolyte, affecting both ionic conductivity and oxygen reduction rates. Compared with flooded cells, the VRLA batteries have a very small amount of sulfuric acid electrolyte relative to active mass.

If the electrolyte is immobilized with SiO_2, the battery is called a gel battery, and in case of immobilization with absorbent glass mat the battery is called an AGM battery. AGM valve release lead-acid batteries (VRLABs) enable shorter distances between the electrodes than gel VRLABs and therefore better power, which is required for SLI batteries. On the other hand, active mass is more stable in gel batteries; it contains more electrolyte and normally does not undergo premature drying. This battery has a better lifetime and is used mostly for stationary applications.

The electrode is designed to minimize the gas generation during charge and overcharge by using a PbCa alloy for the grid instead of antimony. This grid material

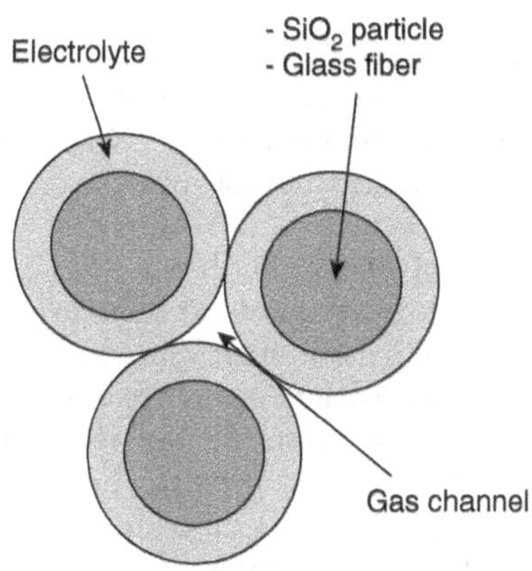

FIGURE 43 Cross-section of SiO_2 particle and glass fiber with absorbed electrolyte.

does not affect hydrogen and oxygen overvoltages, but the absence of antimony leads to week connection of the positive active mass to the grid and battery lifetime reduction at deep discharge. For this reason, the DOD of VRLA cells in cycling applications is mostly restricted to 60%.

During the life of a battery, hydrogen is also generated in a cell but cannot be recombined as in the case of oxygen due to extremely slow reaction. Although hydrogen is evolved only in small quantities, a pressure builds up in the cell over time and VRLB cells are equipped with a gas release valve in the lid of the cell to release a pressure if larger than 50 kPa.

Another method is to use so-called recombination vent plugs in which oxygen and hydrogen react chemically on a precious metal catalyst to form water. The gas mixture has to be stoichiometric, otherwise one of the gases will be in excess.

$$2H_2 + O_2 \rightarrow 2H_2O \tag{35}$$

The VRLA batteries are sealed and do not contain any free liquid acid that could be spilled in case of misuse or damage to a battery. Even in case of battery rupture, only a small amount of acid can escape from the battery because the electrolyte is immobilized. Contrary to flooded LABs, the low risk of acid leakage classifies the VRLB battery as nonhazardous cargo by both air and sea.

The VRLA batteries cannot be made as dry-charge battery type, which would be activated by the addition of electrolyte, similar to dry-charged flooded batteries. Because of the immobilized electrolyte, VRLA batteries have a short shelf life of only about 2 years.

Lead-Acid Batteries for Automotive Applications

LABs used in automotive applications include SLI-flooded type, EFB type, valve-regulated AGM type, and valve-regulated gel type. Operating parameters for these four battery types are compared in Table 11.

SLI-Flooded Type

Flooded LABs have been overwhelmingly used for SLI applications for nearly a century, which makes this battery the most-produced battery in the world, with annual revenues of US $23 B (2015).

The single cells are constructed from one negative and one positive flat plates, with grid collectors and divided by a separator. These 2-V cell blocks are then placed in plastic containers and further assembled into six-cell stacks to form a battery monoblock case of 12 V (Figure 44). The capacity of these stacks is usually between 40 Ah and 110 Ah for cars, while for trucks two 12-V batteries are connected in series to form a 24-V system with capacity of 100 to 230 Ah.

Normally, the 12-V stacks or monoblocks contain a central gas volume in which the gases of all six cells are collected and then vented to the surroundings through a flame arrestor in order to protect the battery against external ignition.

The most important considerations for flooded LABs used in SLI applications are lifetime and high discharge rate (i.e., current). The lifetime is expected to be 5–7 years; however, it depends strongly on the temperature of operation. The discharge rate is particularly important at low temperature and when measured at 0°F (–18°C), it is called a cold cranking rate.

Parameter	Flat-Plate Flooded	Flooded Advanced EFB	Flat-Plate VRLA AGM	Spiral Cells VRLA AGM	Flat-Plate VRLA Gel
Typical weight, kg	17–19	20–21	20.5–21.5	21	21
Typical capacity, Ah	66–72	70	70	55	60
Specific power, W/kg	220–290	270–310	300–310	350–400	170–190
Specific energy, Wh/kg	44–46	40–42	40–42	33	34
Internal resist., mOhm	3.8–4.7	3.0–3.4	3.2–3.5	2.8	5.0
Self-discharge, %/month	3–6	3.3	3.0	2.0	2.0
Temperature range, °C	–30, 75	–30, 75	–30, 60	–30, 60	–30, 75
Cold cracking at –18°C	Good	Very good	Very good	Excellent	Weak
Lifetime, years	5–7	5–7	5–7	5–7	5–7
Cost, $/kWh	80–200	× 1.3	× 1.6	× 2.5	× 2.6
Application	SLI	Microhybrid	Microhybrid	Special veh.	Deep cycle

(Adapted from M. Gelbke, C. Mondoloni, Flooded Starting-Lighting-Ignition (SLI) and Enhanced Flooded Batteries (EFBs): State-of-the-Art, in *Lead-Acid Batteries for Future Automobiles*, ed. by J. Garche, E. Karden, P. T. Moseley, D. A. J. Rand, Elsevier, 2017.)

Table 11 Operating Parameters for Different Automotive Battery Types

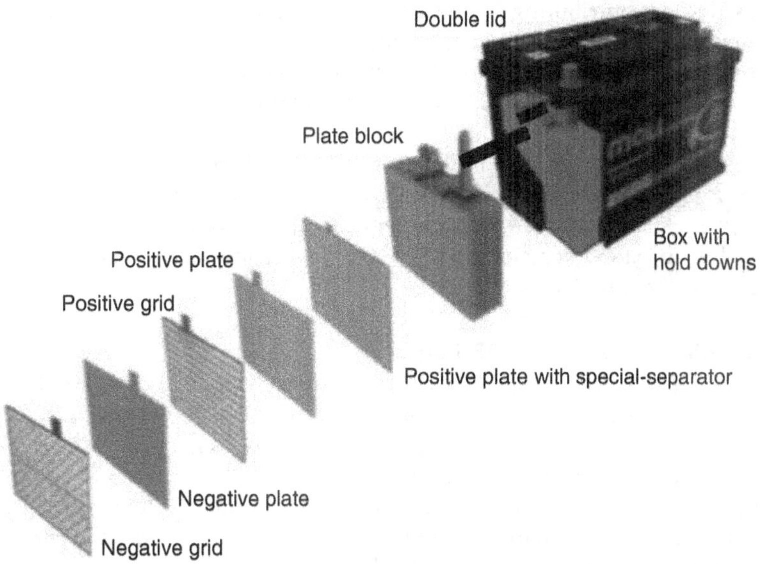

Figure 44 Exploded view of a modern SLI battery. (Courtesy of Akkumulatorenfabrik MOLL GmbH + Co. KG.)

EFB Type

The enhanced flooded battery (EFB) is an improved version of the SLI battery made in particular to meet the requirements of microhybrid EVs, which need to provide start-stop and brake energy recovery functionality as well as electrical motor assistance. The start-stop function involves deeper discharge of a battery than normal discharge in the start and ignition functions, while regenerative braking requires better high charge acceptance by the battery.

These requirements are not normally met by the conventional SLI batteries. AGM valve-regulated LABs are often able to provide deeper discharge and higher charge current acceptance, but they come at about 1.5 times the cost. Hence, the improvements have been made with the enhanced flooded battery such as active mass stability and thinner electrodes to provide higher power.

Another requirement for batteries used in microhybrid EV (electrical vehicle) applications is that the battery maintains continuous partial-state-of-charge (PSOC) to enable charge acceptance when braking. However, this leads to increased sulfation of the negative electrode because acid stratification is developed under the PSOC operating conditions. While stratification is normally reduced by the gas evolution in the full state-of-charge, certain additives (e.g., carbon) to the negative electrode active mass can also reduce the sulfation problem.

Valve-Regulated Adsorbed Glass Matt (VRLAB AGM) Type

The AGM VRLABs are characterized by reduced distance between the electrodes through the use of electrolyte immobilized in glass mat and lowering the electrolyte resistance. The use of electrolyte immobilized in glass mat also enables extension of electrodes since the electrolyte reservoir above the plates, used to store electrolyte and compensate for water loss, can be eliminated. Acid stratification is also reduced with this design contributing to reduced sulfation and longer lifetime. The cell plates are assembled under pressure, which inhibits shedding of active material and formation of lead sulfate dendrites. The AGM batteries also have a better charge acceptance. These characteristics make them ideal for microhybrid vehicle applications with regenerative braking charging.

The cells are available in flat plate and spiral wounded designs. They have a large surface area of the electrodes. Stack compression reduces the internal resistance and helps cycle life and good low temperature rate or cranking.

The AGM battery has captured a relatively low part of the LAB market primarily because of the high cost.

Valve-Regulated Gel (VRLAB Gel) Type

The gel VRLA battery uses a separator between the electrodes and because of that has higher internal resistance (compared with AGM-type batteries), which means lower power and limited automotive applications. However, the battery has excellent lifetime because electrolyte stratification has been eliminated and offers a reliable solution for applications that require good lifetime and light traction (e.g., golf carts, airport transport vehicles, wheelchairs, forklifts, etc.), as well as for energy storage applications such as emergency power systems and solar PV systems.

Gel-type batteries are available only with flat electrodes.

Lead-Acid Batteries for Industrial Applications

The most important performance characteristic for industrial applications is the lifetime of a battery. The factors that affect lifetime are grid corrosion and positive mass shedding because of acid stratification. Furthermore, undesirable reactions in flooded batteries such as electrolyte decomposition and drying out can be solved by refilling with water or completely avoided by use of VRLA batteries of gel type.

Vented, VRLA AGM, and gel-type batteries are used in industrial applications.

Stand-By Applications

Stand-by applications, for example, telecommunications and uninterruptible power supplies (UPSs), require high reliability over a long lifetime. These batteries are expected to deliver energy in an emergency situation, which could occur relatively frequently or not at all in the lifetime of a battery. The assumed DOD is low, and, in most cases, there would be only a limited number of discharge episodes longer than 1 hour. Under the condition of low DOD, batteries would have a high cycle life. One of the important requirements for this application is the expectation of the battery SOC (state-of-charge) of 100% at all times, which means that "float" charge is necessary to compensate for the self-discharge.

The lifetime of batteries used for stand-by and emergency applications is largely determined by the grid corrosion reactions. Additives play an important role in minimizing the rate of corrosion reaction and to that point, antimony alloys are avoided in favor of pure Pb, PbCaSn, and PbSn alloys, which provide better stability toward corrosion reactions but compromise mechanical characteristics of the electrodes. The lifetime for these batteries is also affected by the grid thickness, and it is generally assumed that 4–6-mm grid thickness is required to ensure 10–12-year lifetime at 20°C.

The LAB types considered for stand-by and emergency applications include vented batteries of Planté type (thick electrodes and pure lead), which provide up to 25 years lifetime; flat-pate and tubular electrode cells; VRLA AGM- and gel-type cells; and a so-called UltraBattery, comprising a negative hybrid electrode (i.e., carbon for power and lead for energy) and a conventional PbO_2 electrode (Figure 45).

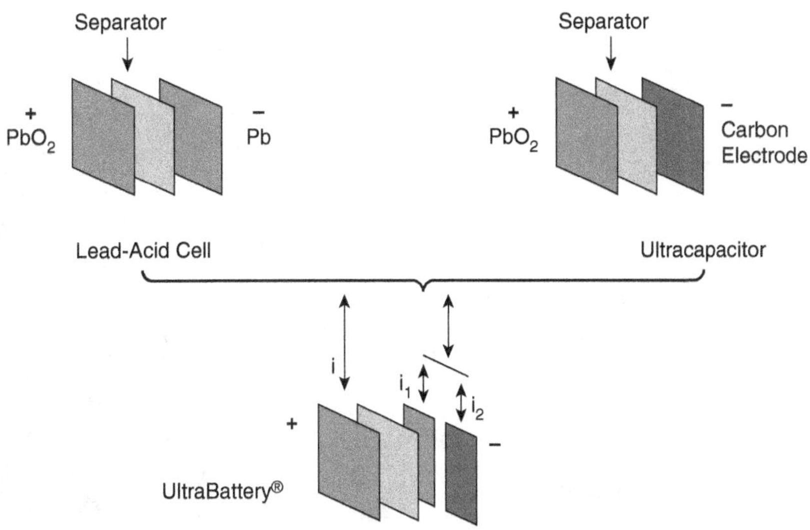

FIGURE 45 Expanded view of UltraBattery, as a combination of conventional LAB and supercapacitor. The negative electrode is a hybrid electrode of lead and carbon.

Performance Factors

Capacity and Energy

The theoretical specific capacity of LABs is 83 Ah g^{-1} and the specific energy is 167 Wh kg^{-1}. In practical cells, the theoretical values are reduced by limited mass utilization and inactive cell components (Figure 46).

The practical specific energy depends on the LAB type and application. For stationary LABs, the energy is roughly about 25 Wh kg^{-1} and for SLI LABs it is 40 Wh kg^{-1}. Besides the battery type, the specific capacity and energy depend strongly on current and temperature (Figure 47).

Two key principles of battery performance are shown in the graph. The higher temperature and lower discharge current result in higher battery capacity. In turn, higher capacity at lower discharge rate (i.e., longer discharge time) results in higher consumption of H_2SO_4 and at the end of charge the acid density is lower and the freezing point of water is higher. This has the consequence of ice formation and the discharge process is finished.

The effect of discharge current on the capacity is given by the previously discussed Peukert equation and Peukert coefficient in the range of 1.2–1.4. The mechanistic reason for the reduced capacity at higher discharge currents is the reduced active mass utilization, which is proportional to $PbSO_4$ concentration at the end-of-discharge.

It is observed from Figure 48 that the concentration of $PbSO_4$ in the electrode is uniform for low discharge currents (long discharge times), but as the discharge current increases it is evident that the discharge is too fast and there is not enough time for the active mass conversion to $PbSO_4$ to take place in the center of the electrode before the outer regions of the electrode are converted and the current cannot be supported, causing end-of-discharge.

It can be envisioned mechanistically that the beginning of discharge of a porous electrode depends on the relation between the ionic conductivity of the electrolyte in

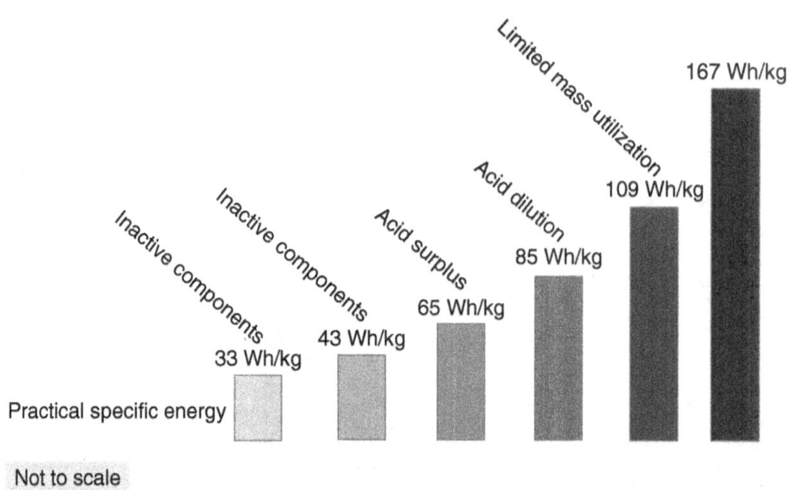

Figure 46 Average values for specific energy reduction and reduction factors in LAB systems.

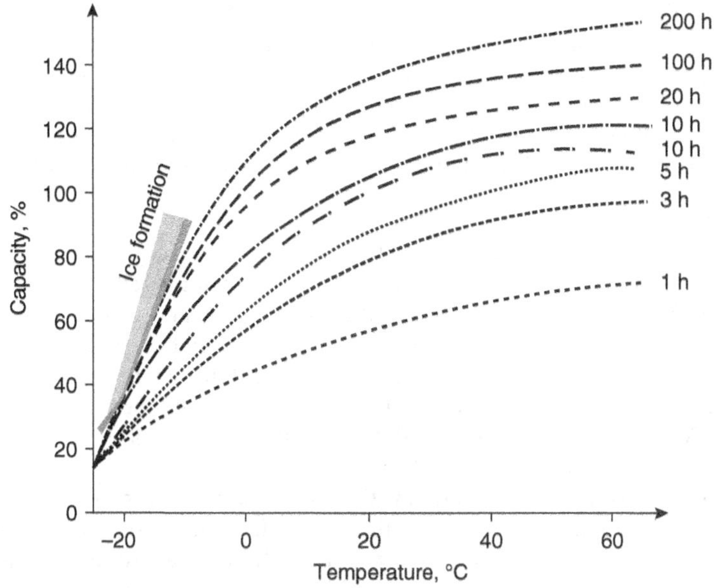

FIGURE 47 Available capacity of an SLI battery (12 V, 80 Ah @ C/20) at different discharge times from 1 to 200 h (1 h, 80 A; 10 h, 8 A; 100 h, 0.8 A) and for temperature range of –20 to 60°C.

the pores (σ_i) and the electronic conductivity of the active mass (σ_e). If the ionic conductivity in the pores is greater ($\sigma_i > \sigma_e$), then the discharge starts at the grid in the middle of the electrode; and if the electronic conductivity is greater ($\sigma_e > \sigma_i$), then the discharge starts at the electrode surface. This phenomenon appears to be universal but has been observed with certainty at the LAB electrodes.

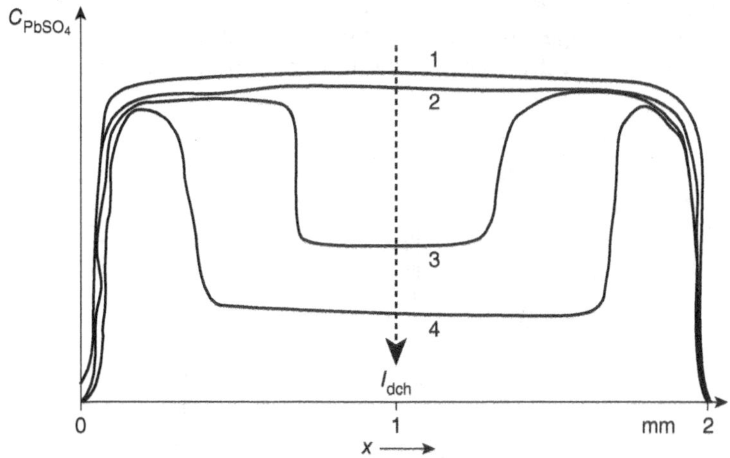

FIGURE 48 $PbSO_4$ concentration distribution across the positive plate thickness (2 mm) at the end-of-discharge, for different discharge currents and times: (1) I_1 = 1.5 mAcm^{-2}, 20 h, (2) I_2 = 6 mAcm^{-2}, 5 h, (3) I_3 = 30 mAcm^{-2}, 1 h, and (4) I_4 = 180 mAcm^{-2}, 10 min. (Reproduced from H. Bode, H. Panesar, E. Voss, *Die Naturwissenschaften*, 55, 1968, 541–542.)

For high discharge currents, large amounts of $PbSO_4$ are formed at the surface, in particular at the pore openings; and since lead sulfate molecules are larger than Pb or PbO_2, the result is that the pores become blocked, triggering the end-of-discharge. As for the effect of temperature, the reduced capacity at lower temperatures is caused mostly by the reduction of the acid diffusion rate. This causes a drop in specific ionic conductivity, ohmic resistivity increase and overall lower energy of the battery.

Power

The specific power of LABs (W kg^{-1}) depends on the overall internal resistance, which includes electrode polarization (due to charge transfer reaction) and ohmic resistance. Both of these phenomena, opposing the current flow and causing voltage drop (when resistance is multiplied by the current), are affected by the cell design. Cells with tubular plates have a longer lifespan, but hindered acid diffusion and their specific power in the range from 150 to 200 W kg^{-1}. Conventional SLI batteries feature specific energy from 600 to 800 W kg^{-1}; and special design batteries such as spiral-wounded AGM batteries are capable of reaching 2000 W kg^{-1}.

The battery power depends on the electrode thickness and more specifically on the amount of active mass close to the electrode surface, in good contact with electrolyte and immediately available for reaction. Thinner electrodes have a higher ratio of surface area of the active mass exposed directly to electrolyte to the amount of active mass. This also means that the average diffusion length of $PbSO_4$ particles into the active mass is shorter and that the battery has high active mass utilization and fast reaction, resulting in a higher battery power. Based on the active mass conversion to $PbSO_4$, a thin electrode, for example, 0.4 mm (Figure 49), exhibits high active mass utilization over the whole thickness, in contrast to a 2 mm electrode (Figure 48).

The thin electrodes are designed for high-power applications where energy content is less important. One long-term disadvantage of thin-design electrodes is a reduced lifetime caused by particle shedding. Cells made with these electrodes are sometimes called thin metal foil cells (TMF). Commercial battery examples of these cells include REBEL, INSPIRE, and BOULDER (Figure 50).

The construction entails coating a 0.05 mm active mass layer on a lead-foil and using the whole upper edge of the foil for electrical contact to improve electronic conductivity from the active mass to the terminal (Figure 51). Because of the structural

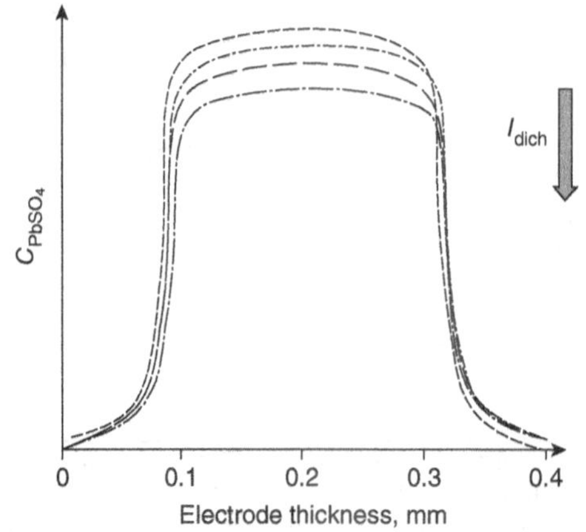

FIGURE 49 $PbSO_4$ distribution profile for thin design electrode.

Figure 50 TMF 12-V starting battery by REBEL.

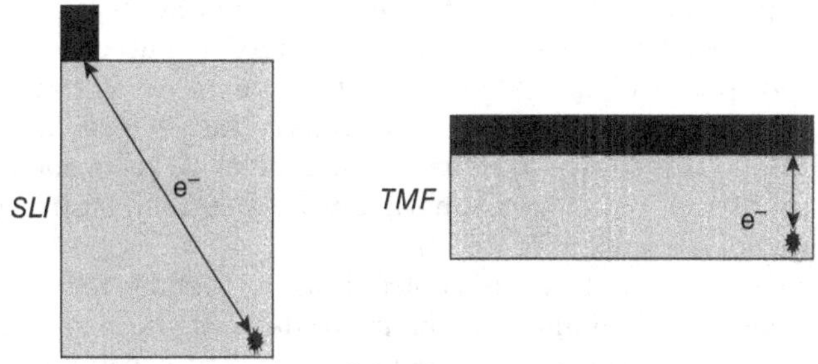

Figure 51 Schematic comparison of TMF and traditional SLI electrode designs showing the longest electron path from the active mass to contact tab or terminal.

advantage, such cells are smaller than the traditional SLI batteries and can be discharged at very high currents. But the lifetime of these batteries is shorter because of insufficient corrosion resistance of lead foil.

Self-Discharge

Battery cells with thermodynamic instabilities are predisposed to self-discharge because of the reaction between the active masses and the electrolyte. Typical self-discharge for LABs at 25°C is between 3% and 5% per month. The rate of the self-discharge reactions depends on the temperature, the composition of the grid material, and the state-of-charge. Grids made of PbSb alloy show a higher self-discharge rate because of the lower hydrogen overvoltage and lower SOC, which determines the H_2SO_4 concentration (Figure 52).

For most batteries, charging is necessary if the capacity falls between 60% and 80% of the full capacity, but if a battery is discharged to less than 60% through self-discharge the battery may have irreversible capacity loss.

Lifetime

As with other batteries, the calendar life of LABs is measured in years or months and for cycling applications also as the number of cycles before the capacity drops below 80% of the nominal capacity. For certain LAB applications, the usable life may be extended until capacity falls to less than 60% of the nominal.

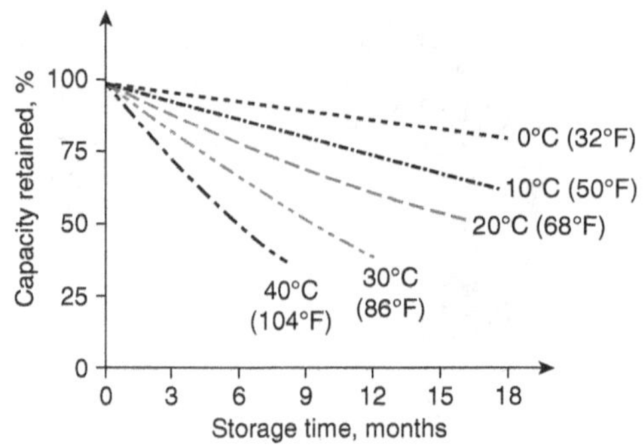

FIGURE 52 Self-discharge of VRLA battery. (Courtesy of POWER-SONIC.)

The lifetime depends on the type of battery, application type, temperature, and DOD. The effect of DOD on average number of cycles is given for a number of different batteries and types of applications (Figure 53).

It is apparent from the graph that the consumer SLI batteries are not designed for long cycling operations. Years ago, the SLI battery was used in developing countries as a battery for photovoltaic systems achieving only a small number of cycles. In stark contrast to that, an advanced industrial VRLA battery in renewable energy applications (10–70% DOD) was shown to achieve a lifetime of 17 years and over 3000 cycles. The lifetime of automotive batteries is normally 5–7 years and is strongly dependent on the temperature of operation (Figure 54).

The most important factors contributing to lifetime reduction in LABs are grid corrosion and plate sulfation. Secondary or derived effects include formation of dendrites and electrical shorts, electrolyte loss (in VRLA batteries), loosening of active materials, and passivation effects on electrodes. The lifetime of LABs is in general shorter than for other battery types because of the fundamental underlying reaction mechanism called solution-precipitation causing active materials to repeatedly dissolve in electrolyte and form solids with electrolyte components. This cycling process leads to mechanical, structural, and interfacial problems in electrodes; and over time causes active material detachment and capacity loss. The $PbSO_4$ is precipitated on the electrode during discharge from Pb^{2+}-ions generated on both electrodes and sulfate ion from the solution, SO_4^{-2} (Figure 55).

$$Pb^{2+} + SO_4^{2-} => PbSO_4 \tag{36}$$

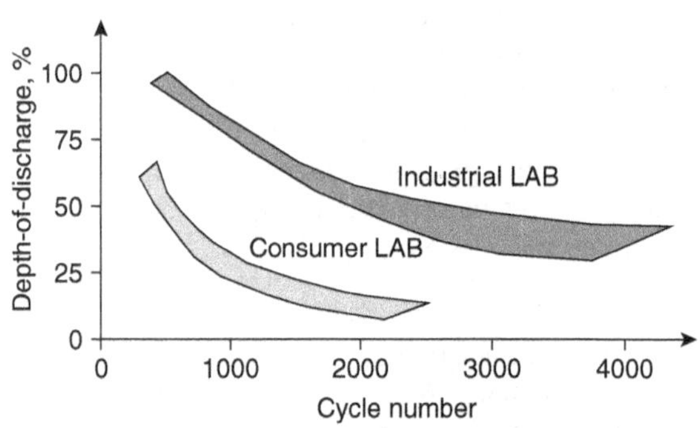

FIGURE 53 Average cycle number of consumer (mostly SLI) and industrial LABs as a function of DOD.

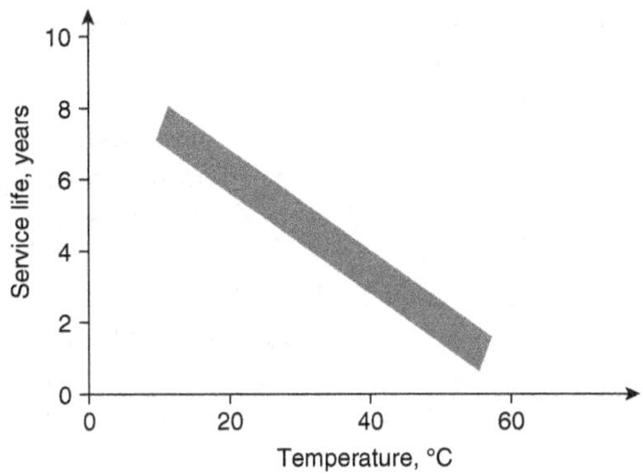

Figure 54 Average values for lifetime of SLI batteries as a function of temperature.

Promising improvements for lifetime extension include better charging process control; negative plate additives such as lignin or carbon; and positive plate design methods such as high-density active mass, plate compression, and corrosion resistant grids.

Safety

Under normal operating conditions, there are little or no safety problems with LABs, but improper charging or discharging as well as manufacturing defects may create disorder and departure from safe operation. The most noteworthy safety concerns are chemical hazard of sulfuric acid, toxicity of lead, and the risk of toxic gas (H_2S or SO_2) generation during charge.

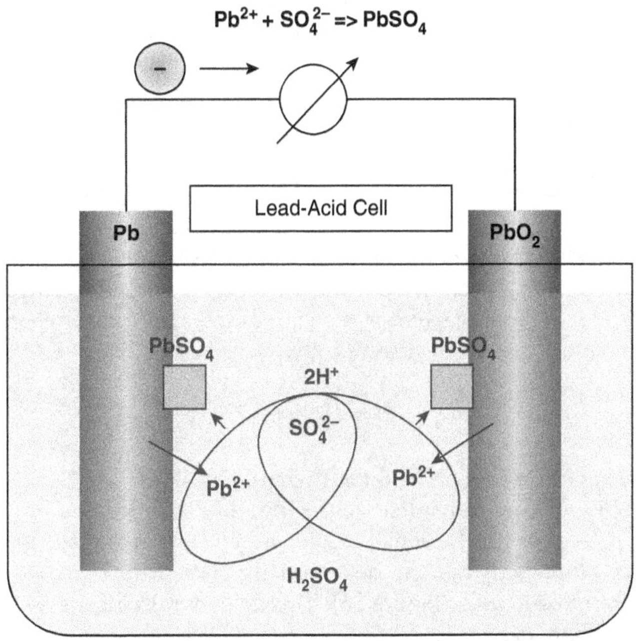

Figure 55 Illustration of the solution-precipitation reactions in LABs.

A less likely hazard is the formation of hydrogen and oxygen in the overcharge phase and subsequent explosive recombination. Because of the health safety concerns, LABs face scrutiny by the regulatory agencies. For example, EU regulations in the car industry prohibit the use of products containing toxic substances (2000/53/EC) and exceptions are allowed only if no other comparative products exist. The use of LAB in the car industry in EU countries is reviewed every 5 years.

Another type of safety risk is a thermal runaway, which occurs if the rate of heat generation is larger than heat dissipation and initiates temperature increase. The consequences may include container softening (at around 100°C), electrolyte evaporation (around 125°C), container melting and rupture (160°C), and even plastic container material fire along with the release of combustible gases. For acrylonitrile butadiene styrene (ABS), this can occur above 300°C and for polypropylene (PP) above 380°C. The thermal runaway is observed mainly in old AGM batteries during overcharge phase due to generation of gases in low electrolyte conditions. The effect in LABs is less pronounced than in the case of LIBs.

Cost

The cost of LABs is somewhat difficult to accurately establish because of the variety of battery types and different manufacturers. The cost for conventional SLI batteries is in the range from \$60 to \$170 kWh^{-1}. The cost of advanced batteries is naturally higher, for EFB about 1.3 times, for AGM VRLA about 1.6 times, and for gel VRLA about 2.8 times the cost for conventional. The more robust and reliable industrial batteries are also more expensive and in the range of \$200 to \$260 kWh^{-1}.

Charging

The charging of LAB should be as fast as possible without parasitic side reactions, at low charging voltages, and without reduction of the lifetime. To meet these three boundary conditions is difficult, especially at high charge currents.

Parasitic side reactions are the electrolyte electrolysis with gas generation (Equations 37 and 38).

$$2H^+ + 2e^- \rightarrow H_2 \qquad E° = 0\text{ V} \tag{37}$$

$$H_2O \rightarrow \frac{1}{2}O_2 + 2H^+ + 2e^- \qquad E° = 1.23\text{ V} \tag{38}$$

The total charge current (I_t) at the positive electrode is composed of the active mass charge current (I^+_{AM}) (Equation 39) and the oxygen evjolution current ($I^+_{O_2}$). At the negative electrode, the total current is the sum of the active mass charge current (I^-_{AM}) and the hydrogen development current ($I^-_{H_2}$).

$$I_t = I^+_{AM} + I^+_{O_2} = I^-_{AM} + I^-_{H_2} \tag{39}$$

At the beginning of charge for a discharged cell, that is, 0% SOC, the charge current is going only toward active mass charging, that is, $I_t = I^+_{AM} = I^-_{AM}$. With the increasing SOC, the potentials for the active mass charge reaction and the gas evolution reactions are getting closer and the gas development current increases while the active mass charge current decreases (Figure 56). This is shown in Figure 55 via the charge acceptance rate of both electrodes.

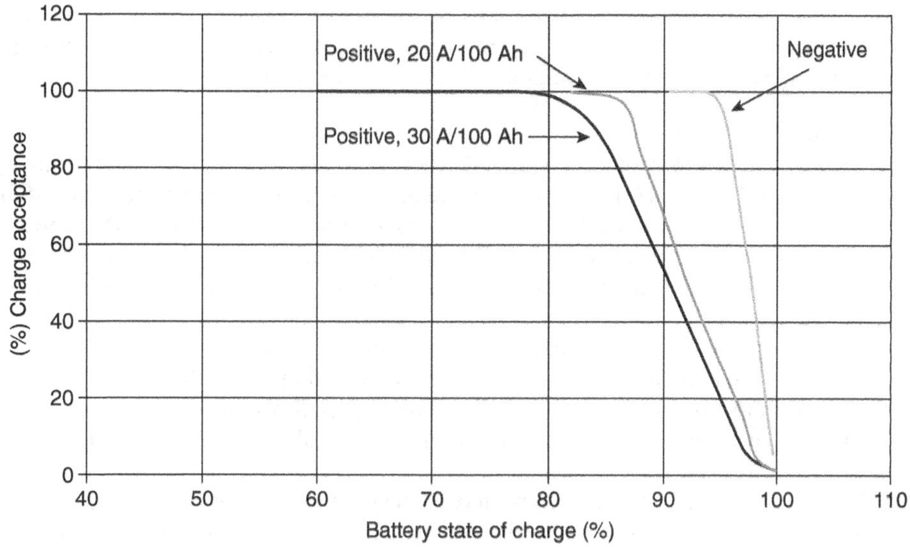

FIGURE 56 Charge acceptance of an LAB at two charge rates (20 A 100 Ah^{-1} and 30 A 100 Ah^{-1}) for the positive and the negative electrode at different SOCs. (Reproduced from S. S. Misra, Charging, in *Encyclopedia of Electrochemical Power Sources*, ed. by J. Garche et al., Elsevier, Amsterdam, 2009, Vol 4, p. 765.)

The charge acceptance rate is defined as $(I_t - I^+_{AM})/I_t$ for the positive electrode and $(I_t - I^-_{AM})/I_t$ for the negative electrode. If the charge acceptance is 100%, the total charge current is going into the active mass charge. Above 80% SOC, the charge acceptance is reduced and some of the current is used for gas evolution of oxygen at the positive electrode. At about 95% SOC at the negative electrode, the hydrogen evolution starts and consumes a portion of the current.

At the positive electrode, the beginning of the oxygen generation depends on the charge current. At lower specific charge current, for example, 20 A current for a 100-Ah battery capacity, the oxygen generation starts at about 85% SOC. Further reduction in the specific charge current shifts the point of oxygen generation to higher SOC as the active mass charge potentials decrease.

On the other hand, the hydrogen evolution on the negative electrode starts independent of the charge current.

As the gas evolution currents $(I^+_{O_2}, I^-_{H_2})$ increase, the active mass charging currents decrease and at some presumed full charge the active masses are still not fully charged, and overcharge is necessary to compensate for the charge loss.

The most common charge method comprises three stages: constant current, constant voltage, and a combination of constant voltage and constant current.

The first phase is performed at a constant current until a critical voltage of 2.35–2.4 V is reached. The second phase continues at the constant voltage (at the critical voltage) until a critical low current value is reached, typically 5% of the capacity, for example, 5 A for 100 Ah^{-1} (C/20). In the third stage, a small current, typically C/20, is applied while voltage increases. The third phase is also used for capacity equilization of different cells in a battery.

The VRLA batteries are charged using the same method, but the total charge current (I_t) is divided into active mass charge current (I^-_{AM}), hydrogen evolution current ($I^-_{H_2}$), and oxygen reduction current at the positive electrode ($I^+_{O_2}$).

Outlook

The advantages of LABs are low costs, simple manufacturing, relatively high specific power, good efficiency, large temperature operating range, good safety, and established recycling (ca. 95% of all LABs). The drawbacks, when compared to other modern batteries, are low specific energy, short lifetime, low charge currents (i.e., long charge time), need for storage at charged conditions, need for maintenance for flooded LABs, and use of toxic lead.

The balance between advantages and disadvantages is still in favor of LABs 130 years after the initial commercialization, and they securely dominate the market despite more advanced battery systems. Further applications where growth is expected for LABs include micro-HEVs (microhybrid electrical vehicles), especially if additional reduction in cost can be achieved.

The improvements can be expected especially through development of UltraBattery, based on the use of carbon additive and accomplishing a supercapacitor effect.

Further Reading

- D. Bernd, *Maintenance Free Batteries*, Research Studies Press, 2003.
- M. Broussely, G. Pistoia, *Industrial Applications of Batteries*, Elsevier, 2007.
- K. Bullock, A. Salkind, chap. 17, Valve Regulated Lead-Acid Batteries, in *Linden's Handbook of Batteries*, ed. by Th. Reddy, D. Linden, McGraw Hill, 4th ed., 2011.
- J. Garche, Ch. K. Dyer, P. T. Moseley, Z. Ogumi, D. A. J. Rand, B. Scrosati, *Encyclopedia of Electrochemical Power Sources*, Elsevier, 2009.
- J. Garche, E. Karden, P. T. Moseley, D. A. J. Rand, *Lead-Acid Batteries for Future Automobiles*, Elsevier, 2017.
- D. Pavlov, *Lead-Acid Batteries—Science and Technology*, Elsevier, 2011.
- D. A. J. Rand, P. T. Moseley, J. Garche, C. D. Parker, *Valve-Regulates Lead-Acid Batteries*, Elsevier, 2004.
- A. Salkind, G. Zguris, chap. 16, Lead-Acid Batteries, in *Linden's Handbook of Batteries*, ed. by Th. Reddy, D. Linden, McGraw Hill, 4th ed., 2011.

CHAPTER 3

Nickel Batteries

Nickel-Cadmium Batteries

Introduction

The battery with alkaline electrolyte was patented in 1899 by E. A. Junger. The original invention included use of alkaline electrolyte such as KOH or NaOH that does not change during cycling into insoluble oxides of finely "divided" metals. Two years later, in 1901, Junger patented the nickel-cadmium battery (Ni-Cd or NiCd), only two weeks before the nickel-iron (Ni-Fe) battery was patented in the United States by T. Edison.

In 1906, Junger established a factory close to Oskarshamn, Sweden, to produce flooded Ni-Cd batteries. In 1947, G. Neumann patented a sealed alkaline cell and established the foundation for small portable sealed Ni-Cd cells. For many years, Ni-Cd was the preferred secondary battery especially for consumer applications. With the introduction of Ni-MH cells and a ban on cadmium use in batteries by the EU, the production of Ni-Cd cells has been reduced.

Operating Principle

Cadmium is the negative active mass (anode) in Ni-Cd batteries and it reacts during the discharge process with hydroxyl ions, OH^-, from the electrolyte to form cadmium hydroxide, $Cd(OH)_2$ (Equation 40).

$$Cd + 2OH^- \rightarrow Cd(OH)_2 + 2e^- \qquad E^0 = -0.81\ V \tag{40}$$

On the positive electrode (cathode) the active mass is nickel oxyhydroxide, NiOOH, which is converted during the discharge to nickel hydroxide, $Ni(OH)_2$ (Equation 41).

$$NiO(OH) + H_2O + e^- \rightarrow Ni(OH)_2 + OH^- \qquad E^0 = +0.49\ V \tag{41}$$

The overall reaction shows cadmium and nickel oxyhydroxide as reactants during discharge that results in cadmium hydroxide and nickel hydroxide as products (Equation 42). Note that cadmium and nickel oxyhydroxide never get in a direct contact. Rather, each substance reacts on an electrode in an electrochemical reaction and the two half-cell reactions are tied together by the hydroxide ions and water. The OH^--ions take part in the reactions on the electrodes but in equal numbers that cancel each other, so OH^--ions are not showing in the overall reaction. The reactions are completely reversed during the charge process.

$$Cd + 2NiOOH + 2H_2O \rightarrow Cd(OH)_2 + 2Ni(OH)_2 \qquad E^0_{cell} = 1.30\ V \tag{42}$$

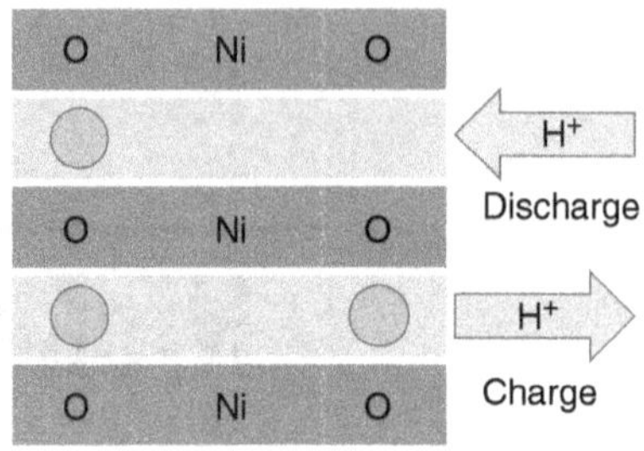

Figure 57 Schematic of proton intercalation during discharge and deintercalation during charge at the $NiOOH/Ni(OH)_2$ electrode.

The mechanisms of charge and discharge reactions on the electrodes are different on a molecular level. On the negative Cd electrode, the reactions proceed through solution-precipitation sequence, while those reactions on the positive Ni electrode have foundation in the intercalation process, a reversible inclusion or insertion of ions between NiO_2 layers (Figure 57).

The application of the Nernst equation on Ni-Cd battery reactions gives the cell voltage in the equilibrium at 25°C in a 5 M KOH solution and assuming the activity of water as 0.90 (Equation 43).

$$E = E^0 + \frac{RT}{nF}\log\frac{1}{[H_2O]^2} = [1.3\ +0.059\log 1.235]\ \mathrm{V} \tag{43}$$

During the charging process (Equation 42), water is consumed and its activity decreases making the term $1/[H_2O]$ and its logarithm larger, while the cell voltage increases. Near the end-of-charge it approaches a voltage at which water decomposition with gas evolution takes place. The overpotentials for hydrogen and oxygen generation on the Ni-Cd battery electrodes in KOH are smaller than in the case of the same reactions in lead-acid batteries (LABs) and, as a result, the reactions can occur at much lower cell voltage. This means that near the end-of-charge (roughly at 90% state of charge), water decomposition generating oxygen and hydrogen takes place in parallel with the main battery reactions. Once active mass is fully charged, the overcharge phase follows, and the only reactions, at pH = 14, are those of gas evolution (Equations 44 and 45).

Negative electrode:

$$4H_2O + 4e^- \rightarrow 2H_2 + 4OH^- \qquad E^\circ = -0.83\ \mathrm{V} \tag{44}$$

Positive electrode:

$$4OH^- \rightarrow O_2 + 2H_2O + 4e^- \qquad E^\circ = 0.40\ \mathrm{V} \tag{45}$$

Two important characteristics of Ni-Cd batteries, self-discharge and sequence of gas evolution, demonstrate the complexity of reactions that occur on the electrodes.

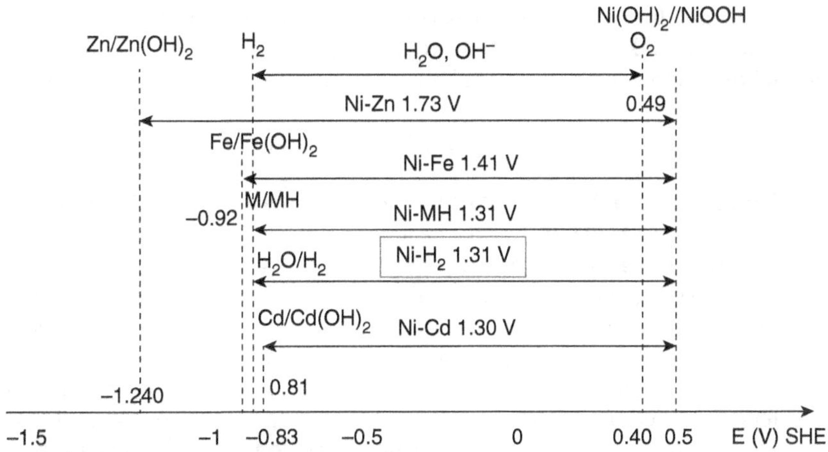

Figure 58 Thermodynamic potentials of electrodes (vs. SHE) in alkaline electrolyte of alkaline batteries. (Reproduced from P. Bernard, Nickel-Cadmium: Sealed, in *Encyclopedia of Electrochemical Power Sources*, ed. by J. Garche et al., Elsevier, Amsterdam, 2009, Vol 4, p. 459.)

Self-Discharge

The Ni-Cd batteries have the propensity for self-discharge reactions in the rest phase (without being connected to a load). The reasons for high rate of self-discharge can be discerned from the relative position of the reaction voltages for the main battery reactions on the electrodes and for the side reactions (Figure 58).

The thermodynamic or theoretical potential for hydrogen reaction in alkaline solution is, at −0.83 V, slightly more negative than the potential of the Cd electrode (−0.81 V), but the practical potential is about 200 mV more negative because of the overvoltage. As a result, the Cd electrode is thermodynamically stable in alkaline solution and no hydrogen evolutions occur during the rest phase.

On the other hand, the potential for oxygen oxidation and evolution in alkaline solutions (5 M KOH) is at 0.40 V, lower than the potential of the NiOOH/Ni$(OH)_2$ electrode (0.49 V), making the positive electrode thermodynamically unstable in alkaline electrolytes and resulting in side reaction of oxygen production and nickel reduction (Equation 46).

$$2NiO(OH) + H_2O \rightarrow 2Ni(OH)_2 + \frac{1}{2}O_2 \tag{46}$$

A local electrochemical couple forms on the positive electrode from the oxygen evolution reaction and nickel reduction, but no electricity is produced, and no power is transmitted to a load (Equations 34 and 38). The rate of self-discharge is low because of the high oxygen overpotential, but nonetheless typically reaches 20–30% per month at room temperature.

Oxygen Cycle in Sealed Ni-Cd Batteries

The second important characteristic of Ni-Cd batteries is the sequence of gas evolution and recombination in the overcharge phase. The reactions of gas evolution result in pressure increase in batteries and prevent sealing of conventionally designed, flooded-electrolyte Ni-Cd batteries. During the charge phase, the main reaction on the negative electrode is cadmium oxidation to $Cd(OH)_2$ (Equation 40), but to produce gas, hydrogen oxidation also takes place close to full charge (Equation 44). The third reaction is the reduction of oxygen gas produced on the positive electrode (Equation 47).

$$\frac{1}{2}O_2 + H_2O + 2e^- \rightarrow 2OH^- \tag{47}$$

The ability of the negative electrode to support oxygen reduction (or recombination) reaction has been utilized to make sealed Ni-Cd batteries. It is based on a simple principle of oversizing the cathode active mass (by about 30%). The disproportion in the amount of active mass means that the charge process ends on the cathode and the oxygen evolution starts before the charge process is finished on the anode and hydrogen can be evolved. Oxygen produced on the cathode diffuses through the interspaces and the electrolyte to the anode, where it is reduced or recombined. The process sequence is called oxygen cycle and it is the principle of sealed Ni-Cd batteries.

The treatment of the oxygen reduction in sealed Ni-Cd batteries is similar to lead-acid batteries. The ratio of oxygen reduction rate to current, $v_{O_2\text{-red}}/I_{O_2\text{-red}}$, is limited by the diffusion of oxygen and the configuration (size and orientation) of the pores in the separator (Equation 48).

$$I_{O_2\text{-red}} = 4F\frac{D_{O_2}}{\delta}K_S\, p_{O_2}\, S \tag{48}$$

D_{O_2} is oxygen diffusion coefficient, K_s is Henry's solubility constant, p_{O_2} is oxygen partial pressure, S is the reaction surface area, and δ is the thickness of the diffusion layer. The reaction area and the diffusion layer thickness are strongly influenced by the gas pores in the immobilized electrolyte. A larger three-phase boundary creates favorable conditions for the oxygen reduction by enabling a diffusion layer thickness of about one-third of the radius of oxygen bubbles, which is roughly 100 times lower than the thickness in a liquid electrolyte (Figure 29).

The oxygen generation rate ($v_{O_2\text{-gen}}$) at the positive electrode in the overcharge phase is proportional to the charge current (I_{ch}). As long as the rate of the oxygen generation is lower than the rate of oxygen reduction, $v_{O_2\text{-gen}} \leq v_{O_2\text{-red}}$, no gas pressure increase occurs in the cell.

The presence of additional Cd in the negative electrode active mass ensures that Cd formed from $Cd(OH)_2$ during the charge is continuously converted back to $Cd(OH)_2$ in the chemical reaction with oxygen. In that way, Cd electrode is never fully charged, which prevents hydrogen evolution.

There is one additional condition for ensuring problem-free performance of a sealed Ni-Cd battery. Once all NiOOH is converted in the overdischarge phase, hydrogen evolution on the positive electrode can occur (Equation 44).

$$4H_2O + 4e^- \rightarrow 2H_2 + 4OH^- \tag{44}$$

The solution has been found in adding a small amount, about 15% of the full capacity, of $Cd(OH)_2$, a so-called antipolar mass to the positive electrode (Equation 49). In the presence of the antipolar mass, after the end-of-charge stage, no hydrogen is generated because the current is used to reduce $Cd(OH)_2$ back to Cd, which takes place at a less negative potential (Figure 59).

$$Cd(OH)_2 + 2e^- \rightarrow Cd + 2OH^- \tag{49}$$

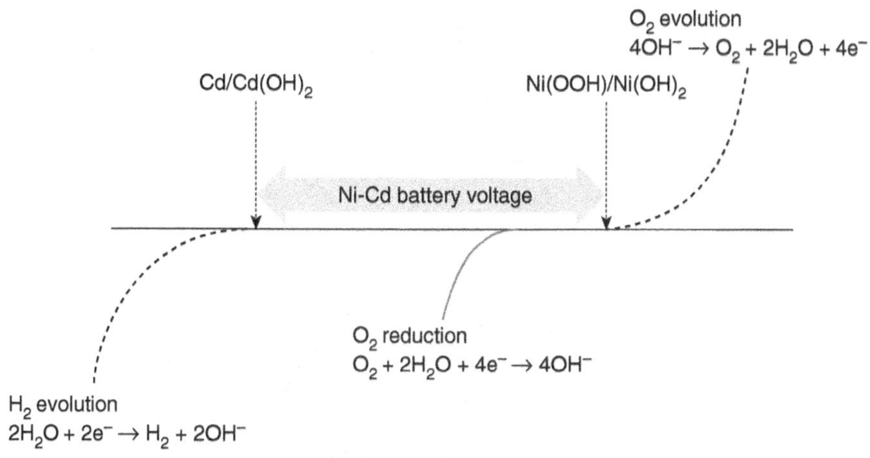

FIGURE 59 Relative potentials for electrochemical reactions taking place in Ni-Cd batteries.

In practice, the antipolar mass is used in today's Ni-Cd batteries only in hermetically sealed coin cells where hydrogen generation has the potential to destroy the cell, while larger sealed cells have a vent for hydrogen release in case of pressure rise.

Components

Positive Electrode

The active mass of the positive electrode is NiOOH in the charged state and $Ni(OH)_2$ in the discharged state. The positive electrode's theoretical capacity is 289 mAh g^{-1}. The mechanism of conversion during charge and discharge reactions involves the exchange of one electron.

Nickel electrode undergoes rather complex changes in the crystallographic phases during the charge/discharge reactions. While the elemental composition is always the same, either NiOOH or $Ni(OH)_2$, a crystallographic phase, α, β, or γ, can quite dramatically change its behavior and performance based on the differences in atomic distances and orientation.

The mechanism of reaction on the cathode involves ion intercalation or insertion between the layers, which critically depends on the interlayer distance (Figure 60). In turn, the rate of reaction depends also on the phase and interlayer distance, whereby the smaller distance leads to lower capacity.

The crystallographic phase has further effect on the mechanical properties of electrodes. It has been shown that phase transition from β-NiOOH to γ-NiOOH during overcharge contributes to electrode swelling and mechanical stress, and in turn reduces electrode lifetime.

The nickel electrode active mass in the discharged state, $Ni(OH)_2$, has very low electronic conductivity for certain electrode types such as foam or plastic-bonded and has obviously a diminished performance. Additives such as cobalt are used to improve the conductivity.

There are different types of electrodes based on the current collector and the way active mass is impregnated into the electrode.

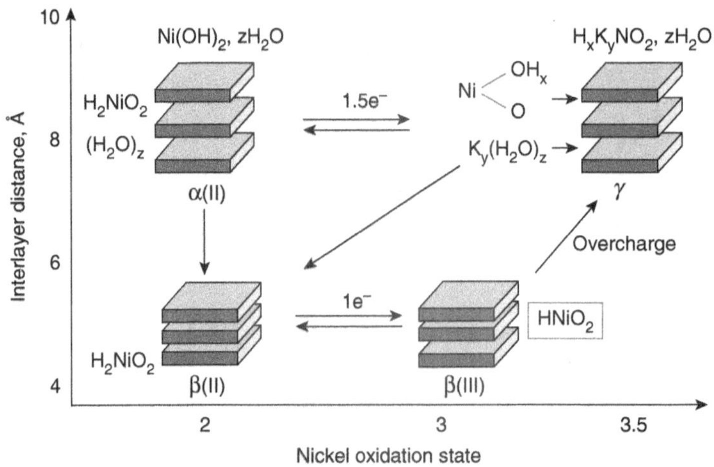

FIGURE 60 Variations of interlayer distance for oxyhydroxide and hydroxide nickel phases. (Adapted from P. Bernard, Nickel-Cadmium: Sealed, in *Encyclopedia of Electrochemical Power Sources*, ed. by J. Garche et al., Elsevier, Amsterdam, 2009, Vol 4, p. 459.)

"Pocket" plates were used in the early days of Ni-Cd production. The active mass was surrounded by pockets made from perforated nickel-plated steel sheets. But the "pockets" could not put enough pressure on the active mass to enable mechanical and electronic consistency and long service life (Figure 61).

Tubular plates were introduced by Thomas Edison in 1908. To extend the electrode cycle life during deep discharge, the active mass was filled into the perforated nickel-plated mild-steel tubes. The tubular plate technology is no longer in use because of the high cost.

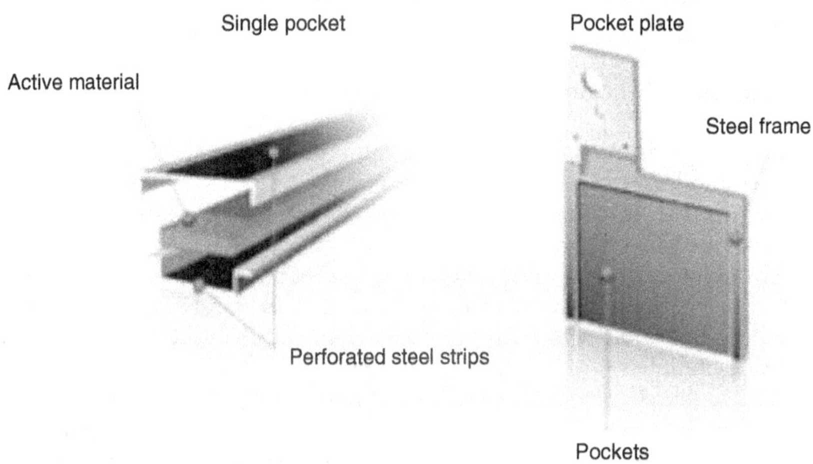

FIGURE 61 Illustration of "pocket" plate design. (Reproduced from P. Bernard, M. Lippert, chap. 14, Nickel-Cadmium and Nickel-Metal Hydride Battery Energy Storage, in *Electrochemical Energy Storage for Renewable Sources and Grid Balancing*, ed. by P. Moseley, J. Garche.)

Sintered-type electrodes are made by coating an aqueous paste of nickel powder onto a perforated nickel-plated steel sheet and treating (i.e., sintering) it at 800–1000°C. In addition to nickel salt, in the form of $Ni(NO_3)_2$, the paste contains $Ni(OH)_2$ to enable precipitation of nickel in the pores.

This sintered current collector, that is, the matrix, has pores in the range of 5–20 μm in diameter and a porosity of 80%. It has excellent electronic conductivity for high-power applications. One disadvantage of sintered nickel electrodes compared to most non-sintered electrodes is the low weight ratio of the active material to the matrix, about 0.4–0.5.

Foam electrodes are prepared by electroless nickel plating on a polyurethane foam substrate and its subsequent removal at high temperature. The nickel foam has a porosity of about 95% and pore sizes in the range of 300–500 μm. The larger pore size leads to lower power density because of a longer electronic pathway for conduction. However, preparation is simplified because of easier pore filling with the active mass, which is accomplished by simple pasting.

The fiber electrode type, known as FNC (fiber-nickel-cadmium), is manufactured by electroless nickel plating on synthetic fiber material and subsequent sintering under compression at 800°C and in a hydrogen atmosphere. The final metalized fiber matrix has 90% free volume and thickness in the range of 0.6–10 mm. The impregnation of the porous structure with active material is a straightforward process. The fiber electrodes have the advantages over the conventional sintered electrodes because of their compactness and structural integrity, better specific energy and life, and lower cost.

Plastic-bonded electrodes are made by adding a polymer binder, such as Teflon (PTFE), to the active mass and pressing it onto an expanded metal sheet or mesh made of nickel. These electrodes contain a conductive additive as well, usually graphite. The electrodes are low cost and easy to prepare; and these are usually made in a thicker form for energy applications.

Negative Electrode

The active mass of the negative electrode is metallic cadmium. Its theoretical capacity is 477 Ah kg^{-1}, but in practice much lower capacities are achieved because of the parasitic oxygen recombination and loss of metallic cadmium. The electrode electronic conductivity is reduced with the loss of metallic cadmium because $Cd(OH)_2$ is nonconductive.

The discharge reaction of the active mass from Cd to $Cd(OH)_2$ proceeds in alkaline solution via a sequence of soluble intermediate complex hydroxide ions $Cd(OH)^+$, $Cd(OH)_3^-$, and $Cd(OH)_4^-$. The soluble Cd^{2+}-ions can migrate from the bulk electrode toward the surface and further into the separator and nickel electrode. Precipitation of $Cd(OH)_2$ in the separator blocks the pores and reduces free electrolyte flow and oxygen diffusion through the pores, resulting in lower oxygen recombination efficiency in sealed cells. Furthermore, the surface area of electrochemical active mass is reduced and a layer of inactive $Cd(OH)_2$ is formed on the electrode surface because of the pore blockage at the electrode/separator interphase. During the charge, $Cd(OH)_2$ in the separator pores converts to metallic Cd and enables formation of dendrites, self-discharge, and a short-circuit danger. The solubility of complex hydroxide ions is further reduced at low temperature leading to limiting discharge current and lowering performance. The effect is negligible for temperatures as low as –20°C and good performance is maintained until about –40°C.

Same electrode manufacturing techniques, previously described for the positive electrode, are used to prepare negative electrodes. The difference is of course that $Cd/Cd(OH)_2$ active mass is used to impregnate the current collectors.

Electrolyte

The electrolyte used in Ni-Cd cells is 5–9 M KOH solution. The addition of NaOH and LiOH improves high-temperature performance of the positive electrode but has negative impact on the low-temperature behavior of the negative electrode. Concentration of NaOH above 9 M supports higher capacity of the positive electrode because of crystalline phase shift from γ-NiOOH to α-$Ni(OH)_2$, but is associated with a significant lifetime reduction.

The amount of electrolyte is also important, especially in sealed cells. If the amount of electrolyte is too low, the internal resistance increases. If it is too high the separator pores fill with electrolyte and hinder oxygen gas transport, which increases gas pressure. The optimal electrolyte amount is related to battery capacity in the range of 1.5–2.5 mL Ah^{-1}.

Although there is no change in KOH electrolyte concentration in Ni-Cd cells during charge and discharge, there is localized imbalance when water formation (at a rate of 0.67 mL/Ah) takes place during discharge and localized dilution inside the active mass pores occurs. With time, the difference between the KOH concentration in the bulk and inside the pores is equalized by the means of diffusion, but since diffusion rates are reduced at lower temperatures the KOH concentration inside the pores drops due to dilution below the electrolyte freezing point. An opposite effect of imbalanced electrolyte concentration between the bulk electrolyte and inside the active mass pores happens during discharge at high current when the concentration increase in the pores causes precipitation of solid KOH. Both of these negatively impact the electrical performance of the cells.

Separator

Two types of separator materials used in Ni-Cd batteries are polyamide and polyolefin. Both types are nonwoven, bonded by meshing short and long fibers in chemical, mechanical, heat, or solvent treatments. These materials are characterized by randomly distributed pores and no pinholes (Figure 62).

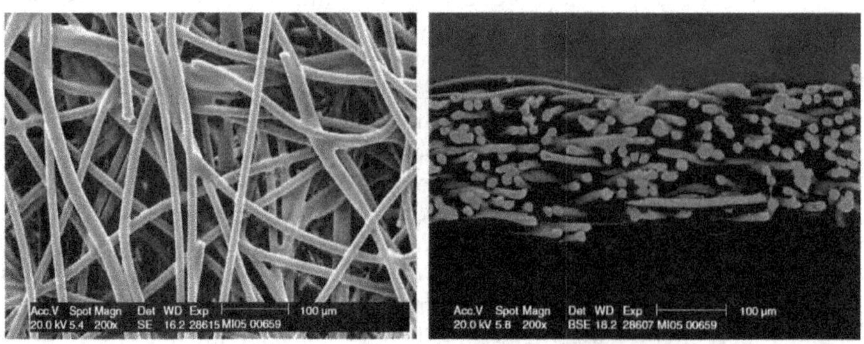

Figure 62 Scanning electron micrographs of nonwoven polyamide separators used in Ni-Cd cells. (Reproduced from J. Cook, P. Kritzer, Alkaline Battery Separators, in *Encyclopedia of Electrochemical Power Sources*, ed. by J. Garche et al., Elsevier, Amsterdam, 2009, Vol 4, p. 424.)

Separator materials must meet several requirements. First, they must be stable in alkaline electrolytes at elevated temperatures of up to 120°C, a condition that can happen inside a cell during high-currents discharge. Polyethylene is not suitable under those conditions, therefore polyolefin or alkaline-resistant polyamide is typically used. At moderate battery temperatures of approximately 55°C, a reaction of a separator material with the electrolyte can occur, making a hydrolysis reaction possible. In those cases, carbon-nitrogen (C-N) bonds in the polymer are weakened, leading to formation of ammonia and amines. This, in turn, leads to high self-discharge and reduction of battery lifetime. The formation of carbonates has also been reported as a result of separator decomposition due to hydrolysis. In addition, while the overall water amount does not change in Ni-Cd cell reactions, water does take part in the reactions on electrodes and the water equilibrium affects the separator.

Cell Types

Vented Cells

Vented, also described as wet or flooded, cells are used for cells with greater capacity and high discharge rate. They are safe and light because they do not need sealing accessories. In most designs, flooded cells are equipped with low-pressure release valves for venting gases generated during charge or discharge.

When gases escape through the vents, they carry some electrolyte and over time electrolyte loss occurs, which must be periodically (a few months to a year) replaced through maintenance. Vented cell Ni-Cd batteries have a long lifetime of over 20 years and can operate at extreme temperatures (−40 to 70°C). They are used in aviation, rail and mass transit, backup power for telecom, and engine starting for backup turbines.

Sealed Cells

Sealed cells rely on the oxygen cycle during overcharging, a mechanism to prevent hydrogen generation while oxygen evolved on the positive electrode is reduced on the negative electrode. In contrast to lead-acid batteries, sealed Ni-Cd cells do not require special gel electrolytes because of the ability of nonwoven separators to support transport of gaseous oxygen. In addition, oxygen transport between the electrodes is simplified by the short distance and limited electrolyte volume, enabled by the fact that Ni-Cd battery reactions do not consume electrolyte.

A variation of a sealed cell is a Ni-Cd battery based on fiber structure (FNC), where two negative electrodes are attached with a special porous fiber-nickel recombination structure (Figure 63). This interelectrode structure is not filled with electrolyte, rather the electrolyte is present only at the interface with the electrodes and allows oxygen reduction in the gas phase over the whole surface of both adjacent Cd electrodes.

Cylindric Cells

In a sealed cylindric Ni-Cd cell, the positive and negative electrodes are rolled together with the separator (so-called jelly roll) and transferred into a cylindrical, nickel-plated steel container. The positive electrode is connected with the top cover and the negative electrode with the body of the container. The container contains a safety vent, which opens at about 2 MPa. After releasing the pressure, the operation continues similarly to a vented cell, with water loss and reduced lifetime. Recently, the vent has been substituted by a resealing valve.

The cylindric Ni-Cd cells are available in a variety of sizes.

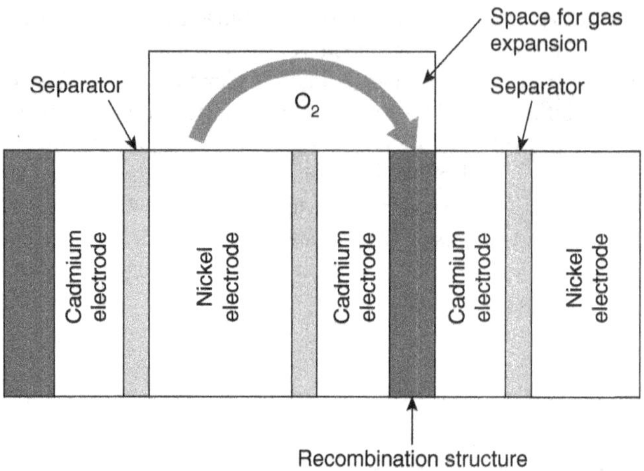

Figure 63 Scheme of the oxygen recombination in FNC cells.

Ni-Cd Battery Performance Parameters

Energy and Power

In general, Ni-Cd batteries have better power characteristics and low-temperature behavior than lead-acid and lithium batteries. They are available in the range from 0.5 Ah to 1500 Ah. The theoretical specific energy of Ni-Cd cells is 244 Wh kg^{-1}, while their practical specific energy depends on the cell design and electrode type:

- Robust industrial batteries: 15 to 30 Wh kg^{-1} and 25 to 50 Wh L^{-1}.
- The common sealed cylindric cells: 45 Wh kg^{-1} to 125 Wh L^{-1}.
- Special cylindric, high energy cells: 55 Wh kg^{-1} and 170 Wh L^{-1}.

The specific energy and energy density for Ni-Cd batteries depend critically on the discharge current and temperature of operation (Figure 64).

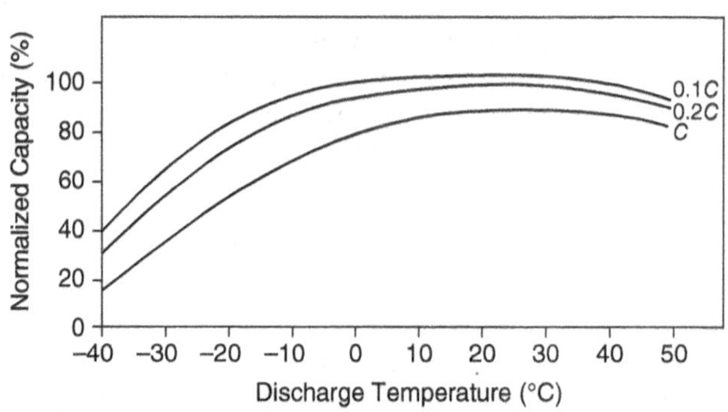

Figure 64 Normalized capacity for different discharge rates and discharge temperatures for typical sealed Ni-Cd cells after charging at 20°C. (Reproduced from A. K. Shukla, S. Venugopalan, B. Hariprakash, *J. Power Sources* 100, 2001, 125–148.)

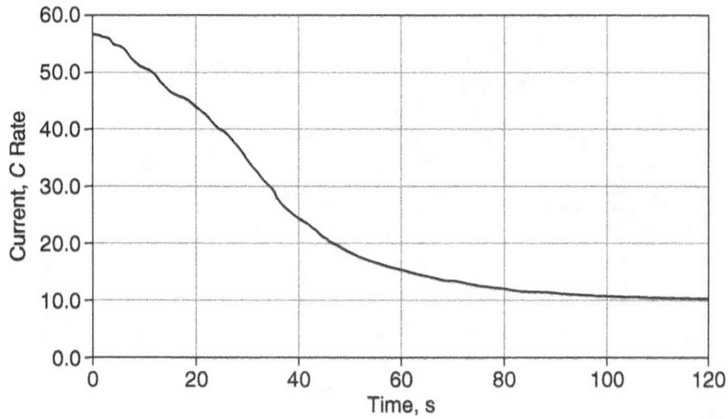

Figure 65 Constant 0.6 V discharge of a vented sintered-plate Ni-Cd cell at 25°C. (Reproduced from R. D. Lucero, chap. 20, Vented Sintered-Plate Nickel-Cadmium Batteries, in *Linden's Handbook of Batteries*, ed. by Th. Reddy, D. Linden, McGraw Hill, 4th ed., 2011.)

The specific power of Ni-Cd is high and can reach 1000 W kg^{-1} without significant capacity loss. The power characteristic of a vented, sintered-plate Ni-Cd battery is displayed as current versus time, for constant voltage discharge. It is important to note that the starting discharge current, which is a measure of power, is extraordinarily high at 60 C (not to be confused with degree Celsius) while after 2 minutes it is still at a remarkable rate of 10 C (Figure 65).

Efficiency

The efficiency of Ni-Cd cells is relatively low for fully charged cells. As for other batteries, it is dependent on the currents and temperature.

For partial cycling, from SOC = 0% to SOC = 50%, the energy efficiency reaches about 85%, while the Faraday efficiency is nearly 100%.

For cycling between SOC = 50% and SOC = 100% the energy efficiency is much lower, because at higher SOC the charge acceptance is lower and a larger portion of the current is used for side reactions, such as gas development. For a slow charge (e.g., C/10), full charge is reached in 14–16 h, while Faradaic efficiencies are around 70%.

Self-Discharge

The capacity loss caused by self-discharge for Ni-Cd batteries is relatively high compared to other battery types, and it is largely recoverable, that is, non-permanent. The reactions contributing to self-discharge are taking place mainly on the nickel electrode. While the Cd-electrode is stable under OCV (open-circuit voltage) conditions in the alkaline electrolyte, the Ni-electrode undergoes decomposition of NiOOH with oxygen formation (Equation 46).

$$2NiOOH + H_2O \rightarrow 2Ni(OH)_2 + \frac{1}{2}O_2 \qquad (46)$$

The self-discharge rate is higher at the beginning of storage and typically 10% capacity loss takes place during the first day of storage. The rates of the chemical reactions contributing to self-discharge depend strongly on the storage temperature

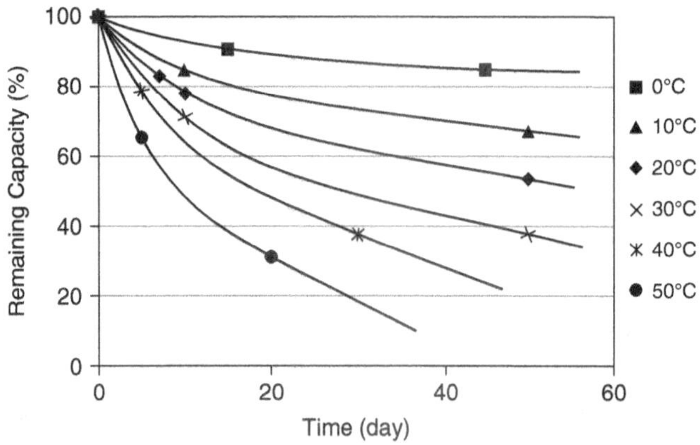

FIGURE 66 Typical self-discharge curves for sealed Ni-Cd cells as a function of storage temperature. (Reproduced from P. Bernard, Nickel-Cadmium: Sealed, in *Encyclopedia of Electrochemical Power Sources*, ed. by J. Garche et al., Elsevier, Amsterdam, 2009, Vol 4, p. 459.)

because all reaction rates, including unwanted side reactions, increase with rising temperatures. For example, after one month of storage at 50°C, a Ni-Cd battery has a remaining capacity of only about 20% of the original capacity (Figure 66). The rate of self-discharge typically doubles with every 10°C (18°F) of temperature increase.

The cell and electrode designs have also a significant impact on the self-discharge reactions. In cells with high surface area electrodes designed for high current applications, the NiOOH decomposition is accelerated because of the larger amount of active mass in direct contact with the electrolyte. Similarly, in cells with narrow electrode spacing, the diffusion pathways of ions and gases driven by thermal motion are shorter and self-discharge reactions faster. As a result, the self-discharge for sealed cells is faster than for vented cells.

Cycle Life

The lifetime of Ni-Cd batteries, measured by the number of cycles before a battery capacity drops below a certain level (usually 80% of the original value), is excellent compared to other battery types. Modern cells, with sintered positive and "plastic" negative plates, can easily reach 500 cycles on full depth-of-discharge (DOD), and routinely reach 1000–2000 cycles (Figure 67), depending on the specific cell and electrode type and cycling conditions such as temperature, current, and DOD. A shallow discharge results in longer cycle life.

The factors affecting Ni-Cd battery lifetime are dominated by the processes on Cd electrode: formation of passivating $Cd(OH)_2$ surface layer, agglomeration of the active mass, and Cd dendrite formation. The rate of active mass agglomeration is determined by the temperature and the charge/discharge currents, for example, low charge/discharge currents and high temperatures lead to a stronger agglomeration, while high currents and low temperature reduce the rate of agglomeration.

Memory Effect

Memory effect, also known as the voltage depression effect, is an aging effect which is observed after repeated partial discharge. The cells "remember" that only a part of their

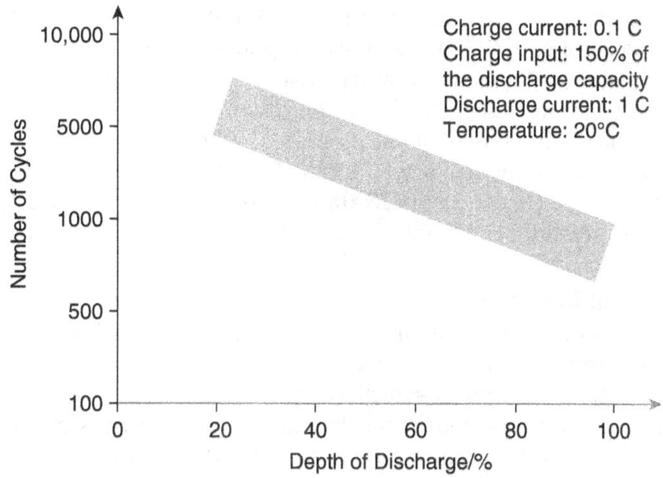

FIGURE 67 Cycle number vs. DOD for SANYO N-type, sealed Ni-Cd cell. (Adapted from D. Berndt, *Maintenance-Free Batteries*, 3rd ed., 2003, p. 362 [6].)

capacity is needed, which then leads to a reduced voltage, premature end-of-discharge voltage, and reduced capacity. In Ni-Cd batteries, both electrodes undergo reactions described as the memory effect.

The memory effect is reversible if not older than on average 2 months, and after a few full charge/discharge cycles at low current the effect is removed. The exact reasons for the memory effect are not yet known. Several hypotheses have been explored, such as formation of intermetallic phases between Cd and the sintered Ni collector or a proton rearrangement in the nickel lattice. Both effects lead to a reduced conductivity and therefore to a reduced discharge voltage. A simpler explanation is the recrystallization of the active mass, which reduces the electrochemically active surface. This recrystallization occurs preferentially in the open-circuit (rest) phase, at higher temperatures and low charge/discharge currents. During the partial discharge, the portion of the active mass not used is never electrochemically reformed. In a good cell, the crystals of the active mass remain small and surface area large. In affected cells, memory effect causes growth of electrochemically unused active mass crystals with less exposure to electrolyte. In extreme cases, the sharp edges of the crystals penetrate the separator, causing high self-discharge or even an electrical short.

Crystal growth causing "memory effect" takes place if a battery is left on the charger for days or if repeatedly recharged without a periodic full discharge to 1 V per cell. This step is essential in preventing memory effect and on average it should be performed once a month for a battery in a stand-by mode. If a battery shows signs of the memory effect loss it should be treated with a slow, deep discharge to 0.4 V per cell. However, after 6–12 months in the memory condition, the treatment is usually not effective anymore.

Safety

Safety problems in Ni-Cd batteries come as a result of exposure to alkaline electrolyte (eye contact, skin irritation, and electrolyte fume inhalation) and cadmium, which is toxic and carcinogenic. Normally a user does not come into contact with cadmium, but in case of improper disposal cadmium compounds could be released in the groundwater after corrosion of the cell container. Also, in case of fire, cadmium compounds and sublimated cadmium are released in the environment as small particles.

Based on the high-level of risk associated with the toxicity of cadmium, the production of Ni-Cd consumer cells, which can often end up in the household waste, has been forbidden in the EU countries in 2000 and in 2006 completely banned for EV applications. Today, they are authorized only for special applications.

Charging and Discharging

It is recommended to charge new Ni-Cd batteries with a low current for 16–24 hours before the first use. The slow charge equalizes the charge level of cells in a battery caused by different rates of self-discharge. Furthermore, during long storage, electrolyte gravitates to the bottom of the container and the initial slow charge eliminates dry spots on the separator by uniform electrolyte redistribution.

The main problem in charging Ni-Cd cells is the Ni electrode. From the theoretical (thermodynamic) point of view, the $Ni(OH)_2$ oxidation reaction is possible only because of the high oxygen overvoltage. The electrode potential during the charge increases with SOC and as it approaches oxygen evolution potential the rate of oxygen generation increases as well. However, as the overpotential decreases with increasing temperature, it becomes impossible to charge nickel electrode above 70°C. Charging is still possible if oxygen evolution potential is increased by adding cobalt to nickel hydroxide active mass or LiOH (or NaOH) to the electrolyte.

Charging Ni-Cd batteries below 1.43 V, which is defined as the thermoneutral voltage, produces a cooling effect (rather than warming) and compensates for the resistive heat generation or Joules heat, as long as the active mass charging is the dominant reaction. As a result, there is no temperature stress during charging and very high currents can be used without any significant reduction in capacity or safety concerns. However, if a sealed cell reaches oxygen generation stage (Figure 68), pressure increases and temperature sharply rises, while the cell voltage drops due to oxygen reaction on the negative electrode.

Vented cells are not based on the oxygen cycle and exhibit only a slow temperature increase even at high rates of charge, that is, high current. For sealed cells, full-charge detection possibilities are derived from the temperature increase and voltage drop (Figure 68).

The temperature method for sensing full charge used in low-cost chargers is not very accurate because the measured temperature at the cell container surface is lower than in the inside of the cell, which may lead to overcharge. Typically, the chargers use 50°C as a cutoff temperature value. Advanced chargers, on the other hand, do not use a fixed temperature threshold but the "delta temperature method," where a critical dT/dt value, usually 0.5-1 K min^{-1}, triggers the end-of-charge command. This method, which requires a significant charge current, is usually combined with a temperature cutoff set to 60°C to terminate the charge.

Another charging method seeks to establish the end-of-charge through negative voltage change, that is, "negative delta V" (NDV) or a voltage drop caused by the reaction of oxygen on the cadmium electrode. Compared with the temperature method to determine end-of-charge, the NDV method has a faster response and is used when higher charge currents are used. To enhance end-of-charge point detection, most NDV

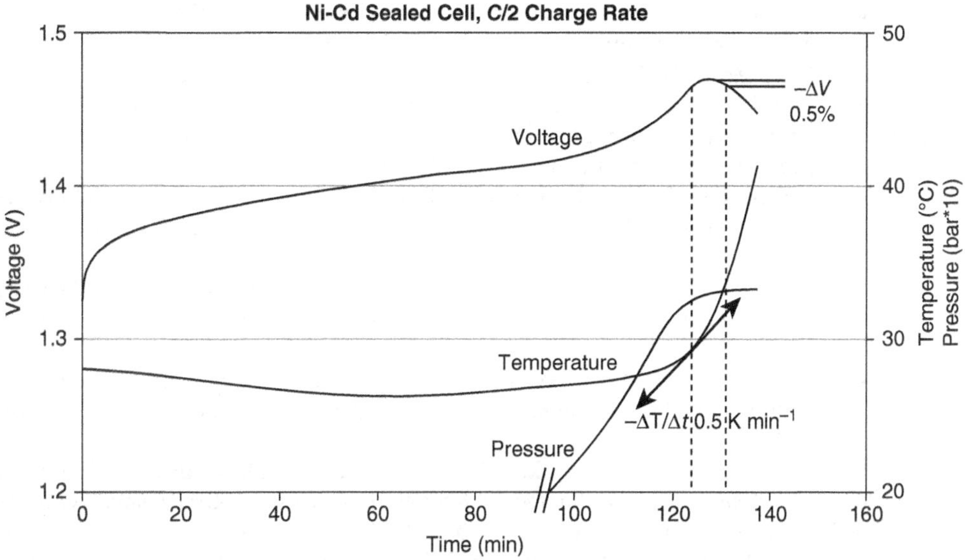

Figure 68 Voltage, temperature, and internal pressure during constant current charge (C/2) for a sealed Ni-Cd cell. (Reproduced from P. Bernard, Nickel-Cadmium: Sealed, in *Encyclopedia of Electrochemical Power Sources*, ed. by J. Garche et al., Elsevier, Amsterdam, 2009, Vol 4, p. 459.)

chargers also use identification of the short voltage plateau between voltage increase and voltage drop.

Advantages and Disadvantages

The advantages of Ni-Cd batteries are high cycle life and good calendar life, high power, ultrahigh charging currents, and very good low-temperature performance.

The limitations are relatively low specific energy, memory effect (which requires periodical full discharge), toxicity of cadmium, low electrical efficiency, high self-discharge, relatively high cost, and low cell voltage of 1.20 V.

Because of the cadmium toxicity, the use of Ni-Cd batteries is banned in some countries and for most applications. Exceptions are allowed in cases where good low-temperature behavior and very fast charging are needed. In general, the market share for Ni-Cd batteries has been reduced as Ni-MH and lithium-ion (Li-ion) batteries captured large segments of the market.

Further Reading

- P. Bernard, M. Lippert, chap. 14, Nickel-Cadmium and Nickel-Metal Hydride Battery Energy Storage, in *Electrochemical Energy Storage for Renewable Sources and Grid Balancing*, ed. by P. Moseley, J. Garche, Elsevier, 2015, p. 223.
- P. Bernard, Nickel Systems—Nickel-Cadmium: Sealed, in *Electrochemical Energy Storage for Renewable Sources and Grid Balancing*, ed. by P. Moseley, J. Garche, Elsevier, 2015, p. 459.
- D. Bernd, *Maintenance Free Batteries*, Research Studies Press, 2003.

- M. Broussely, G. Pistoia, *Industrial Applications of Batteries*, Elsevier, 2007.
- J. A. Carcone, Portable Sealed Nickel-Cadmium Batteries, in *Linden's Handbook of Batteries*, ed. by Th. Reddy, D. Linden, McGraw Hill, 4th ed., 2011.
- J. Cook, P. Kritzer, Nickel Systems—Alkaline Battery Separators, in *Electrochemical Energy Storage for Renewable Sources and Grid Balancing*, ed. by P. Moseley, J. Garche, Elsevier, 2015, p. 424.
- J. K. Erbacher, chap. 19, Industrial and Aerospace Nickel-Cadmium Batteries, in *Linden's Handbook of Batteries*, ed. by Th. Reddy, D. Linden, McGraw Hill, 4th ed., 2011.
- R. D. Lucero, chap. 20, Vented Sintered-Plate Nickel-Cadmium Batteries, in *Linden's Handbook of Batteries*, ed. by Th. Reddy, D. Linden, McGraw Hill, 4th ed., 2011.
- Y. Sato, Nickel Systems—Memory effect, in *Electrochemical Energy Storage for Renewable Sources and Grid Balancing*, ed. by P. Moseley, J. Garche, Elsevier, 2015, p. 534.
- A. K. Shukla, B. Hariprakash, Nickel Systems—Electrodes: Cadmium, in *Electrochemical Energy Storage for Renewable Sources and Grid Balancing*, ed. by P. Moseley, J. Garche, Elsevier, 2015, p. 412.
- A. K. Shukla, B. Hariprakash, Nickel Systems—Electrodes: Nickel, in *Electrochemical Energy Storage for Renewable Sources and Grid Balancing*, ed. by P. Moseley, J. Garche, Elsevier, 2015, p. 404.
- A. K. Shukla, S. Venugoplan, B. Hariprakash, Nickel Systems—Nickel-Cadmium: Overview, in *Electrochemical Energy Storage for Renewable Sources and Grid Balancing*, ed. by P. Moseley, J. Garche, Elsevier, 2015, p. 452.

Nickel–Metal-Hydride Batteries

Introduction

Research on nickel–metal-hydride (Ni-MH) batteries started in 1967 in response to a need for better specific energy and elimination of toxic materials present in Ni-Cd batteries. Metal hydride (MH) alloys stable in alkaline electrolytes were developed as the cathode material and new Ni-MH batteries became a substitute for alkaline primary cells and for Ni-Cd batteries.

In addition, another battery type with hydrogen as the cathode was developed.

Operating Principle

The negative electrode (anode) in Ni-MH batteries uses hydrides of metal alloys (MH). The cell voltage for Ni-MH cell is 1.31 V. The discharge reaction on the negative electrode involves oxidation of MH to metal and water, while the reactions on the positive electrode (cathode) is identical to that of Ni-Cd cells (Equation 41).

Negative electrode:

$$MH + OH^- \rightarrow M + H_2O + e^- \qquad E° = -0.82 \text{ V} \qquad (50)$$

Positive electrode:

$$NiO(OH) + H_2O + e^- \rightarrow Ni(OH)_2 + OH^- \qquad E° = +0.49\ V \qquad (41)$$

Cell (or overall reaction):

$$MH + 2NiOOH \rightarrow 2Ni(OH)_2 + M \qquad E°\ (cell) = +1.31\ V \qquad (50a)$$

The hydrogen stored in the MH electrode is in equilibrium with the gas phase hydrogen (Equation 51):

$$MH + MH \rightleftharpoons 2M + H_2 \qquad (51)$$

The equilibrium potential of the MH electrode is determined by the hydrogen pressure. For example, the calculations using the Nernst equation show that hydrogen pressure between 0.01 bar and 10 bars results in equilibrium potential of –0.82 to –0.77 V versus standard hydrogen electrode.

Unlike the underlying mechanism for the negative cadmium electrode in Ni-Cd batteries based on solution-precipitation reaction, the reactions proceed in the negative MH electrode through the proton-intercalation mechanism. This makes the Ni-MH battery similar to Li-ion battery, as the two battery systems share a so-called rocking chair mechanism (Figure 69).

There are unwanted gas evolution reactions that take place at the end-of-charge and in the overcharge phase (Equations 44 and 45):

Negative electrode (hydrogen evolution):

$$4H_2O + 4e^- \rightarrow 2H_2 + 4OH^- \qquad E° = -0.83\ V \qquad (44)$$

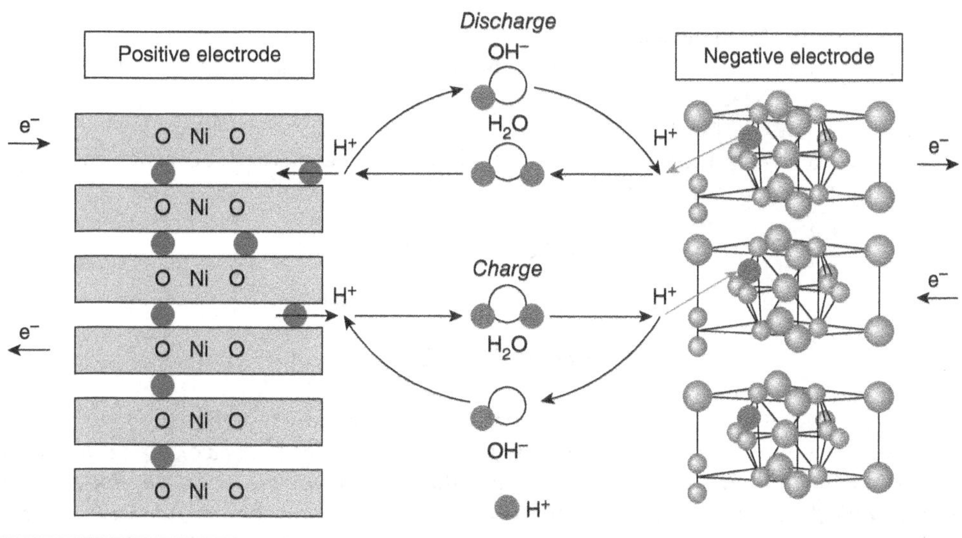

Figure 69 Schematic representation of the "rocking chair" mechanism reaction of the Ni-MH battery whereby ionic charge carriers, OH^--ions, alternate between two host materials at the positive and negative electrodes.

Positive electrode (oxygen evolution):

$$4OH^- \rightarrow O_2 + 2H_2O + 4e^- \qquad E° = 0.40 \text{ V} \qquad (45)$$

Self-Discharge Reactions

Reactions in the open-circuit voltage state are examined. The MH electrode reaction is thermodynamically stable in alkaline solutions and takes place at around -0.82 V. The exact potential depends on the type of MH alloy and hydrogen pressure. The overall reaction on the anode in the open-circuit state leads to hydrogen evolution (Equation 52) and it is composed of anodic and cathodic parts (Equations 53 and 54).

$$MH \rightarrow M + \frac{1}{2} H_2 \qquad (52)$$

$$MH + OH^- \rightarrow M + H_2O + e^- \qquad (53)$$

$$H_2O + e^- \rightarrow \frac{1}{2} H_2 + OH^- \qquad (54)$$

The hydrogen evolution overpotential is large and therefore there is no observable hydrogen generation in the open-circuit voltage (OCV) state.

The reaction on the cathode involves oxygen generation that proceeds with high overpotential (Equation 45) and NiOOH reaction with electrolyte (Equation 55), resulting in low-rate self-discharge reaction producing $Ni(OH)_2$ and oxygen (Equation 56).

$$4OH^- \rightarrow O_2 + 2H_2O + 4e^- \qquad (45)$$

$$NiO(OH) + H_2O + e^- \rightarrow Ni(OH)_2 + OH^- \qquad (55)$$

$$2NiO(OH) + H_2O \rightarrow 2Ni(OH)_2 + \frac{1}{2} O_2 \qquad (56)$$

Oxygen and Hydrogen Cycles

The oxygen gas evolution reaction leads to pressure increase, similar to reactions in lead-acid and Ni-Cd batteries. Oxygen formed at the end-of-charge and during the overcharge at the positive electrode diffuses to the negative electrode where it is reduced. There are three reactions taking place at the negative MH during charge process (Equations 57, 54, and 58).

$$M + H_2O + e^- \rightarrow MH + OH^- \qquad (57)$$

$$H_2O + e^- \rightarrow \frac{1}{2} H_2 + OH^- \qquad (54)$$

$$\frac{1}{2} O_2 + H_2O + 2e^- \rightarrow 2OH^- \qquad (58)$$

The most likely of the three reactions is the oxygen reduction (Equation 58) because of the most positive potential. This reaction proceeds for as long as oxygen is available and such condition is met by oversizing the capacity of the negative electrode by typically 30%. In that way, the positive electrode reaches full charge and starts generating oxygen before the negative electrode is fully charged. Oxygen then diffuses from the positive to the negative electrode, where it is captured or recombined (i.e., reduced), thereby preventing the negative electrode from reaching full charge.

The ratio between the oxygen recombination rate and cell current ($\nu_{O_2\text{-red}}/I_{O_2\text{-red}}$) is diffusion limited, which means that it depends on the oxygen gas pressure and size of separator pores that enable flow of gas. The oxygen generation rate ($\nu_{O_2\text{-gen}}$) at the positive electrode in the overcharge phase is proportional to the charge current (I_{ch}) and must be lower than that of oxygen recombination ($\nu_{O_2\text{-red}} > \nu_{O_2\text{-gen}}$) to prevent cell pressure increase.

In the discharge process on the positive electrode, when all of NiOOH is reduced to form $Ni(OH)_2$, hydrogen generation (Equation 45) starts before negative MH electrode reaches full discharge. The hydrogen gas is then transported inside the cell and through the separator to the negative electrode where it is oxidized, completing the hydrogen cycle, while preventing the negative electrode from ever reaching the over discharge stage (Equation 59).

$$H_2 + 2OH^- \rightarrow 2H_2O + 2e^- \tag{59}$$

Components

Positive Electrode

Positive electrode in Ni-MH batteries is the same as in Ni-Cd batteries.

Negative Electrode

There are three basic families of hydrogen metal alloys:

- AB_5: LaCePrNdNiCoMnAl. The specific capacity for these alloys is 290 to 320 mAh/g.
- A_2B_7: LaCePrNdSmMgNiCoMnAlZr. The specific capacity for this family of alloys is 335 to 400 mAh/g.
- AB_2: VTiZrNiCrCoMnAlSn. The specific capacity for this family of alloys is 385 to 450 mAh/g.

Despite the fact that the specific capacity for AB_5 alloys is considerably lower than for the other two families, this group of alloys is at the moment predominantly used in commercial technologies because of the lower material cost, simpler material activation and cell formation, and greater processing flexibility. The performance of the negative active mass based on AB_5 alloy can be partially tuned by varying the ratio La:Ce, adjusting thereby the discharge capacity, cycle life, and power.

The cells with A_2B_7 alloy have been lately used for applications requiring high energy and low self-discharge. AB_2 alloys are seldom used because of the high rate of self-discharge caused by the solubility of vanadium oxide in the electrolyte.

The hydrogen bonding mechanism in the alloy during the charge involves its progressive dissolution in a hydrogen-poor solid phase, called α-phase, until saturation, followed by formation of hydrogen-rich β-phase once constant hydrogen pressure is achieved. As the hydrogen concentration in the alloy increases, the α-phase disappears and transforms to β-phase, while hydrogen pressure is increasing. The hydrogen equilibrium pressure depends on the MH composition and temperature (Figure 70).

During battery charge and discharge, AB5 retains its hexagonal crystalline structure, but undergoes significant volume changes of approximately 25%. The expansion and contraction lead to cracks in the active mass, followed by crumbling to small particle

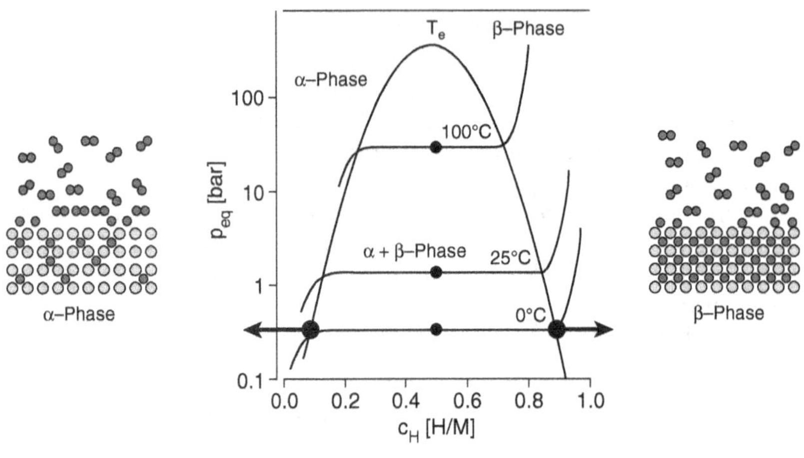

FIGURE 70 Pressure-composition isotherms for a hypothetical MH with α, α + β, and β phase. The pressure plateau is essential for device functionality.

size and creation of surface susceptible to electrochemical corrosion. In addition, pure $LaNi_5$ has equilibrium hydrogen pressure of 1.7 bar at room temperature, while pressure up to 0.1 bar is considered suitable for use in Ni-MH cells. As a result, pure $LaNi_5$ cannot be used in practical Ni-MH cells without elemental substitutions that help stabilize the pressure plateau and improve lifetime by reducing volume changes. The intermetallic cell volume is directly proportional to the logarithm of plateau pressure. Several elements are used as substitutes for Ni and La including Al, Mn Co, and 4f lanthanides (called 4f because of incomplete 4f shell), as well as natural alloys of rare earth elements called "mish-metals."

The function of MH active mass in the negative electrode is not only to store hydrogen but also to catalyze the proton reduction and hydrogen oxidation reaction. The catalytic rate is affected by formation of passivating layers during corrosion reactions.

Unlike most other electrode materials, the electrode conductivity of MHs is nearly independent of the state of charge and no electronic conductivity enhancers are necessary.

The negative electrode can be prepared by using a dry process to compact MH powder directly onto a Ni-mesh, Cu-mesh, expanded Ni, or Ni-foam without the use of a binder. Alternative method involves wet pasting of MH slurry, containing binder and additives, onto the nickel-plated perforated stainless steel.

The MH electrode, unlike the cadmium electrode in Ni-Cd systems, does not have the high load capability and good low temperature performance. These performance limitations can be traced back to limited rate of the hydrogen desorption during discharge.

Separator

The separator in Ni-MH batteries is thinner than that in Ni-Cd cells because it does not need to act as a donor or absorber of electrolyte. The separator is designed simply to ensure electronic isolation and sufficient electrolyte penetration to maintain oxygen recombination during overcharge.

The separator materials used must be resistant to hydrolysis by the electrolyte at operating temperature above 55°C, a condition that excludes polyamide and favors polyolefins such as polypropylene and polyethylene. Additional treatment using sulfuric acid is applied to polymer fibers to form hydrophilic groups and create a permanently wettable surface.

Electrolyte

Electrolyte used in Ni-MH cells is almost exclusively 30% KOH. At this concentration, it has a good balance between the conductivity and freezing temperature. A small amount of LiOH (1.5 g/L) is added for better low temperature performance. For improved corrosion resistance at high temperature, NaOH is added. No other additives are used in mass produced Ni-MH cells.

Cell Types

Vented Cells

In Ni-MH cells, hydrogen gas is in equilibrium with metal hydride ($MH \rightarrow M + \frac{1}{2}H_2$). In vented cells, hydrogen gas is released on the negative electrode during the charge process, disrupting thereby the equilibrium conditions and leading ultimately to capacity decrease. As a result, vented cells have poor cycle life and low practical market relevance.

Sealed Cells

Sealed nickel-based cells and batteries are fabricated in various shapes and sizes, from small button cells to cylindrical, large prismatic, and monoblock cells (Figure 71 and Table 12).

Consumer cells in AA and AAA formats are produced in billions of units per year. The AA cells are commercially available in capacity of 1800, 2000, 2300, 2500, and 2700 mAh. In the period from 1993 till 2006, a sequence of batteries with progressively larger capacity was developed in the same form factor as a result of improved manufacturing methods (Figure 72).

Similar improvement in performance was demonstrated in Ni-MH batteries for the hybrid electrical vehicle, Toyota Prius (Table 13).

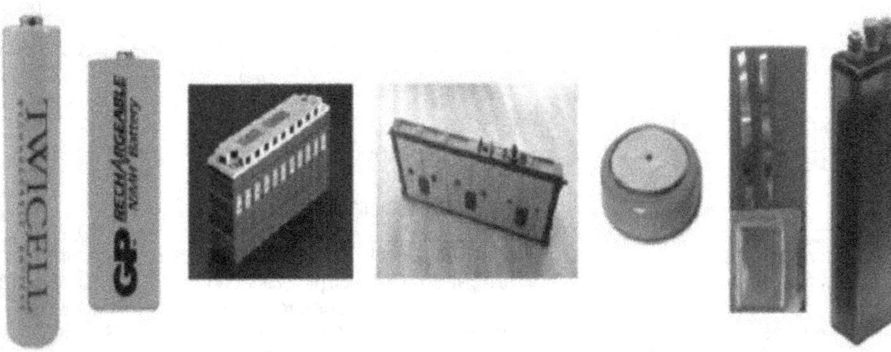

(a) (b) (c) (d) (e) (f) (g)

Figure 71 Ni-MH cells in (a) cylindrical, (b) stick, (c) metallic prismatic, (d) plastic prismatic, (e) button, (f) pouch, and (g) flooded configuration. (Reproduced from K.-H. Young, S. Yasuoka, Capacity degradation mechanisms in nickel/metal hydride batteries, *Batteries* 2016, 2.)

Shape	Case Material	Sealed	Manufacturability	Cost	Energy Density	Head Dissipation	Abuse Tolerance
Cylindrical	Metal	Yes	Easy	Low	High	Easy	High
Stick	Metal	Yes	Medium	Low	High	Easy	High
Prismatic	Metal	Yes	Medium	High	Low	Easy	Medium
Prismatic	Plastic	Yes	Medium	High	Low	Hard	Medium
Button	Metal	Yes	Easy	Low	Low	Easy	Low
Pouch	Al foil	Yes	Easy	Low	High	Easy	Low
Cylindrical/ prismatic	Plastic/ flooded	No	Easy	Low	Low	Hard	High

(Reproduced from K.-H. Young, S. Yasuoka, Capacity degradation mechanisms in nickel/metal hydride batteries, *Batteries* 2016, 2.)

Table 12 Characteristics of Ni-MH Cells

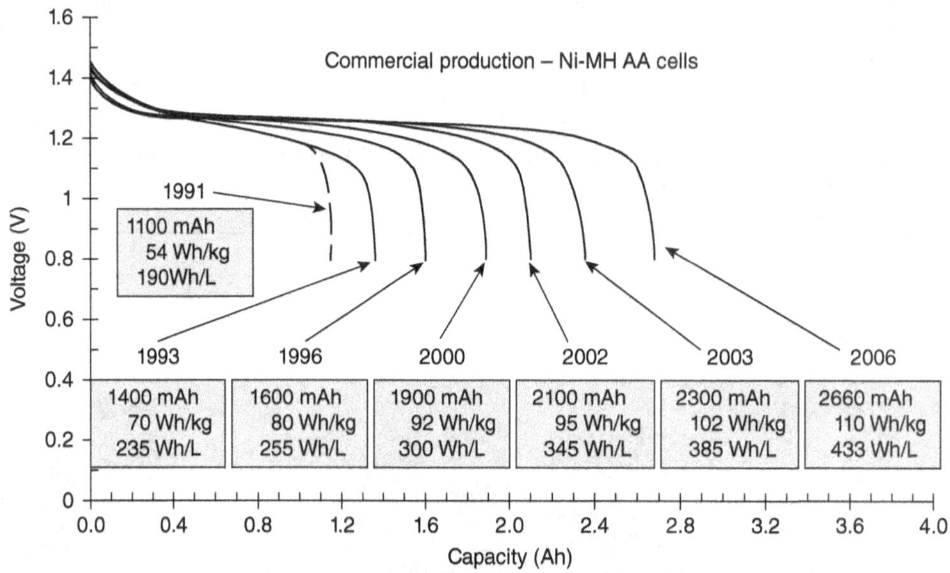

Figure 72 Voltage as a function of capacity graph is used to show chronological progression of battery parameters from 1993 till 2006, for commercial NiMH AA cells. (Reproduced from M. Fetcenko, J. Koch, M. Zelinsky, chap. 6, Nickel-Metal Hydride and Nickel-Zinc Batteries for Hybrid Vehicles and Battery Electrical Vehicles, in *Advanced in Battery Technologies for Electrical Vehicles*, ed. by B. Scrosati, J. Garche, W. Tillmetch, Woodhead, 2015, p. 103.)

Parameter	97 Prius (Generation I), Japan Only	00 Prius (Generation II)	04 Prius (Generation III)	2010 Prius (Generation IV)
Form factor	Cylindrical	Prismatic	Prismatic	Prismatic
Cells (Modules)	240 (40)	228 (38)	158 (28)	168 (28)
Nominal voltage	288.0 V	273.6 V	201.6 V	201.6 V
Nominal capacity	6.0 Ah	6.5 Ah	6.5 Ah	6.5 Ah
Specific power	800 W kg^{-1}	1000 W kg^{-1}	1300 W kg^{-1}	1310 W kg^{-1}
Specific energy	40 Wh kg^{-1}	46 Wh kg^{-1}	46 Wh kg^{-1}	44 Wh kg^{-1}
Module weight	1090 g	1050 g	1045 g	1040 g
Module dimensions	35 (diam) × 384 (L)	19.6 × 106 × 275	19.6 × 106 × 285	19.6 × 106 × 285

TABLE 13 Progress in Ni-MH Battery Performance for Toyota Prius

Performance Factors

Capacity and Energy

The theoretical specific energy of Ni-MH cells is 278 Wh kg^{-1}, while the practical specific energy depends on the temperature, discharge current, cell design, electrode types, and MH alloy used. Industrial Ni-MH batteries used, for example, for railways and hybrid electrical vehicles, have energy density of around 60 Wh kg^{-1} and 100 Wh L^{-1}, while the consumer cells have from 80 to 120 Wh kg^{-1} and 250 to 400 Wh L^{-1}.

The size of commercially available Ni-MH batteries covers a very wide range, from 0.5 Ah to 100 Ah.

Power

The specific power of Ni-MH cells is in the range from 250 to 500 W kg^{-1} and 950 to 1250 W L^{-1}, for continuous 3 C discharge. For special cells used in hybrid electrical vehicles, the specific power of 1500 W kg^{-1} has been demonstrated. The specific power is significantly higher for short current pulses.

Efficiency

The energy efficiency of Ni-MH batteries is roughly 70% at room temperature and mainly determined by the low charge efficiency of the positive electrode. The nickel hydroxide electrode features nearly complete charge acceptance at room temperature, until it reaches about 80% charge and a competing oxygen evolution reaction starts. At elevated temperatures, the oxygen evolution starts earlier and compromises the charge efficiency.

Self-Discharge

There are several reactions that contribute to self-discharge in Ni-MH batteries. The reaction of charged nickel electrode (NiOOH) with the electrolyte, which produces $Ni(OH)_2$ and O_2, is accelerated by MH corrosion products produced at the negative electrode that lower oxygen generation overpotential. Next, oxygen produced is transported to the MH electrode where it reacts with MH. Both oxidation and reduction reactions are taking part on the negative electrode and MH decomposition is driving oxygen reduction by producing the electrons (Equations 53 and 58).

Anodic reaction:

$$MH + OH^- \rightarrow M + H_2O + e^- \tag{53}$$

Cathodic reaction:

$$\frac{1}{2}O_2 + H_2O + 2e^- \rightarrow 2OH^- \tag{58}$$

Overall reaction on the negative electrode:

$$\frac{1}{2}O_2 + 2MH \rightarrow 2M + H_2O \tag{58a}$$

The result of these reactions is reduced capacity of both positive and negative electrodes through so-called oxygen shunt.

The next reaction that contributes to self-discharge is initiated by the evolution of the hydrogen gas from the MH electrode (Equation 51) and its diffusion to the positive nickel electrode where it reacts with NiOOH (Equation 60).

$$MH + MH \rightleftharpoons 2M + H_2 \tag{51}$$

$$\frac{1}{2}H_2 + NiO(OH) \rightarrow Ni(OH)_2 \tag{60}$$

Similar to the reaction mechanism at the negative electrode, there are both anodic and cathodic reactions that proceed on the positive electrode (Equations 59 and 55):

Anodic reaction (Equation 59):

$$\frac{1}{2}H_2 + OH^- \rightarrow H_2O + e^- \tag{59}$$

Cathodic reaction (Equation 55):

$$NiO(OH) + H_2O + e^- \rightarrow Ni(OH)_2 + OH^- \tag{55}$$

Overall reaction on the positive electrode is shown in Equation 61:

$$MH + 2NiOOH \rightarrow 2Ni(OH)_2 + M \tag{61}$$

Another reaction potentially contributing to self-discharge in Ni-MH batteries is related to redox couples that form between the negative electrode and trace amounts of impurities from the metal alloys, for example, vanadium in AB_2. Another redox couple NO_2/NH_3 can be formed from trace amounts of NO_3^- ions after precipitation of $Ni(OH)_2$ in the positive electrode. These reactions result in slow rate of MH consumption, which is the self-discharge. Some organic compounds from the decomposition of separator material can also form redox couples. The self-discharge from redox couples can be reduced by recently developed separator materials containing chemical grafting agents that bind corrosion products and trace amounts of impurities.

All self-discharge reactions are accelerated at higher temperature (Figure 73).

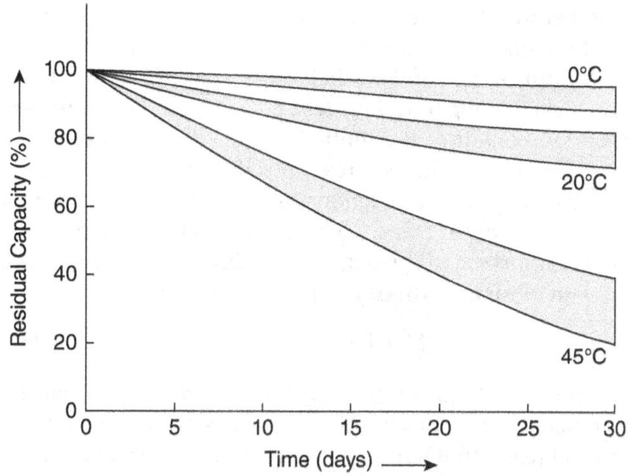

Figure 73 Self-discharge of Ni-MH cells as a function of time, for several temperatures. (Reproduced from P. H. L. Notten, M. Latroche, Nickel-Metal Hydrides, in *Encyclopedia of Electrochemical Power Sources*, ed. by J. Garche et al., Elsevier, Amsterdam, 2009, Vol 4, p. 502.)

Lifetime

The lifetime of Ni-MH batteries expressed as the number of cycles of charge and discharge is dependent on the DOD (Figure 74). The cycle life is usually considered the point or number of cycles when the capacity drops to 80% of the original value.

The extreme effect of DOD on the number of cycles is very apparent and points to an extraordinary mechanism of capacity degradation. While cycle life for 100% DOD is

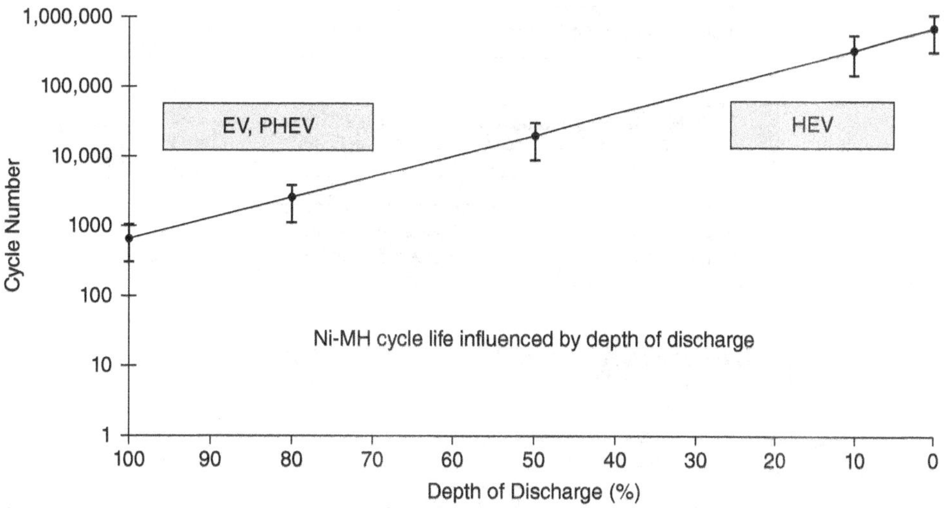

Figure 74 Relationship between DOD and cycle life for Ni-MH batteries. (Reproduced from M. Fetcenko, J. Koch, chap. 12, Nickel-Metal-Hydride Batteries, in *Linden's Handbook of Batteries*, ed. by Th. Reddy, D. Linden, McGraw Hill, 4th ed., 2011.)

typically just below 1000, it increases steeply for more shallow cycling, thus, for example, for 50% DOD is 10,000 and for 20% DOD is 100,000. This particular data has been collected for EV applications of the MH battery where DOD for a typical pattern is 80%. In general, the cycle life depends, besides DOD, on the temperature and discharge current.

Another component of the lifetime is the calendar life, which depends not on the number of cycles, but on the side reactions that take place in the battery and eventually degrade the capacity. These reactions, related to aging, include swelling of the nickel electrode, corrosion of MH, and separator degradation (Figure 75).

The most important MH reaction contributing to aging of the battery active material is corrosion of metal with oxygen (Equation 62).

$$2.15\ \mathrm{M} + 1.15\ \mathrm{H_2O} \rightarrow \mathrm{M(OH)}_{1.15} + 1.15\ \mathrm{MH} \tag{62}$$

This reaction leads to a loss of active material due to electrolyte drying out and subsequent increase in internal resistance as a result of water consumption. In addition, the volume increase that takes place as metal hydroxide forms from metal causes mechanical breakage of the active mass and electronic detachment from the electrode, which results in capacity loss. The obvious design strategy to reduce this effect is to select MH alloys with low volume change property ($\Delta V/V$). In addition, the hydroxide surface has inferior catalytic activity for MH oxidation reaction, which lowers the discharge current.

Methods to improve calendar life of Ni-MH batteries used in off-grid applications include increase in electrolyte volume to prevent drying out and reducing the surface corrosion of MH by adding yttrium oxide.

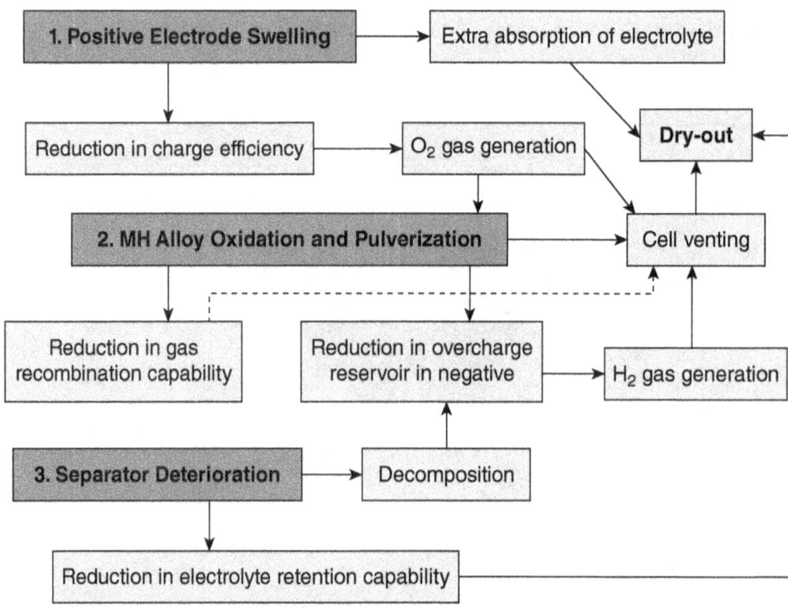

FIGURE 75 Chart depicting failure modes, associated reactions, and effects in Ni-MH batteries. (Reproduced from K.-H. Young, S. Yasouka, Capacity degradation mechanisms in nickel/metal hydride batteries, *Batteries* 2016, 2, 3; 10.3390/batteries2010003.)

Safety

Ni-MH batteries are the least safe among batteries with aqueous electrolytes. The hydrogen evolution that occurs from the MH alloy (Equation 52) results in the cell pressure increase (Figure 68), a hazardous condition compounded by the rise in temperature above 60°C and leading to potential explosion. The factors that contribute to rise in temperature include overcharging and discharging with too high current, mechanical shock to the battery, separator degradation, and high external temperature. Most Ni-MH cells of larger size are protected by the overpressure release valve, which opens when temperature reaches 120°C.

Cost

The cost of Ni-MH batteries is determined mainly by the cost of active materials, NiOOH and MH. When they were first introduced, the cost of Ni-MH batteries was above $1500/kWh, but through increased manufacturing volumes and improvements in material performance, the cost of Ni-MH batteries for electrical vehicle applications decreased to $600–$800/kWh and for high-volume consumer cells to $200/kWh.

Charging

The charging process for Ni-MH cells is similar to that for Ni-Cd and it is usually carried out using 0.1–0.2 C currents. Unpredictably at first, because of inefficiency, the charge endpoint would be reached only when about 150% of charge based on capacity is reached. At the end of charge, with an oversized MH electrode, the charging current is proportional to the rate of oxygen evolution on the positive electrode and reduction on the negative. This reaction mechanism supports oxygen cycle in sealed batteries, ensuring that pressure inside the sealed battery never increases. The principle of operation based on oxygen cycle assumes that both hydrogen generation from MH and internal temperature are kept low.

Detection of the end-of-charge in Ni-MH batteries is critically important, but also more challenging to detect than in the case of Ni-Cd batteries. Because lower charging currents are used (less than 0.5 C), the negative delta method (NDV) is more difficult to implement. Consequently, in addition to the NDV method, advanced chargers use a combination of additional techniques to reliably detect end-of-charge point. These include voltage plateau, delta temperature (dT/dt), temperature threshold (usually 60°C), and timed charge. These five techniques, comprising NDV as well, secure reliable detection of the end-of-charge through a multiple-trigger setup, where any of the sensors can initiate disruption in current and end-of-charge.

During the charging process, heat is generated in Ni-MH cells at a faster rate than in Ni-Cd cells. The higher temperature limits the current during float and trickle charge steps to about 0.05 C (compared to 0.1 C in Ni-Cd cells). Consequently, Ni-Cd chargers cannot be used to charge Ni-MH cells, while the opposite is possible, i.e., Ni-MH chargers can be used to charge Ni-Cd cells.

Advantages and Disadvantages

The main advantage of Ni-MH batteries is up to 40% higher specific capacity compared with Ni-Cd batteries. In addition, they have relatively high cycle and calendar life, reasonably high power, large operating temperature range, and are environmentally friendly.

The limitations of Ni-MH batteries compared to nonnickel-based systems, such as lead-acid or Li-ion, are in high rate of self-discharge, low efficiency, relatively high cost, sensitivity to overcharge, and low cell voltage of 1.2 V. Comparison of Ni-MH batteries with other battery systems has been previously shown in Table 3.

The prospects for Ni-MH batteries remain favorable. Despite their partial displacement in many applications by Li-ion batteries, the complete transition and elimination of Ni-MH batteries did not take place. This has been a result of certain performance issues with Li-ion batteries, primarily their safety, but also an apparent progress in Ni-MH performance. As part of the Advanced Research Projects Agency for Energy (ARPA-E) program, a specific energy of 127 Wh kg^{-1} has been demonstrated on a cell level and the target has been subsequently increased to 148 Wh kg^{-1}. This level of performance is comparable with the Li-ion battery. Furthermore, in another development direction, a new material, amorphous silicon, was proposed to replace MH alloy, and this could lead to storage capacity of 3650 Ah kg^{-1}, which is about 10 times higher than A_2B_7 alloys. As these improvements on the cell level get scaled up and commercialized, a resurgence of nickel batteries with hydride anode could be expected.

Further Reading

- P. Bernard, M. Lippert, chap. 14, Nickel-Cadmium and Nickel-Metal Hydride Battery Energy Storage, in *Electrochemical Energy Storage for Renewable Sources and Grid Balancing*, ed. by P. Moseley, J. Garche, Elsevier, 2015, pp. 223.
- D. Bernd, *Maintenance Free Batteries*, Research Studies Press, 2003.
- M. Broussely, G. Pistoia, *Industrial Applications of Batteries*, Elsevier, 2007.
- M. Fetcenko, J. Koch, chap. 22, Nickel-Metal Hydride Batteries, in *Linden's Handbook of Batteries*, ed. by Th. Reddy, D. Linden, McGraw Hill, 4th ed., 2011.
- M. Fetcenko, J. Koch, M. Zelinsky, chap. 6, Nickel-Metal Hydride and Nickel-Zinc Batteries for Hybrid Electric Vehicles and Battery Electric Vehicles, in *Advances in Battery Technologies for Electric Vehicles*, ed. by. B. Scrosati, J. Garche, W. Tillmetz, Woodhead, 2015, p. 103.
- B. Hariprakash, A. K. Shukla, S. Venugoplan, Nickel-Metal Hydride: Overview, in *Encyclopedia of Electrochemical Power Sources*, ed. by J. Garche et al., Elsevier, Amsterdam, 2009, Vol 4, p. 49402.
- P. H. L. Notten, M. Latroche, Nickel-Metal Hydride: Metal Hydrides, in *Encyclopedia of Electrochemical Power Sources*, ed. by J. Garche et al., Elsevier, Amsterdam, 2009, Vol 4, p. 502.
- K.-H. Young, S. Yasuoka, Capacity degradation mechanisms in nickel/metal hydride batteries, *Batteries* 2016, 2, doi:10.3390/batteries2010003.

Nickel-Hydrogen Batteries

Introduction

The nickel-hydrogen battery (Ni-H_2) is a hybrid system comprising a conventional positive battery electrode and a negative electrode similar to those in fuel cells. It emerged as a viable option to replace secondary batteries in the 1970s because of concerns associated with long-term prospects of MH anodes. The first application of this battery was in the US Navy's navigation technology satellite and from that time the Ni-H_2 system has nearly completely replaced Ni-Cd batteries for use in space but has not gained any significant terrestrial markets due to high cost.

At present, the Ni-H_2 batteries serve a niche market of about 1000 satellites, such as ISS, Mercury Messenger, Mars Odyssey, or the Mars Global Surveyor. They are recently facing strong competition from Li-ion batteries; however, higher cycle life of Ni-H_2 batteries has slowed down the substitution process.

Operating Principle

The reaction on the positive electrode (cathode) in Ni-H_2 batteries is the same as in Ni-Cd and Ni-MH batteries (Equation 55).

$$NiO(OH) + H_2O + e^- \rightarrow Ni(OH)_2 + OH^-$$

On the negative electrode (anode), hydrogen is oxidized during the discharge reaction (Equation 61).

$$H_2 + 2OH^- \rightleftharpoons 2H_2O + 2e^-$$

The overall cell reaction shows formation of nickel hydroxide on the positive electrode while hydrogen is consumed on the negative electrode (Equation 63).

$$H_2 + 2NiOOH \rightleftharpoons 2Ni(OH)_2 \tag{63}$$

The process shows that water is generated at the hydrogen electrode and consumed at the nickel electrode; consequently, the concentration of the electrolyte does not change throughout the reaction.

Side reactions at the end-of-charge and in the overcharge phase take place only on the positive electrode because there is no active mass on the negative electrode. The most important side reaction is oxygen evolutiòn (Equation 45).

$$4OH^- \rightarrow O_2 + 2H_2O + 4e^-$$

Oxygen generated at the positive electrode in the overcharge phase can be reduced at the negative electrode electrochemically or in a direct chemical reaction (Equation 64).

$$O_2 + 2H_2 \rightarrow 2H_2O \tag{64}$$

The viability of this reaction as means to regulate internal cell pressure depends on the charge current, oxygen reduction catalysis on the negative electrode, and temperature. To avoid explosive mixtures between hydrogen and oxygen and resulting micro explosions, called "popping," that can damage the cell, oxygen concentration, that is, partial pressure, must be kept low. The controllable rate of oxygen generation is achieved by moderate charging current and using gas permeable separator materials to allow unhindered oxygen transport to the negative electrode.

Components

The positive electrode is identical to electrodes used in Ni-Cd and Ni-MH batteries. The negative electrode is similar to fuel cell electrodes and consists of a platinum black catalyst, supported on a photo-etched grid. Platinum black, its name coming from black color, is a finely divided metallic platinum prepared from its salts. It is used in the negative electrode of Ni-H_2 batteries in the loading of about 7 mg/cm^2. A hydrophobic membrane is then pressed to the back side of the grid, forming thereby a three-phase boundary—solid platinum, liquid electrolyte, and hydrogen gas, whereby gas is reactant, electrolyte conducts ionic charge, and solid phase conducts electrons.

Electrolyte in Ni-H_2 batteries is KOH, in concentrations of 26%–31%. The amount of electrolyte required is 1.8–2.2 g per Ah of battery capacity.

Asbestos and ceramic zirconia filaments (manufactured by Zircar) are the most common separator materials. The typical thickness is 250–380 μm. The asbestos separator is a nonwoven fabric, mechanically stable, and characterized by poor oxygen transport rates. The zirconia separator material has excellent chemical and high temperature resistance, but it is brittle and in general has poor mechanical properties. A third material category sometimes used is polyamide. It is, however, susceptible to partial degradation as a result of hydrolysis reaction in the electrolyte.

A gas diffusion screen, typically made of polypropylene, is also applied on the back side of the negative electrode to promote oxygen gas transport.

Cell and Battery Types

The first design of the Ni-H_2 battery included repeat units of gas screen, negative electrode, separator, and positive electrode. In most cases, asbestos, known for relatively poor oxygen transportation rate, was used as a separator material, which led to an explosive ratio between hydrogen and oxygen and caused small explosions or popping. The original design was replaced by the back-to-back design, which used Zircar separators with better oxygen transport properties. A single unit comprised two positive electrodes, two negative electrodes, two separator sheets, and one gas screen. One later iteration developed for NASA involved a bipolar design.

Due to concerns regarding high hydrogen pressure, all Ni-H_2 cells are placed in a protective pressure vessel made of 0.6–0.8 mm thick Inconel 7/18 alloy. A pressure vessel can accommodate one or more cells and withstand maximum pressure of 7 MPa (70 bar), with a burst pressure three times higher. The pressure cells or vessels are connected together to form a battery (i.e., a stack), by mounting the single pressure vessels on a base plate, which also helps dissipate heat evolved due to inefficiencies (Figure 76).

FIGURE 76 Ni-H_2 pressure cells (left) and battery made by SAFT company. (Reproduced from Y. Borthomieu, P. Bernard, Nickel Systems, Nickel-Hydrogen, in *Encyclopedia of Electrochemical Power Sources*, ed. by J. Garche et al., Elsevier, Amsterdam, 2009, Vol 4, p. 482.)

Practical batteries have been made with up to 22 cells in a single pressure vessel, for the Geosynchronous Earth Orbit (GEO) application, with a lifetime of 10–15 years at 50–80% DOD and for Low Earth Orbit (LEO) applications with a lifetime of 3–7 years at 20–40% DOD.

Performance Factors

Ni-H_2 batteries have an open-circuit voltage of 1.55 V and an average voltage during discharge of 1.25 V. They have specific energy between 40 and 75 Wh kg^{-1} and energy density of 60–100 Wh L^{-1}. Their specific power is in the range of 200–250 W kg^{-1}.

Specific energy is shown as a function of capacity and cell size (Figure 77) for the Ni-H_2 battery. It is evident that specific energy critically depends on the design of a cell, in particular the cell diameter.

The energy efficiency of Ni-H_2 cells is around 85% and the Faradaic efficiency is nearly 100%. Their lifetime depends, as in case of all other batteries, on the DOD, and for a particular battery the lifetime is 40,000 cycles at 40% DOD. Calendar life is typically longer than 15 years.

Lifetime can also be expressed in terms of capacity throughput, which is the total amount of energy a battery can store and deliver over its lifetime. The advantage of the Ni-H_2 battery is evident from the graph (Figure 78).

The self-discharge for the Ni-H_2 cell is based on the hydrogen reaction with the active mass of the positive electrode. The rate of self-discharge is extremely high for a particular cell especially at an elevated temperature of 30°C (Figure 79).

Ni-H_2 batteries are prohibitively expensive for applications on Earth. A single cell (used for space applications) can cost thousands of dollars.

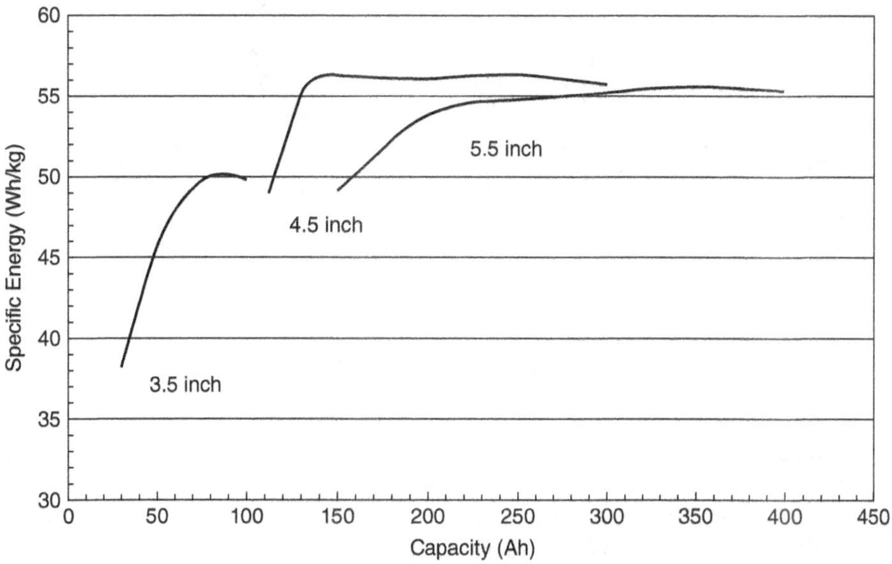

FIGURE 77 Specific energy as a function of capacity and cell diameter for 28 V Ni-H_2 battery. (Reproduced from J. N. Brill, chap. 24, Nickel-Hydrogen Batteries, in *Linden's Handbook of Batteries*, ed. by B. Reddy, McGraw Hill Professional, 4th ed., 2011.)

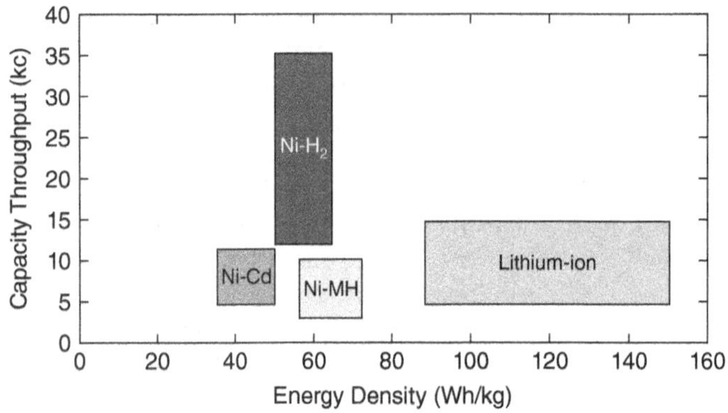

FIGURE 78 A throughput capacity for different space-battery systems. (Reproduced from A. H. Zimmerman, *Nickel-Hydrogen Batteries, Principle and Practice*, The Aerospace Press, El Sequndo, California, American Institute of Aeronautics and Astronautics, Inc., Reston, Virginia, 2009.)

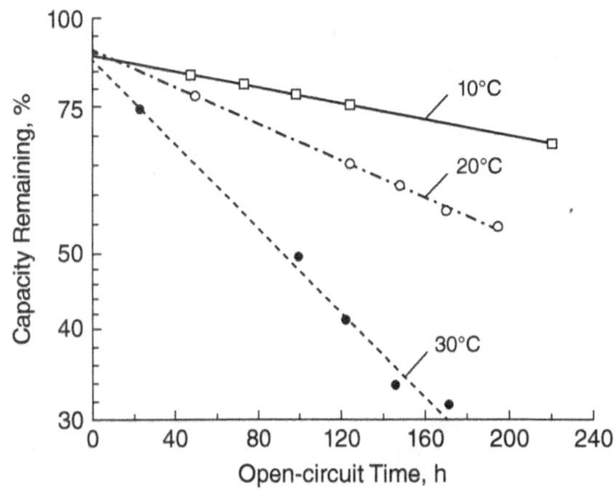

FIGURE 79 Self-discharge for a 50 Ah Ni-H_2 cell as a function of time for different temperatures. (Reproduced from J. N. Brill, chap. 24, Nickel-Hydrogen Batteries, in *Linden's Handbook of Batteries*, ed. by B. Reddy, McGraw Hill Professional, 4th ed., 2011.)

Safety of Ni-H_2 batteries is related to high oxygen content in hydrogen and potential explosive reaction if oxygen concentration surpasses 2%. These reactions can cause destruction of electrodes, but the cells are contained inside the pressure vessel and no dangerous conditions for the surroundings exist.

Charging

Ni-H_2 batteries exhibit different charge behavior than other batteries with nickel positive electrodes in two aspects. First, the overcharging and polarity reversal can be controlled; and second, hydrogen pressure in the cell provides good indication of the state of charge, a characteristic not available in Ni-Cd and Ni-MH batteries.

There are two main processes that take place during overcharge:

- First, the temperature increases during the oxygen recombination reaction and leads to heat and water vapor movement, followed by condensation on cooler parts of the cell. The most consequential result of this process is higher rate of self-discharge, which nevertheless takes place at the end-of-charge. Hence, a trickle-charge is required to maintain good state of charge. Temperature increase of more than 10 K becomes a concern and requires measures to impede the heat evolution process.
- As mentioned previously, oxygen cycle in the cell, its evolution at the positive electrode and recombination on the platinum catalyst of the negative electrode, can produce strong recombination or "popping" if the oxygen concentration is more than 2%.

It is apparent that cell control must be used to limit the rate of oxygen evolution and this is usually accomplished by reducing the charging voltage, which exponentially affects the oxygen evolution rate. The voltage limit must also be temperature-compensated, or the end-of-charge can be detected based on temperature-compensated cell pressure.

Advantages and Disadvantages

The most important advantages of Ni-H_2 batteries are extremely long service life, that is, cycle life and remarkable performance at temperatures down to –30°C. In addition, these batteries have very good calendar life, a result of low rates of corrosion reactions.

The disadvantages include very high cost, high rate of self-discharge, and low energy density compared with modern lithium batteries. Specific energy is approximately 1/3 and energy density 1/10 of the lithium batteries. Based on these comparative values, Ni-H_2 batteries are being largely substituted by lithium batteries.

Further Reading

- Y. Borthomieu, P. Bernard, Nickel Systems: Nickel-Hydrogen, in *Encyclopedia of Electrochemical Power Sources*, ed. by J. Garche et al., Elsevier, Amsterdam, 2009, Vol 4, p. 482.
- J. N. Brill, chap. 24, Nickel-Hydrogen Batteries, in *Linden's Handbook of Batteries*, ed. by B. Reddy, McGraw-Hill Professional, 4th ed., 2011.
- L. H. Thaller, A. H. Zimmerman, *Overview of the Design, Development, and Application of Nickel-Hydrogen Batteries*. The Aerospace Corporation, El Segundo, California, NASA/TP—2003-21190, 2003. https://ntrs.nasa.gov/archive/nasa/casi.ntrs.nasa.gov/20030062140.pdf.
- A. H. Zimmerman, *Nickel-Hydrogen Batteries: Principles and Practice*, The Aerospace Press, El Segundo, California, American Institute of Aeronautics and Astronautics, Inc., Reston, Virginia, 2009.

Nickel-Zinc Batteries

Introduction

A rechargeable Ni-Zn battery was initially patented by Edison in 1901. Throughout the 20th century, there were no commercial Ni-Zn batteries because of the inability to improve their cycle life. Recently, these batteries have found certain applications and have been produced in AA format and prismatic format for micro-HEVs by the manufacturers such as EnZinc and ZincFive (formerly PowerGenix).

Operating Principle

The reaction on the negative electrode (anode) in a Ni-Zn battery during discharge is oxidation of zinc to zinc hydroxide in the presence of a KOH electrolyte. On the positive electrode (cathode), a well-known NiOOH reduction to $Ni(OH)_2$ takes place (Equation 65).

Negative electrode:

$$Zn + 2OH^- \rightarrow Zn(OH)_2 + 2e^- \qquad E° = -1.24\ V \tag{65}$$

Positive electrode:

$$NiO(OH) + H_2O \rightleftharpoons Ni(OH)_2 + 2e^- \qquad E° = +0.49\ V$$

Overall reaction:

$$Zn + 2NiO(OH) + 2H_2O \rightleftharpoons Zn(OH)_2 + 2Ni\ (OH)_2 \qquad E° = 1.73\ V \tag{66}$$

Side reactions in an alkaline electrolyte also take place at the end-of-charge and in the overcharge phase: oxygen is evolved on the positive electrode and hydrogen on the negative electrode (Equations 67).

$$\begin{gathered} 2H_2O + 2e^- \rightarrow H_2 + 2OH^- \\ 4OH^- \rightarrow O_2 + 2H_2O + 4e^- \end{gathered} \tag{67}$$

Self-Discharge

It has been previously described for Ni-Cd and Ni-MH batteries that NiOOH positive active mass is not stable in the KOH electrolyte. The reaction that ensues leads to oxygen generation at a low rate because of high oxygen overpotential. Similarly, the Zn electrode is not stable in the electrolyte either and low rate self-discharge reaction produces hydrogen.

Oxygen Cycle

Oxygen generated on the positive Ni electrode in the overcharge phase can be reduced at the negative electrode in the electrochemical reaction (Equation 58) or direct chemical reaction with Zn. The reaction product is $Zn(OH)_2$, which is then reduced back to zinc (Equation 68). In that way, the oxygen cycle is completed, and no pressure buildup occurs in the cell. The reactions that take place are:

$$\frac{1}{2}O_2 + H_2O + 2e^- \rightarrow 2OH^- \tag{58}$$

$$Zn(OH)_2 + 2e^- \rightarrow Zn + 2OH^- \tag{68}$$

The third reaction that can take place on the negative electrode (during charge) is hydrogen evolution (Equation 54).

$$4H_2O + 4e^- \rightarrow 2H_2 + 4OH^-$$

The oxygen reduction reaction has the most positive potential, which means that it will occur first, as long as oxygen is available. This prevents the hydrogen evolution, which would create a problem because the hydrogen oxidation rate on the positive electrode is very low.

Components

The positive electrode is identical to electrodes used in all other nickel batteries. The negative zinc electrode undergoes more complex reactions than shown above. The electrochemical reaction (Equation 65) is followed by two chemical reactions (Equations 69 and 70).

$$Zn + 2\,OH^- \rightarrow Zn(OH)_2 + 2e^- \qquad (65)$$

$$Zn(OH)_2 + 2\,OH^- \rightarrow [Zn(OH)_4]^{2-} + 2\,H_2O \qquad (69)$$

$$[Zn(OH)_4]^{2-} \rightarrow ZnO + 2\,H_2O + 2\,OH^- \qquad (70)$$

The complex set of reactions, resulting in a number of problems with soluble reaction products, includes:

- Shape change. The soluble zincate ions (ZnO_2^{2-}) migrate to the lower part of the electrode, where they deposit and create a convex growth.
- Dendrite formation. The soluble zincate ions (ZnO_2^{2-} and/or ZnO_4^{2-}) migrate into the separator where they are reduced (i.e., they "plate") during the subsequent charge step and form random, tree-like growths, called dendrites. Dendrites grow randomly and can also propagate directly through the separator and touch the positive electrode, causing a short circuit.
- Zinc agglomeration. All solids try to reduce their surface energy by increasing their particle size. This phenomenon is known as Ostwald ripening and is observed in solid solutions or liquid solutions. The inhomogeneous structures change over time by dissolving small crystals and re-depositing as larger crystals. This process is favored if well-soluble species are present such as zincate ions and leads to agglomeration and reduction in surface area. This, in turn, means lower current capability. The dissolution and re-crystallization as larger crystals also leads to electrical disconnection and isolation of parts of the active mass that results in capacity reduction.
- Passivation. The products of the precipitation reaction on the electrode surface, ZnO and $Zn(OH)_2$, create passivation layers. The thickness of these passivation layers depends on the concentration of Zn species, current density, and solubility of reduction products (which is affected by the temperature).

The processes described above are caused by the high solubility of zincate ions. This is amplified by the formation of reaction products inside the pores of a porous, 3D structure and subsequent lowering of the solubility limit. Additives to the active mass and electrolyte can reduce the solubility of zincate ions and stir the reaction away from harmful products such as dendrites, agglomeration, or reactions such as passivation. The additives used include calcium hydroxide, calcium zincate, calcium titanite, fluoride, phosphate, lead oxide, bismuth oxide, and others. For example, the addition of calcium hydroxide results in bonding of ZnO and preventing its redeposition (Equation 71).

$$Ca(OH)_2 + 2\,ZnO + 4\,H_2O + 4\,OH^- \rightleftharpoons Ca[Zn(OH)_3]_2 \cdot 2\,H_2O \tag{71}$$

An additional problem with the zinc electrode is that it is susceptible to corrosion. Since zinc is not stable in aqueous solutions it corrodes, along with hydrogen evolution. Because of high hydrogen overpotential on zinc the overall reaction is slow, but it is accelerated in the presence of cobalt, which is often found in trace amounts on the nickel electrode. Additives that can reduce corrosion rate include various alloys of Bi, In, and Pb, as well as some organic corrosion inhibitors.

The current collector for the negative electrode is made of copper-perforated foil, expanded metal, or wire mesh. The active mass is pressed as powder on a current collector or deposited using electrodeposition or pasting. Another type of electrode is a so-called plastic-bonded electrode where a polymer like Teflon is added to the active mass slurry to improve mechanical properties of electrode.

Electrolyte in Ni-Zn cells is 20 wt% potassium hydroxide solution. Higher KOH concentration would improve the nickel electrode capacity, but it would also increase the solubility of zincate. Additives such as KF and K_2CO_3 are used to reduce the solubility of zincate.

The main challenge when selecting a separator is to find an optimum between a small pore size that prevents dendrite protrusion and larger pore size necessary for fast oxygen transport in the overcharge phase. Separators initially used were made of celluloses and had a small pore size that successfully prevented dendrite protrusion but were susceptible to oxidation from oxygen produced in the charge and overcharge phase. A more stable class of separators was found in polypropylene materials with micropores although attention must be paid to sufficient wettability that allows electrolyte penetration into the pores. To minimize the risk of dendrite protrusion, multilayer separators with a larger pore size are often used.

Cell and Battery Types

Ni-Zn cells are usually sealed to prevent ingress of CO_2 from the atmosphere into the cell and reaction with electrolyte that produces carbonates, which subsequently precipitate. Zinc also reacts with CO_2 creating a charge imbalance between the electrodes.

The batteries are available in cylindrical and prismatic form factor. Cylindrical design is used for capacities of maximum 20 Ah and prismatic cells for higher capacities. The AAA cells (700 mAh) and AA cells (roughly 1500 mAh) are used in consumer applications as a substitute for alkaline primary cells. Both types of batteries are built as monoblocks or as single cells connected in a standard manner. Monoblocks have usually 12 V nominal voltage based on 7 cells connected in series. Ni-Zn batteries are also used in power tools and in a 12 V configuration for microhybrid EVs. They provide a higher rate of regenerative breaking than LABs. Ni-Zn batteries were also demonstrated in full HEVs.

New developments include flow Ni-Zn batteries with $NiOOH/Ni(OH)_2$ suspension.

Performance Factors

The nominal voltage for Ni-Zn cells is 1.65 V, while fully charged cells have 1.85 V open-circuit voltage.

The specific energy of Ni-Zn batteries is 70–110 Wh kg^{-1}, energy density is 130–350 Wh L^{-1}, specific power is 280–2500 W kg^{-1}, and power density is 420–7000 W L^{-1}. The capacity of a battery depends critically on the discharge rate and temperature (Figure 80).

The lifetime of Ni-Zn batteries is between 300 and 900 cycles, and depends on the DOD (Figure 81).

The self-discharge of Ni-Zn batteries is about 20% per month at 25°C.

As with all other nickel batteries, hydrogen and oxygen are generated in the overcharge phase. The rate of oxygen reduction on the negative electrode must be sufficiently high to prevent hydrogen evolution and pressure buildup. For safety reasons, cells are equipped with a release valve in case of pressure increase.

The cost of Ni-Zn batteries is roughly 25% lower than Ni-MH batteries and is controlled mainly by the price of Ni.

Charging

A two-stage charging protocol, consisting of constant current followed by constant voltage charging, is used for Ni-Zn cells. The constant current (CC) phase is terminated when a preset, temperature-compensated voltage is reached. The constant voltage (CV) phase is terminated when the current falls below a certain value or maximum charge

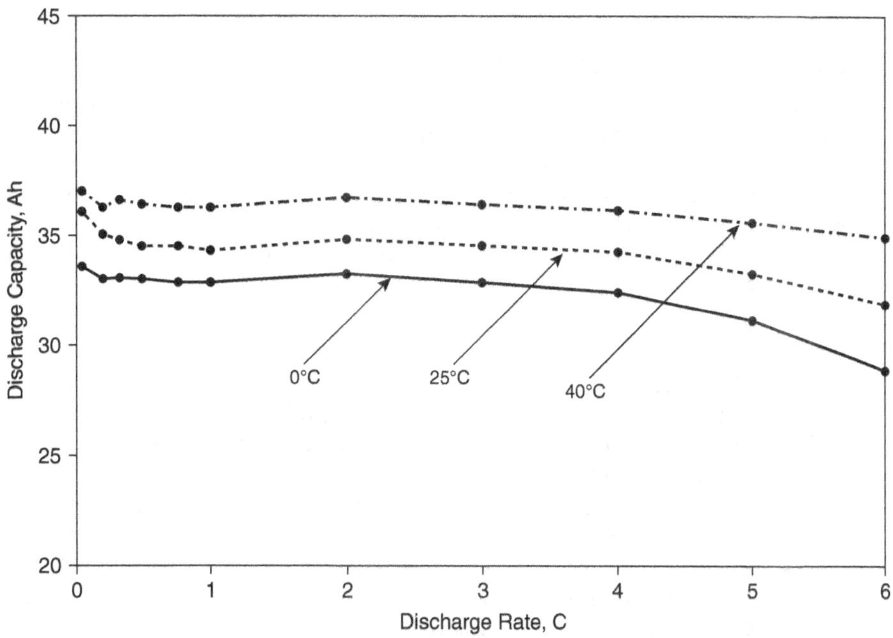

Figure 80 Discharge capacity of a 30 Ah Ni-Zn cell as a function of discharge rate for different temperatures. (Reproduced from D. Coates, A. Charkey, chap. 31, Nickel-Zinc Batteries, in *Linden's Handbook of Batteries*, ed. by B. Reddy, McGraw Hill Professional, 3rd ed., 2002.)

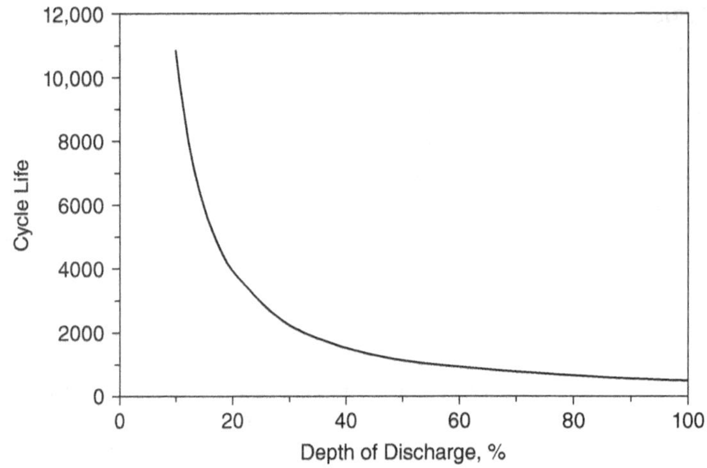

FIGURE 81 Ni-Zn cell cycle life as a function of DOD. (Reproduced from D. Coates, A. Charkey, chap. 31, Nickel-Zinc Batteries, in *Linden's Handbook of Batteries*, ed. by B. Reddy, McGraw Hill Professional, 3rd ed., 2002.)

time is reached. Ni-Zn cells can be fast charged in roughly 2.5 h for smaller systems, such as bicycle batteries.

Advantages and Disadvantages

The advantages of Ni-Zn batteries are fast recharge capability, good specific energy, low environmental impact, high safety, relatively low cost compared with Ni-Cd or Ni-MH batteries, and superior cycle life in comparison to lead-acid.

Their disadvantages are relatively low energy density, lower lifetime than Ni-Cd or Ni-MH batteries, and higher cost than LABs.

Ni-Zn batteries are likely to be replaced in most applications by Li-ion batteries as their cost further decreases.

Further Reading

- E. J. Cairns, Y. Borthomieu, P. Bernard, Nickel Systems: Nickel-Zinc, in *Encyclopedia of Electrochemical Power Sources*, ed. by J. Garche et al., Elsevier, Amsterdam, 2009, Vol 4, p. 582.
- D. Coates, A. Charkey, chap. 31, Nickel-Zinc Batteries, in *Linden's Handbook of Batteries*, ed. by B. Reddy, McGraw-Hill Professional, 3rd ed., 2002.
- M. Fetcenko, J. Koch, M. Zelinsky, chap. 6, Nickel-Metal Hydride and Nickel-Zinc Batteries for Hybrid Electric Vehicles and Battery Electric Vehicles, in *Advances in Battery Technologies for Electric Vehicles*, ed. by. B. Scrosati, J. Garche, W. Tillmetz, Woodhead, 2015, p. 117.
- J. Phillips, S. Mohanata, chap. 23, Nickel-Zinc Batteries, in *Linden's Handbook of Batteries*, ed. by B. Reddy, McGraw-Hill Professional, 4th ed., 2011.

Nickel-Iron Batteries

Introduction

The Ni-Fe battery was invented by Junger in 1899 and later produced by Svenska Ackumulatore Aktibolaget Junger. Edison worked on a Ni-Fe system as well and filed patents in 1901. He produced this battery in the Edison Storage Battery Company in East Orange, NJ, from 1903 to 1972, when the company was sold to Exide Battery Corporation. The production of this battery was discontinued in 1975.

This battery is characterized by very long lifetime and high resistance toward physical stress (e.g., temperature, vibration, etc.). The ability to deliver long life has played the major role in the recent resurgence of Ni-Fe batteries for stationary energy storage. Still, there are serious shortcomings of this system, mostly due to low energy efficiency attributed to the iron electrode.

Operating Principle

The negative electrode of the Ni-Fe battery undergoes a reaction of elemental iron with electrolyte to form iron hydroxide (Equations 72 and 73), while the reaction on the positive electrode is identical to electrode reactions in other nickel batteries (Equation 55).

Negative electrode:

$$\mathrm{Fe} + 2\,\mathrm{OH^-} \rightleftharpoons \mathrm{Fe(OH)_2} + 2\,\mathrm{e^-} \qquad \text{(1st plateau)} \qquad E^\circ = -0.88\ \mathrm{V} \qquad (72)$$

$$\mathrm{Fe(OH)_2} + \mathrm{OH^-} \rightleftharpoons \mathrm{FeO(OH)} + \mathrm{H_2O} + \mathrm{e^-} \qquad \text{(2nd plateau)} \qquad E^\circ = -0.56\ \mathrm{V} \qquad (73)$$

Positive electrode:

$$\mathrm{NiO(OH)} + \mathrm{H_2O} + \mathrm{e^-} \rightleftharpoons \mathrm{Ni(OH)_2} + \mathrm{OH^-} \qquad E^\circ = 0.49\ \mathrm{V}$$

Overall reactions (Equation 74):

$$\mathrm{Fe} + 2\,\mathrm{NiOOH} + 2\,\mathrm{H_2O} \rightleftharpoons \mathrm{Fe(OH)_2} + 2\,\mathrm{Ni(OH)_2} \qquad \text{(1st plateau)} \qquad E^\circ\ \text{(cell)} = 1.37\ \mathrm{V} \qquad (74)$$

$$3\,\mathrm{Fe\ (OH)_2} + 2\,\mathrm{NiOOH} \rightleftharpoons \mathrm{Fe_3O_4} + 2\,\mathrm{Ni(OH)_2} + 2\,\mathrm{H_2O} \qquad \text{(2nd plateau)} \qquad E^\circ\ \text{(cell)} = 1.05\ \mathrm{V} \qquad (75)$$

As with other nickel batteries, there are side reactions that lower the efficiency or increase self-discharge rate. Both electrodes are involved in these unwanted reactions of hydrogen evolution on the negative electrode and oxygen evolution on the positive electrode, especially near the end-of-charge and in the overcharge phases.

Negative electrode:

$$2\,\mathrm{H_2O} + 2\,\mathrm{e^-} \rightarrow \mathrm{H_2} + 2\,\mathrm{OH^-}$$

Positive electrode:

$$4\,\mathrm{OH^-} \rightarrow \mathrm{O_2} + 2\,\mathrm{H_2O} + 4\,\mathrm{e^-}$$

Because of the lower hydrogen reduction potential on iron, the hydrogen generation on the negative electrode starts earlier (before the end-of-charge) than in the case of other negative electrodes in nickel batteries. This fact significantly lowers the energy efficiency of the Ni-Fe battery.

Self-Discharge

The self-discharge reaction on the positive electrode is identical to reactions in other nickel batteries because NiOOH is not stable in the presence of oxygen and in the electrolyte. In addition, iron electrode is not stable in electrolyte either, enabling a self-discharge reaction.

Components

As expected, the positive electrode is the same as in other nickel batteries. The negative electrode is iron, in which the voltage curve during discharge and charge is showing two plateaus (Figure 82).

The anode current collector is made of steel strips or perforated, nickel-plated ribbon. The strips are formed into pockets, with an open end through which iron active material is filled. The starting negative active mass is a mixture of iron and iron oxide (Fe_3O_4) to which additives are also included to reduce gas evolution at the end-of-charge. After applying the active mass, the pockets are crimped and pressed into openings in the nickel-plated steel frame, to make a design similar to a pouch cell.

Besides this type of electrode, advanced designs include sintered or fibrous electrodes.

Electrolyte used in Ni-Fe cells is 25–30% KOH solution with up to 50 g/L of LiOH, for higher capacity and prevention of capacity loss during cycling. The separator is most commonly a nonwoven polypropylene sheet because of the susceptibility of polyamide fabric material to hydrolysis at temperatures above 55°C.

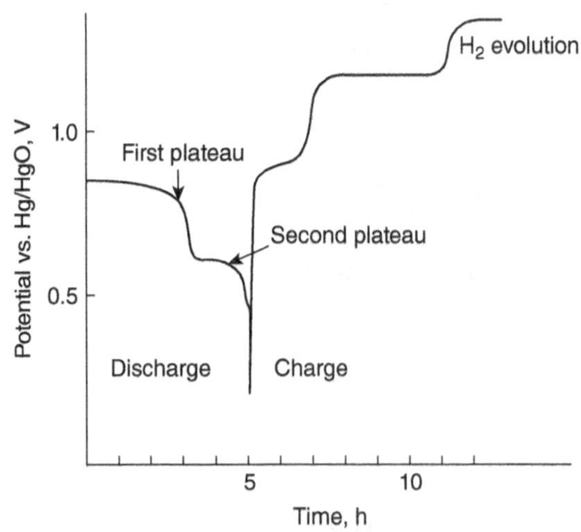

FIGURE 82 Potential of iron electrode as a function of time during discharge and charge of Ni-Fe battery. (Reproduced from G. A. Bayles, chap. 18, Iron Electrode Batteries, in *Linden's Handbook of Batteries*, ed. by B. Reddy, McGraw Hill Professional, 4th ed., 2011.)

Cell and Battery Types

Unlike other nickel batteries, the excessive hydrogen evolution rate on the iron electrode makes the oxygen cycle a less-than-satisfactory solution and, in most cases, precludes the use of sealed cells. However, there is a version of a sealed Ni-Fe cell based on a concept of recombining hydrogen and oxygen to water on a Pt catalyst located in the overhead space of a cell. This approach is safe only if the gases are generated in a stoichiometric ratio 2:1. The single-cell batteries are built in the range from 10 to 1200 Ah. Larger batteries are available in 5-cell configurations to give 6 V, 10 cell (12 V), 20 cell (24 V), and 40 cells (48 V).

Performance Factors

The open-circuit voltage of Ni-Fe cells is 1.4 V and the average discharge voltage is 1.2 V. The theoretical specific energy is 265 Wh kg^{-1}, higher than Ni-Cd batteries (244 Wh kg^{-1}), but lower than Ni-MH (278 Wh kg^{-1}). The practical specific energy is between 20 and 30 Wh kg^{-1} for conventional pouch cells and 40 to 60 Wh kg^{-1} for cells with sintered or fibrous electrodes. The energy density for conventional cells is between 35 and 45 Wh L^{-1} and the specific power is up to 100 W kg^{-1}.

The capacity of Ni-Fe cells depends on the discharge rates and temperature (Figure 83).

Ni-Fe batteries have extraordinary lifetime at temperatures below 30°C of roughly 5000 cycles. The US company Encell reported about 11,000 at 80% DOD and roughly 30 years of use. There are some Ni-Fe batteries in operation for more than 50 years. However, the lifetime is greatly reduced at temperatures above 30°C due to faster rate of carbonate buildup in the electrolyte from atmospheric CO_2. In some cases, it is possible to replace electrolytes to extend the lifetime of a battery.

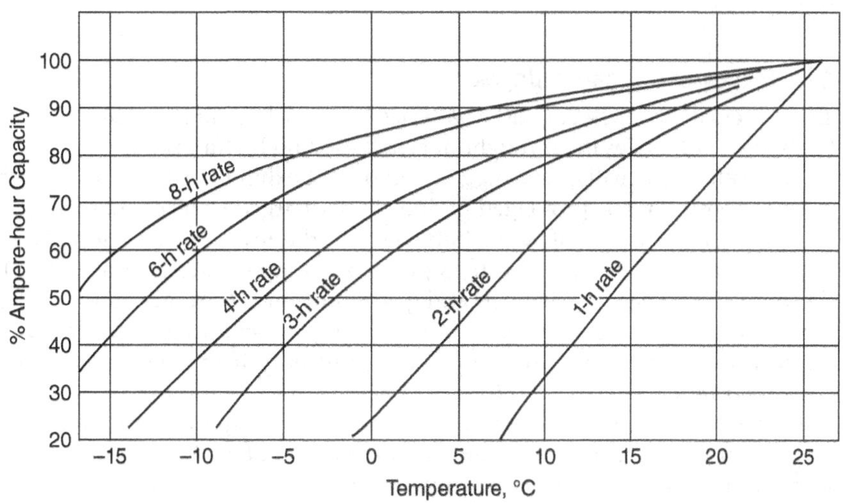

FIGURE 83 Capacity of a Ni-Fe cell as a function of temperature for different discharge rates. (Reproduced from G. A. Bayles, chap. 18, Electrode Batteries, in *Linden's Handbook of Batteries*, ed. by B. Reddy, McGraw Hill Professional, 4th ed., 2011.)

The self-discharge rate of Ni-Fe cells depends on the temperature. At 0°C the capacity loss is roughly 10% per month, while at 45°C the cell is completely discharged after about 15 days.

Typical commercial Ni-Fe batteries have the efficiency of up to 60%, while 80% has been reported recently for some advanced systems.

As previously mentioned, most Ni-Fe batteries are the open or flooded type, with hydrogen evolution on the negative electrode and oxygen on the positive in the overcharge phase. This leads to loss of water from the electrolyte as a result of water electrolysis and requires periodical (every 1–4 months) addition of water to restore the electrolyte.

The main safety concern with Ni-Fe batteries is, as already indicated, hydrogen evolution at advanced stages of the charge process, but also in the rest phase. Unless the battery is equipped to enable recombination of hydrogen and oxygen when in stoichiometric ratio, the battery cells have to be equipped with vents through which the gases escape and prevent explosion. As with all other nickel batteries, the alkaline electrolyte's hazardous and poisonous nature must be taken into account when designing battery safety protections to prevent leakage and possible injuries.

Ni-Fe battery cost is lower than the other nickel batteries, but higher than lead-acid. When the total lifetime is taken into account, the total or sometimes called throughput cost is very low: $0.02–0.04/kWh.

Charging

The optimum charge rate (i.e., current expressed in C-rate) for Ni-Fe cells is C/4, while the maximum rate should not exceed C/2. Sometimes even higher rates are possible as long as the temperature does not exceed 45°C. Anytime during the charge, if the temperature exceeds this temperature or if the cell voltage reaches 1.85 V the charge should be discontinued. While overcharge is not particularly damaging to the cell, the loss of water through electrolysis must be controlled by replenishing water.

Advantages and Disadvantages

The principal advantage of Ni-Fe batteries is a very high lifetime. In addition, the relatively low cost makes the throughput cost (based on lifetime) extremely low.

The cells are usually very rugged and do not exhibit particular sensitivity to overcharge, overdischarge, prolonged periods in discharged state, or short circuiting.

The disadvantages of Ni-Fe batteries are the low specific energy, energy density, and power density; as well as poor low-temperature performance, high self-discharge, low energy efficiency, and gas evolution at the end-of-charge. Renewable energy applications may be one of the primary target applications for Ni-Fe batteries because of their low throughput cost, but only if the energy efficiency can be improved to above 80%. Additional concepts have been proposed, such as collection and use of hydrogen produced in the overcharge and rest phase. It should come as no surprise that some new manufacturers of Ni-Fe batteries have emerged: Zapp Works (USA), Kursk Accumulator Plant (Russia), and Changhong Battery Manufacture (China).

Further Reading

- G. A. Bayles, chap. 18, Iron Electrode Batteries, in *Linden's Handbook of Batteries*, ed. by B. Reddy, McGraw-Hill Professional, 4th ed., 2011.
- Iron Edison Ni-Fe Owner Manual, 2017, https://ironedison.com/images/DOCS/Iron_Edison_Nickel_Iron_Owner%27s_Manual.pdf.
- J. F. Jackovitz, G. A. Bayles, chap. 25, Iron Electrode Batteries, in *Linden's Handbook of Batteries*, ed. by B. Reddy, McGraw-Hill Professional, 3rd ed., 2002.
- A. K. Shukla, B. Hariprakash, Nickel Systems: Nickel-Iron, in *Encyclopedia of Electrochemical Power Sources*, ed. by J. Garche et al., Elsevier, Amsterdam, 2009, Vol 4, p. 522.

CHAPTER 4

Silver Batteries

Introduction

Silver batteries, or more correctly silver oxide batteries based on the cathode material in the charged state, have been developed for high specific energy and high-power applications. They are based on silver oxide cathode, alkaline electrolyte, and a number of different anodes, such as Zn, Fe, Cd, and Cu. These batteries were first introduced as primary batteries, but it was established fairly quickly, around 1900, that rechargeable versions were possible.

Silver batteries are characterized by high cost and low cycle life mainly due to separator oxidation on the cathode side. Early versions were also prone to dendrite buildup, but since the 1940s the use of a semipermeable cellophane membrane as a separator has minimized the problems with short circuiting, and consequently the lifetime has been improved.

Silver batteries have been commercially available in the Ag-Cd version since the 1940s, and currently two additional combinations, Ag-Zn and Ag-Fe, are available.

Operating Principle

The overall reactions in silver batteries with different anodes are shown in Equations 76 to 78. In each case, the discharge reaction on the cathode is reduction of silver from oxide to its elemental form. AgO is not silver(II)-oxide but a mixed oxide $Ag_2O_3{\cdot}Ag_2O$.

$$AgO + Zn + H_2O \rightleftharpoons Zn(OH)_2 + Ag \tag{76}$$

$$AgO + Cd + H_2O \rightleftharpoons Cd(OH)_2 + Ag \tag{77}$$

$$4AgO + 3Fe + 4H_2O \rightleftharpoons Fe_3O_4 + 4H_2O + 4Ag \tag{78}$$

Components

While all possible anode materials are discussed in the text in the context of nickel electrodes, the silver oxide cathode deserves further elucidation. During discharge, silver active mass undergoes reductions that correspond to two potential plateaus (Equations 79 and 80).

$$2AgO + H_2O + 2e^- \rightleftharpoons Ag_2O + 2OH^- \tag{79}$$

$$Ag_2O + H_2O + 2e^- \rightleftharpoons 2Ag + 2OH^- \tag{80}$$

The cathode is made from silver powder by sintering it (coalescing it into a solid by heating and sometimes also compressing it without liquefaction) onto a supporting, woven or expanded (or stretched), silver grid. In some cases, a silver-plated copper grid is used. An alternate, dry process method involves preparation of silver slurry with binders and pressing or pasting it onto the grid. The active mass initially prepared in this way must be then electrochemically treated and converted to silver oxide (AgO). After the preparation is completed, the capacity of the electrode depends on the active mass particle size, and finely divided particles of the active mass can produce nearly a theoretical silver utilization of 2.0 g/Ah.

Electrolyte for silver batteries is a 35–40% KOH solution with LiOH as an additive. Multilayer polyamide or polypropylene separators are used to decrease the risk of dendrite growth. The separator layer on the cathode side has additional functions to serve as an electrolyte reservoir and as a barrier to reduce the oxidation of the main separator layer by preventing a direct contact. It should be noted again that separator oxidation is the main cause of short calendar life in silver batteries.

Cell and Battery Types

Silver (or silver oxide) cells are made in prismatic, cylindrical, and button form factors. In most cases, the cell containers are made of plastic, acrylonitrile-styrene copolymer (SAN), or acrylonitrile butadiene styrene (ABS) to reduce susceptibility to strong combined chemical attacks by the KOH electrolyte and AgO cathode active mass.

The silver battery cells are usually not fully sealed but are equipped with a vent cap or a low-pressure release valve. Because of the short calendar life, they are delivered dry, without electrolyte, in charged or discharged condition and have to be activated before use by adding electrolyte.

An example of a Ag-Zn battery with a number of cells connected in series and in parallel is a 240-V, 850-Ah battery used as a backup by U.S. Navy submersibles.

Performance Factors

Ag-Zn batteries have specific energy of 130 Wh kg^{-1} and energy density of about 500 Wh L^{-1}, and can deliver discharge power of 20 C. The high rate of discharge is possible as long as there is no temperature increase in the cell.

The lifetime of these batteries is only 50 cycles, but the U.S. company ZPower announced the release of cells with more than 100 cycles.

Silver-cadmium batteries have specific energy of 80–100 Wh kg^{-1} and a longer cycle life typical for Ni-Cd batteries. They are used in space applications requiring nonmagnetic properties, such as Giotto Halley Comet intercept spacecraft.

Silver-iron batteries have specific energy of about 100 Wh kg^{-1}, energy density of 160 Wh L^{-1}, and cycle life of over 300 cycles at deep discharge. They are also tolerant to overcharge, deep discharge (very low voltage), and overdischarge. As with other batteries with iron anodes, the concern exists with the necessity for gas and water management.

All three versions of silver battery have a fairly flat discharge curve due to the resistance balancing that takes place. On the cathode, the formation of metallic silver during discharge decreases resistivity balancing thereby resistivity increase that comes from the anode active mass at increased depth-of-discharge.

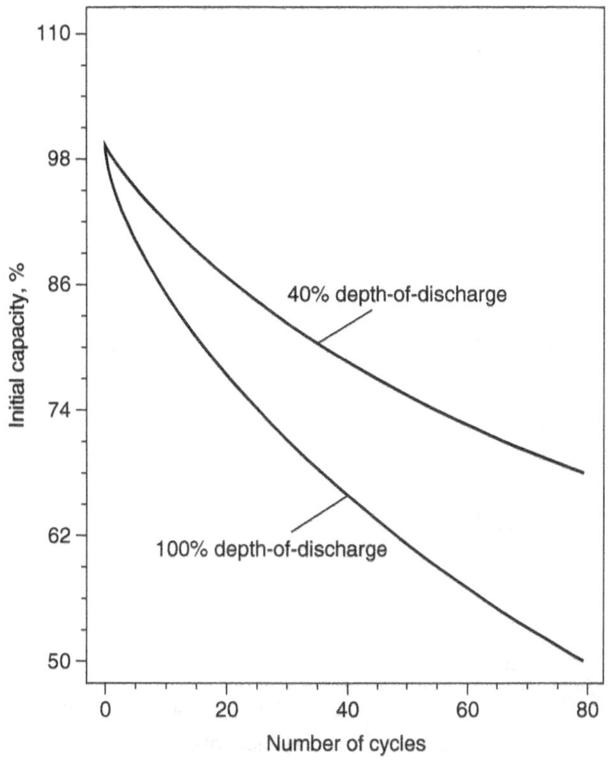

FIGURE 84 Capacity change with the number of cycles for different depths-of-discharge for Ag-Zn batteries. (Reproduced from A. P. Karpinski, St. F. Schiffer, P. A. Karpinski, chap. 38, Silver Oxide Batteries, in *Linden's Handbook of Batteries*, ed. by B. Reddy, McGraw Hill Professional, 3rd ed., 2002.)

Capacity declines with the number of cycles for Ag-Zn (Figure 84) and Ag-Cd (Figure 85) batteries.

Charging

Silver batteries are typically charged at a 10 h or 20 h rate. Charging is terminated when the voltage reaches 2.0 V for Ag-Zn, 1.9 V for Ag-Fe, and 1.5 V for Ag-Cd cells. Constant potential and pulsed charging techniques are sometimes also used.

Overcharge can be problematic and can lead to water loss for Ag-Fe and Ag-Zn cells and to Zn dendrite growth. The Ag-Cd battery is less sensitive to overcharge.

Advantages and Disadvantages

Silver-zinc batteries have high specific energy and energy density, high rate of discharge, low self-discharge, and good safety. Their disadvantages are high cost, low lifetime, poor temperature behavior, and high sensitivity to overcharge.

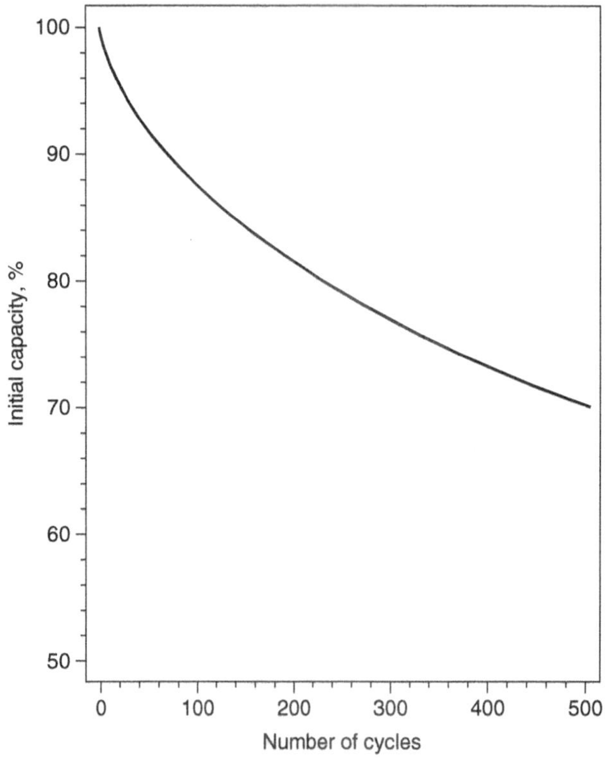

FIGURE 85 Capacity change with the number of cycles at 100% DOD for Ag-Cd batteries. (Reproduced from A. P. Karpinski, St. F. Schiffer, P. A. Karpinski, chap. 38, Silver Oxide Batteries, in *Linden's Handbook of Batteries*, ed. by B. Reddy, McGraw Hill Professional, 3rd ed., 2002.)

Silver-cadmium batteries also have high specific energy and energy density, high rate of discharge, low self-discharge, and good safety. Their disadvantages are high cost, low lifetime, poor temperature behavior, and sensitivity to overcharge.

Silver-iron batteries have high specific energy and energy density, high power capability, good lifetime, good safety, and nonmagnetic properties. Their disadvantages are high cost and poor temperature behavior.

Based on the combination of their advantages and disadvantages, and in particular their high cost, silver batteries are mainly used for niche applications. They are likely to be replaced by Li-ion batteries in those applications in the near future.

Further Reading

- G. A. Bayles, chap. 18.6, Silver-Iron Battery, in *Linden's Handbook of Batteries*, ed. by B. Reddy, McGraw-Hill Professional, 4th ed., 2011.
- P. Karpinski, chap. 25, Silver Oxide Batteries, in *Linden's Handbook of Batteries*, ed. by B. Reddy, McGraw-Hill Professional, 4th ed., 2011.
- P. Karpinski, St. F. Schiffer, P. A. Karpinski, chap. 33, Silver Oxide Batteries, in *Linden's Handbook of Batteries*, ed. by B. Reddy, McGraw-Hill Professional, 3rd ed., 2002.

CHAPTER 5

Aqueous Metal-Air Batteries

Metal-air batteries are hybrid power sources, where one electrode contains active mass as in batteries and the other electrode contains catalysts which is identical to fuel cell electrodes. These batteries comprise a metal anode, a battery-like electrode, and an air cathode, a fuel cell-like electrode (Figure 86). The electrolyte is usually alkaline, although there are some versions with neutral aqueous electrolyte and nonaqueous electrolytes.

There is no cathode active mass because the reactant is oxygen from air, and that is why the cell has very high specific energy. The anode metal should be light, have potential more negative than hydrogen, have a high number of exchanged electrons, be readily available at low cost, and be environmentally friendly. The candidate systems including anodes such as Zn, Fe, Mg, Al, Na, K, and Li can be compared based on properties (Table 14). With the exception of Zn-air batteries, metal-air batteries are at the laboratory-scale development phase.

Reactions on the metal-air battery electrodes can be shown in a generalized form, where M represents any of the anode metals (Equations 81 to 84).

Negative electrode:

$$M + n\,OH^- \rightleftharpoons MO_{n/2} + n/2\,H_2O + n\,e^- \tag{81}$$

$$M + n\,OH^- \rightleftharpoons M(OH)_n + n\,e^- \tag{82}$$

Positive electrode:

$$\frac{1}{2}O_2 + H_2O + 2e^- \rightleftharpoons 2OH^-$$

Overall reaction:

$$M + n/4\,O_2 \rightleftharpoons MO_{n/2} \tag{83}$$

$$M + n/4\,O_2 + n/2\,H_2O \rightleftharpoons M(OH)_n \tag{84}$$

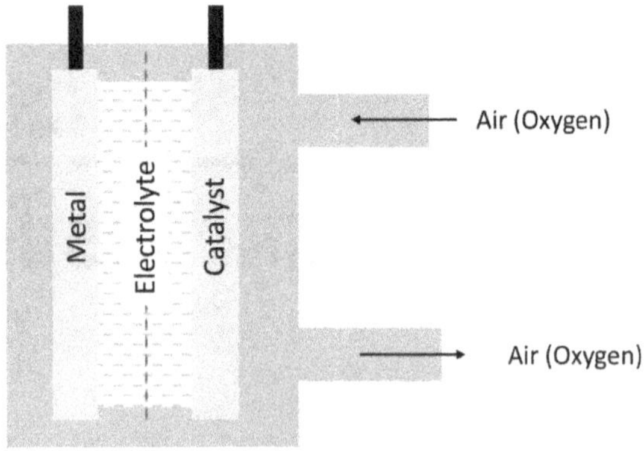

Figure 86 Simple schematic depiction of a metal-air cell, where metal anode is on the left and oxygen cathode is on the right.

Based on their electrochemical potential (or standard reduction potential), Zn and Fe can both be charged and discharged in aqueous solutions. On the other hand, Al and Mg can only be discharged but not charged because of the competition with hydrogen evolution (i.e., reduction) reaction.

The reactions during the discharge phase, affected by self-discharge, are highly exothermic and accompanied by significant entropy change. The metal-air batteries are, because of that, provided with thermal management auxiliary systems. The ratio of heat generation and electrical power varies for different metal-air systems (Table 15). It should be obvious that better thermal management is necessary for systems with higher ratio.

Zinc-Air Batteries

Introduction

The first zinc-air (Zn-air) battery was demonstrated in a primary (i.e., nonrechargeable) version by G. Leclanche in 1868. The battery comprised metal zinc anode and manganese oxide (MnO_2) cathode active material on carbon. The carbon in this active mass mixture was used as a catalyst for oxygen reduction. Then, in 1953, Edison Industries introduced a large primary Zn-air battery with carbon cathode, replacing copper oxide previously used. The battery, named Carbonair, was used for long-term, low-power applications.

Further improvements characterized the next phase of development as the principles of fuel cell electrode performance were used to optimize the thickness of air electrodes and enable construction of small button and prismatic primary cells for medical applications. At the moment, the only commercially available Zn-air batteries are primary systems while rechargeable batteries have been in the demonstration phase.

Battery System	Fe-air	Zn-air	Al-air	Mg-air	Na-air	K-air	Li-air
Year invented	1968	1878	1962	1966	2012	2013	1996
Cost of metals ($ kg^{-1})*	0.40	1.85	1.75	2.75	1.70	~20	68
Theoretical voltage (V)	1.28	1.65	2.71	3.09	2.27	2.48	2.96
Theoretical specific energy (Wh kg^{-1})†	76.3	1086	2796	2840	1106	935	3458
Electrolyte in practical batteries	Alkaline	Alkaline	Alkaline in saline	Saline	Aprotic	Aprotic	Aprotic
Practical voltage (V)	~1.0	1.0–1.2	1.1–1.4	1.2–1.4	~2.2	~2.4	~2.6
Practical specific energy (Wh kg^{-1})	60–80	350–500	300–500	400–700	Unclear‡	Unclear‡	Unclear‡
Primary (P) or electrically rechargeable (R)	R	R	P	P	P	R	R

(Reproduced from Y. Li, H. Dai, *Chem. Soc. Rev.*, 43, 2014, 5257–5275.)
*Data source: http://www.metalprices.com.
†Oxygen inclusive.
‡Reported values in literature were normalized to mass of catalyst.

Table 14 Overview of Several Metal-Air Systems

Electrochemical Couple	Heat Generation (W)/Electrical Power (W), at 40 mA cm^{-2}
Al-air	1.15
Mg-air	1.78
Fe-air	0.61
Zn-air	0.46
Cd-air	0.55

TABLE 15 Overview of Heat Generation and Performance Factors for Several Metal-Air Batteries During Discharge (Calculated from the Thermoneutral Voltage and the Working Voltage at 40 ma/cm^2)

Principle of Operation

The power-producing reactions in the zinc-air battery involve two-stage zinc oxidation to zinc oxide on the anode (Equations 85 and 87) and familiar oxygen reduction on the cathode.

Negative electrode:

$$\mathrm{Zn} + 4\mathrm{OH}^- \rightleftharpoons [\mathrm{Zn(OH)_4}]^{2-} + 2e^- \qquad E° = -1.266\ \mathrm{V} \tag{85}$$

$$[\mathrm{Zn(OH)_4}]^{2-} \rightleftharpoons \mathrm{ZnO} + \mathrm{H_2O} + 2\mathrm{OH}^- \qquad \text{(chemical reaction)} \tag{86}$$

Positive electrode:

$$\frac{1}{2}\mathrm{O_2} + \mathrm{H_2O} + 2e^- \rightleftharpoons 2\mathrm{OH}^- \qquad E° = 0.401\ \mathrm{V\ at\ pH} = 14$$

Overall discharge reaction in the cell proceeds to ZnO on the anode (Equation 87).

$$\mathrm{Zn} + \frac{1}{2}\mathrm{O_2} \rightleftharpoons \mathrm{ZnO} \qquad E° = 1.667\ \mathrm{V\ at\ pH} = 14 \tag{87}$$

Further depiction of the reactions and potentials in the zinc-air battery reveals the transitions from charge to discharge for the battery as well as overcharge phase (Figure 87).

A cross-section of a cell (Figure 88) reveals a two-stage zinc oxidation reaction during discharge, direction of movement of hydroxyl ions and water, and a structure of the cathode with a gas diffusion layer (GDL).

Components

The negative electrode in Zn-air batteries exhibits very similar reactions to zinc electrode as previously discussed in Ni-Zn batteries. The main concerns remain the hydrogen evolution reaction near the end-of-charge and in the overcharge phase and unwanted reactions such as zinc dendrite formation and shape change. To suppress these undesirable reactions, alloys, surface coatings, and additives have been used (Table 16).

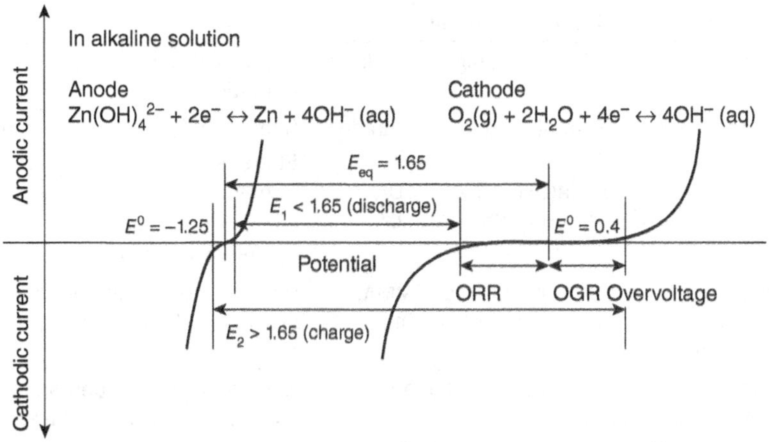

Figure 87 Current-voltage dependence in zinc-air cell shown for the individual electrodes and the cell. ORR is the oxygen reduction reaction and OGR is the oxygen generation reaction.

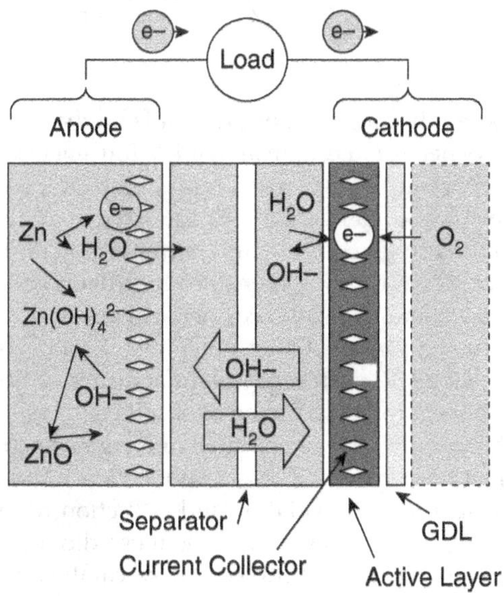

Figure 88 Diagram of zinc-air battery during discharge. Zinc electrode (on the left) undergoes two-stage oxidation of zinc to ZnO and oxygen reduction (on the right) produces hydroxyl ions, which then transport the charge through the electrolyte to the anode. (Reproduced from ReVolt.)

The oxygen reduction reaction on the cathode progresses according to a four-electron mechanism, a two-electron mechanism (Equations 58, 88, and 89), or a combination of the two.

$$O_2 + 2H_2O + 4e^- \rightarrow 4OH^- \quad (58)$$

$$O_2 + H_2O + 2e^- \rightarrow O_2H^- + OH^- \quad (88)$$

$$O_2H^- \rightarrow OH^- + \frac{1}{2}O_2 \quad (89)$$

Modifications to the Zn Electrode	Effects
Alloys with Pb, Cd, Bi, Sn, In, Mg, Al, or Ni	Supress H_2 generation, reduce dendrite formation, improve cycling reversibility
Surface coatings with Al_2O_3 or lithium boron oxide Inorganic additives $Ca(OH)_2$, $B_{i2}O_3$, Tl_2O_3, Ga_2O_3, In_2O_3, $In(OH)_3$, HgO, PbO, CdO, or silicates	Supress H_2 generation and self-discharge Supress H_2 generation, reduce dendrite formation Improve discharge performance and cycling reversibility
Polymer additives: ionomers, PEG, PMMA, polypyrole, polyaniline, polyvinyl acetate, and polycarbonate	Restrict the dissolution of discharge product; reduce dendrite formation and shape change
Surfactant additives: perfluorosurfactants, CTAB, TBABr, tetra alkyl ammonium hydroxides, triethanolamine, or lignosulfonate	Supress H_2 generation and electrode corrosion, reduce dendrite formation

(Reproduced from Y. Li. H. Dai, *Chem. Soc. Rev.*, 43, 2014, 5257–5275.)

Table 16 Electrode Modifications Toward Zn Electrode Performance Improvement

The ratio between the two reactions depends on the catalyst. Nonnoble metal catalysts promote a two-electron mechanism while a four-electron reaction path proceeds at a low rate. Platinum group metals (ruthenium, rhodium, palladium, osmium, iridium, and platinum) are not suitable catalysts for oxygen reduction in zinc-air batteries mainly because of their high cost, but also because they reduce the hydrogen overvoltage at the negative electrode, which enables hydrogen generation and in turn lowers the charging efficiency and intensifies corrosion of zinc (i.e., chemical reaction to oxidize zinc without producing power). Other noble metal catalysts, such as expensive silver and gold, are occasionally (i.e., in some manufacturing solutions) used because they do not have a strong effect on the hydrogen overvoltage.

During the charge process, the potential of oxygen electrode appreciably increases, causing the catalyst to oxidize and lose activity. Hence, it is critical to employ special bifunctional catalysts for both oxidation and reduction reactions. These are, for example, perovskites (e.g., $La_{1-x}Ca_xCoO_3$) or manganese dioxide, usually coated on carbon substrates from permanganate solutions. These catalyst replacements work well for either oxidation or reduction but not for both reactions and do not attain overall necessary activity. Hence, a configuration with two positive electrodes has been examined, whereby one of the electrodes operates during charge and the other during discharge. For instance, a cobalt oxide catalyst on the first positive electrode can be used for an O_2 reduction reaction, and a nickel-iron catalyst on the second positive electrode with CNT (carbon nanotube) support is used for the O_2 evolution.

Since the positive electrode cannot be sealed against the environment, the overall water management in the cell is challenging. High-humidity air leads to too much water in the pores of electrodes and electrode flooding, which lowers the rate of reactions. On the other hand, low humidity causes dehydration, increase in resistivity, and reduction of lifetime. Also, open-cell design makes ingress of CO_2 from air into electrolyte possible. Then, reaction of carbon dioxide with alkaline electrolyte ensues, and carbonates are produced causing conductivity loss in electrolyte and subsequent precipitation.

A concentrated KOH solution is typically used, except for low-temperature applications where NaOH is preferred. Since the evolution of hydrogen in the overcharge phase is the most serious concern, additives to electrolytes such as Pb, Sb, Bi, Cd, and Ga or organic inhibitors are used to inhibit the undesired reaction. The shape change problem with a zinc electrode is averted or minimized by using additives in the electrolyte such as calcium hydroxide, fluoride, phosphate, borate, and carbonate. Other issues affecting the reactions and lowering efficiency include excessive water absorption by hygroscopic KOH and flooding of the electrode pores, which lowers the rate of oxygen diffusion to reaction sites and overall rate of reaction. The carbonization problem of CO_2 reaction in alkaline electrolytes has been pointed out earlier; in order to eliminate such a problem, neutral electrolytes have been investigated, but these are ultimately rejected because of large overpotentials for oxygen reduction on the cathode, resulting in loss of voltage.

The fluctuation of oxygen electrode potential with current depends on the catalysts and electrolyte additives (Figure 89).

Different materials have been considered for the separator, including polyethylene, polyvinyl alcohols, polyolefins, and polypropylene.

Battery Types

The first electrode under consideration is electrically rechargeable. It incorporates, in the most common design, a zinc electrode sandwiched between two air electrodes (Figure 90).

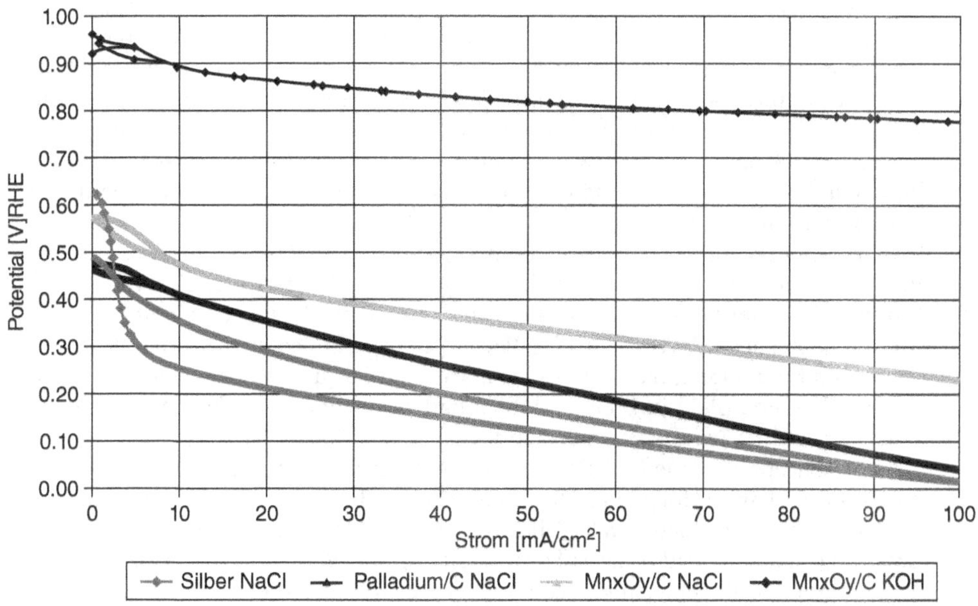

FIGURE 89 Voltage of oxygen electrode as a function of current for different catalysts (Ag, Pd/C, MnO_x) in alkaline (KOH) and neutral (NaCl) electrolytes. (Reproduced from H. J. Kohnke, DECHEMA Weiterbildung Frankfurt/M., Germany, December 4, 2012.)

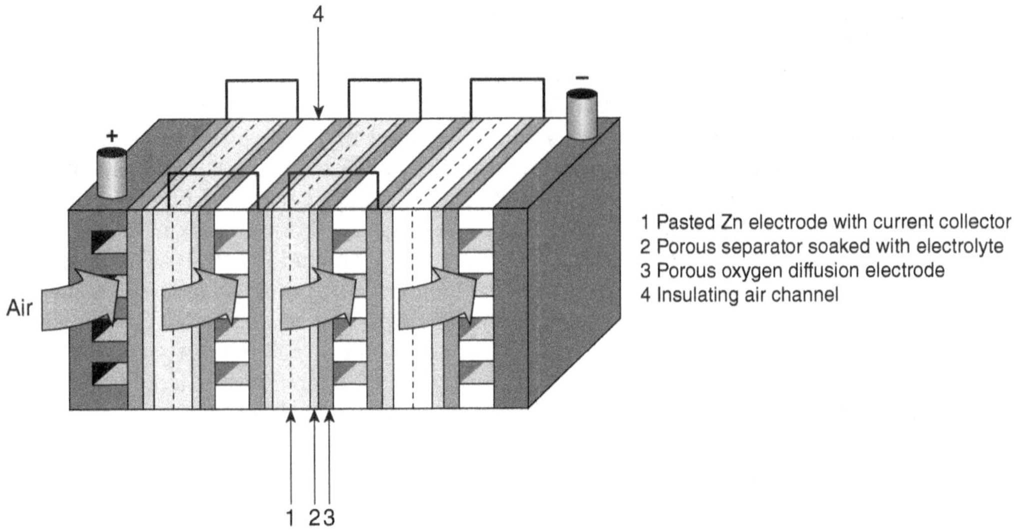

FIGURE 90 Electrically rechargeable zinc-air battery. (Reproduced from O. Haas, J. van Wesemael, Secondary Batteries—Metal-Air Systems: Zinc-Air, Electrical Recharge, in *Encyclopedia of Electrochemical Power Sources*, ed. by J. Garche, Elsevier, 2009, Vol. 4, p. 384.)

The layers shown in light gray represent zinc electrodes, while white sections show air channels of the cathodes. The black plate on the left is the end-oxygen electrode.

The complete stack (also called pack) system contains air supply controllers to balance water and air supply. Also, for high discharge rates (i.e., currents) a pump is used to increase transport of air into the pores, while in the rest phase the air inlets are closed to reduce carbon dioxide diffusion into electrolyte and carbonation reaction as well as loss of water through evaporation and dehydration. Another solution involves the use of CO_2 absorbers such as soda lime, LiOH, or molecular sieves to prevent carbonation. Service requirements for such absorber material depends on air flow rate, temperature, and humidity.

Another concept tested in the 1990s involved mechanical recharging for Zn electrodes in the form of removable cassettes. Such a system was built by electrical fuel (EFL) and tested in German Postal vehicles. Between 1994 and 1998, Deutsche Post had 13 transporters, Mercedes MB410s and Vito Vans, converted to Zn-air batteries. The vehicles were used for a total of 60,000 km. The zinc electrodes in cassettes were reconditioned in a reprocessing plant in Bremen, which had a capacity of 100 kg Zn/h. The driving range for vehicles was between 300 and 350 km.

The concept of a zinc-air battery with mechanical recharging was also used to construct a flow battery for EV applications. The mechanical recharging or reloading of zinc slurry (i.e., zinc active mass) has the potential to shorten the battery recharging times. Due to the hybrid nature of such systems, it is sometimes called a Zn fuel cell.

The best results for a Zn fuel cell included energy density of 110 Wh kg^{-1} and peak power of 80 W/kg, with energy efficiency of 40%. Main development problems involved uniform distribution of zinc particles and charge limitation because of ZnO supersaturation in the electrolyte (280–300 g ZnO/L). Several organizations worked on these systems including SGE (France), Sony, Lawrence Berkeley Laboratory, Metallic Power Inc., Power Air, Revolt, EOS, Zimair Inc., Fluidicenergy, and EFL-Arotech.

Performance Factors

The theoretical cell voltage for a Zn-air cell is 1.65 V, the open-circuit voltage is 1.45 V, and the cell discharge voltage is 1.2–0.9 V. The practical specific energy density is 350–600 Wh kg^{-1} for primary batteries and only 120–180 Wh kg^{-1} for secondary Zn-air batteries. The lifetime for rechargeable cells is 100–400 cycles, depending on the depth-of-discharge, load (discharge current), and temperature. Higher cycle life can be achieved by using neutral electrolytes and ionic liquids. The characteristics of some systems are listed in Table 17.

The electrical efficiency of zinc-air cells is typically low, around 50% as a result of high polarization (i.e., voltage change) of the air electrode during charge and discharge. The efficiency can be increased with unifunctional cathodes with separate charge and discharge functions, but adding the size, weight, and complexity affects specific energy, energy density, and cost. A high cost of ~\$1000 kWh^{-1} (NantEnergy) has been reported for zinc-air cells.

Advantages and Disadvantages

Advantages of zinc-air batteries are high specific energy and energy density, low cost, and good safety. Disadvantages are low power, poor lifetime and cycle life, and low energy efficiency.

Organization	Feature	Status of Technology
EOS	Tri-electrode, horizontal structure, neutral electrolyte	2700 cycles at 2.1 V charge and 0.9 V discharge voltage
Fluidic Energy, Stanford University	Ionic liquids, OER, or ORR catalysts: CoO/NCNT, NiFe LDH/CN	5000 cycles 20 h cycle period, ~0.7 V charge-discharge voltage polarization at 20 mA/cm^2
Revolt Technology EDF (France)	Biofunctional catalysts, alkaline electrolyte Biocatalysts: MnO_2/NiCo Zn/ TiN/Polymer binder, 7M KOH	200 cycles at 100 mA cm^{-2} 200 cycles at 15 mA cm^{-2}
University of Waterloo	Bifunctional catalysts: Co_3O_4 nanowire/stainless steel	100 cycles at 50 mA cm^{-2}
National University of Singapore Daegu Gyeongbuk of Science and Technology	MnO_2/Ag catalysts, 6M KOH, Zinc plate Bifunctional catalysts: $CoMn_2O_4$, N-rGO	270 cycles at charge potential 2.6 V 100 cycles, 0.86 V charge/ discharge polarization at 20 mA cm^{-2}

(Reproduced from P. Pei, K. Want, Z. Mat, *Applied Energy* 128, 2014, 315–324.)

TABLE 17 Characteristics of Secondary Zn-Air Systems

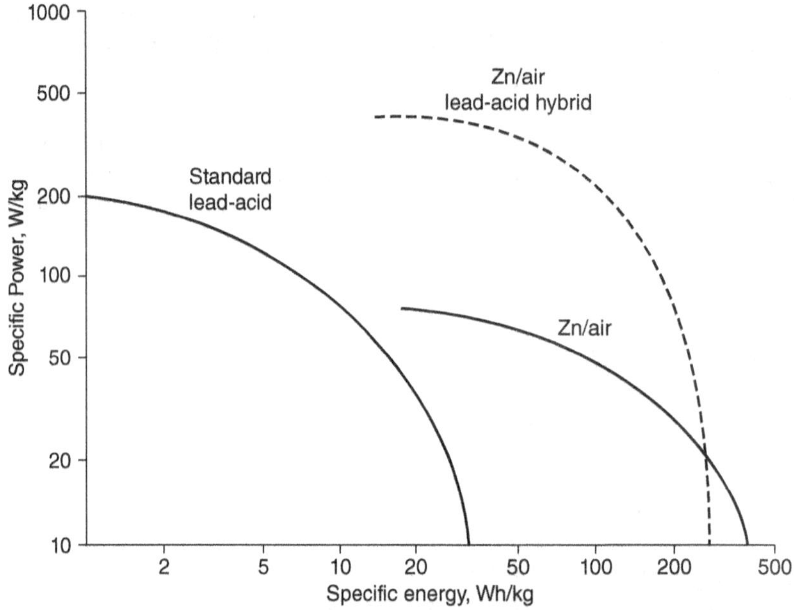

FIGURE 91 Ragone diagram of specific power and energy for zinc-air, lead-acid, and a hybrid system between the two. (Reproduced from T. B. Atwater, A. Dobley, *Linden's Handbook of Batteries*, ed. by T. Reddy, D. Linden, McGraw Hill, 4th ed., 2011.)

The zinc-air battery represents an alternative to popular Li-ion batteries based on excellent safety and low cost if its power and lifetime can be improved. It is also conceivable that higher power can be achieved by constructing hybrid systems with lead-acid, lithium-ion batteries and/or supercapacitors (Figure 91). Improved lifetime can be accomplished by using neutral or ionic liquid electrolytes and through mechanical recharge.

Currently, precommercial systems are available in distributed critical load applications, for example, NantEnergy (formerly Fluidic Energy) installations in Indonesia and Malaysia and in larger demonstration units (300 kWh and 100 kW) for stationary electrical energy storage and grid balancing (e.g., projects by EOS—USA).

Further Reading

- H. Arai, J. Garche, L. Colmenares (editors)—Metal-Air Batteries: Present and Perspectives, Elsevier, Amsterdam 2020.
- H. Arai, M. Hayashi, Secondary Batteries—Metal-Air Systems: Overview (Secondary and Primary), in *Encyclopedia of Electrochemical Power Sources*, ed. by J. Garche et al., Elsevier, Amsterdam, 2009, Vol 4, p. 347.
- T. B. Atwater, A. Dobley, chap. 33, Metal/Air Batteries, in *Linden's Handbook of Batteries*, ed. by Th. Reddy, D. Linden, McGraw Hill, 4th ed., 2011.
- M. Egashira, Secondary Batteries—Metal-Air Systems: Iron-Air (Secondary and Primary), in *Encyclopedia of Electrochemical Power Sources*, ed. by J. Garche et al., Elsevier, Amsterdam, 2009, Vol 4, p. 372.

- O. Haas, J. Van Wesemael, Secondary Batteries—Metal-Air Systems: Zinc-Air: Electrical Recharge, in *Encyclopedia of Electrochemical Power Sources*, ed. by J. Garche et al., Elsevier, Amsterdam, 2009, Vol 4, p. 384.
- L. Jörissen, Secondary Batteries—Metal-Air Systems: Bifunctional Oxygen Electrodes, in *Encyclopedia of Electrochemical Power Sources*, ed. by J. Garche et al., Elsevier, Amsterdam, 2009, Vol 4, p. 356.
- Y. Li, H. Dai, Recent advances in zinc-air batteries, *Chem. Soc. Rev.*, 43 (2014) 5257–5275.
- S. Smedley, X. G. Zhang, Secondary Batteries—Metal-Air Systems: Zinc-Air: Hydraulic Recharge, in *Encyclopedia of Electrochemical Power Sources*, ed. by J. Garche et al., Elsevier, Amsterdam, 2009, Vol 4, p. 393.
- K. Wang, P. Pei, Y. Wang, Ch. Liao, W. Wang, Sh. Huang, Advanced rechargeable zinc-air battery with parameter optimization, *Applied Energy*, 225 (2018) 848–856.

CHAPTER 6

Lithium Batteries

Introduction

Lithium is the best choice for the negative electrode in batteries because of its equilibrium potential of –3.04 V, atomic weight of 6.94 g mol^{-1}, and density of 0.53 g cm^{-3}. Lithium also has the advantage over other alkali metals that it reacts relatively slowly with organic and aqueous electrolytes because of formation of a protective layer on the surface.

Development work began on lithium batteries in the late 1950s, and the first commercial lithium primary batteries were introduced in the early 1970s by Panasonic, Sanyo, and Saft. At first, the cathode material was $(CF_3)_n$ before emergence of other materials, such as CuO, MnO_2, FeS, SO_2, $SOCl_2$, and SO_2Cl_2. The early lithium primary batteries had specific energy of 250 Wh kg^{-1} and energy density of 500 Wh L^{-1}. They were used predominantly in military applications. Later, new cathode materials were developed, such as I_2 and $Ag_2V_4O_{11}$, and new applications in the medical field, such as cardiac pacemakers, began.

The R&D work on secondary or rechargeable Li batteries started in the 1970s by M. St. Wittingham (Exxon), who developed a Li-TiS_2 system. Many other cathode materials were proposed and tested later. At the end of the 1980s, the first commercial Li-MoO_2 battery with organic electrolyte (Li/MoO_2) was released by the Canadian company Moli Energy, but was withdrawn soon thereafter in 1989 due to safety concerns with Li dendrites. The problem was solved by Sony in 1991 through the use of a graphite intercalation electrode instead of metallic lithium, and that started the age of lithium batteries.

The superior performance of lithium-ion batteries has made them the main power source for portable, automotive, and standby power applications. Lithium batteries can be classified by the anode material (lithium metal, intercalated lithium), the electrolyte system (liquid, polymer), or based on the cathode material. Efforts to develop secondary cells with *lithium metal* anodes, however, have shown problems of stability and safety.

Principle of Operation

Lithium is the lightest metal (density 0.534 g cm^{-3} at 20°C) and provides the greatest electrochemical potential (E^0 = –3.045 V). As lithium metal reacts violently with water and can thus cause ignition, modern lithium-ion batteries use carbon negative electrodes (at discharge: the *anode*) and lithium metal oxide positive electrodes (at discharge: the *cathode*). The electrolyte is usually based on a lithium salt in an organic solution. Thin-film lithium-ion batteries use solid oxide or polymer electrolytes.

Two negative electrode types are considered when reviewing the reactions: lithium metal anode and carbon intercalation anode. During the discharge, lithium is oxidized, forming lithium ions (Equations 90 and 91).

$$Li \rightleftharpoons Li^+ + e^- \qquad E° = -3.04\ V \tag{90}$$

$$Li_xC_6 \rightleftharpoons x\,Li+ + x\,e^- + C_6 \tag{91}$$

Lithium ions are then transported to the cathode, where reduction of transition metal oxide takes place during lithium intercalation into the metal oxide, which is generically represented as MO_2 (Equation 92).

$$Li_{1\text{-}x}MO_2 + x\,Li^+ + x\,e^- \rightleftharpoons LiM_{1\text{-}x}O_2 \tag{92}$$

On charge, the process is reversed, and lithium is "plated" back onto the anode.

Overall cell reaction for the lithium metal anode and carbon intercalation anode demonstrates the movement of lithium through discharge and charge (Equations 93 and 94).

$$x\,Li + Li_{1\text{-}x}MO_2 \rightleftharpoons LiM_{1\text{-}x}O_2 \tag{93}$$

$$Li_xC_6 + Li_{1\text{-}x}MO_2 \rightleftharpoons LiM_{1\text{-}x}O_2 \tag{94}$$

Challenges

The process of metal stripping and plating, in the case of a lithium metal anode, destroys at least part of the layer between the electrode and electrolyte called solid electrolyte interphase (SEI), which then has to be reformed consuming lithium and electrolyte. The plated deposit is not smooth, leading to shape change of the negative electrode. The greatly increased surface area leads to higher reactivity of the lithium metal with the liquid organic electrolyte. Dendritic growth can lead to short circuits between the electrodes, which can cause local heating and a violent reaction of the lithium with the electrolyte.

The use of lithium-carbon intercalation compounds in Li-ion cells has eliminated or reduced these issues of the Li metal anode. However, the capacity of a graphite anode is limited to about one-tenth of the metal anode by weight and one-third by volume.

Charging

Charging of lithium batteries is accomplished in two stages: first the constant current phase until predetermined maximum voltage is reached (CC), followed by constant voltage (CV) charge at the maximum voltage (i.e., saturation charge phase).

With charge current of 0.5 C to 1 C, the duration of the CC phase is between 1 and 2 hours; and the duration of the CV phase is 2–3 hours. The full charge is normally reached when the current during the CV phase reaches 3% of the rated capacity expressed as current. Higher current or C rate is possible in the CC phase, but it does not notably reduce the time to reach the full-charge state. Namely, the first CC phase will be reached faster, but the CV phase will then take longer. The only effect of the higher current in the first phase will be that the battery faster reaches about 75% of the full charge.

It is of critical importance for the safety of the battery to discontinue the charging when the maximum charging voltage, usually 4.2 V but not for all cathode materials, is reached. Most chargers are equipped to sense the maximum charging voltage and stop the CC charge. But not all charges are capable to then switch to CV charge at the maximum voltage and this may leave a battery less then fully charged. It should be explained as well that layered cathode materials (e.g., cobalt oxide, manganese oxide, nickel oxide, etc.) exhibit fixed voltages, and charging at higher voltages than those leads to higher capacity but reduced lifetime. However, the most destructive effect of charging beyond maximum voltage (e.g., 4.2 V) is the safety risk based on the danger of lithium not being intercalated inside the graphitic support, but plated as metallic Li on the electrode surface. This dangerous and undesired reaction may lead, under some circumstances, to dendrite growth and electrical short in the cell. In turn, such development might lead to thermal runaway, explosion, and fire. Hence, it is critically important to follow the manufacturer's instructions for charging.

Intentionally lowering the end-of-charge voltage or eliminating the saturation charge brings the outcome of lower capacity on every cycle in such regime, but also a positive effect of higher (i.e., longer) battery life. For the majority of lithium battery cathodes with the end-of-charge voltage of 4.2 V, charging to a lower voltage of 3.8 V will result in state-of-charge (SoC) of 40% at the end of CC phase and 65% at the end of CV phase. Charging to 4.0 V will result in SoC of 70% at the end of CC phase and 80% at the end of CV phase; while charging to a maximum of 4.2 V will give an 85% charged battery at the end of CC phase and 100% charged battery at the end of CV phase.

Occasionally, a battery with unknown history has to be evaluated and then charged; and in such case it is recommended to first charge the battery with a low current of 0.01 C and up to 3.0 V in order to prevent copper dendrite formation (copper foil is used as current collector for negative electrode in lithium batteries). If the battery has been deeply discharged, the negative electrode potential starts rising rapidly to the value where copper is oxidized ($Cu \rightarrow Cu^{2+} + 2e^-$) and copper ions can get involved in forming dendrites and eventual short circuit with the cathode. When this negative outcome has been avoided using a low current in this, so called, mild pre-charge phase, the charging can proceed with typical charging current of 0.5–1.0 C.

For lithium cells with layered-structured cathode materials, the SoC can be estimated after the end of the CV phase by measuring the open circuit voltage (OCV) after the rest time of usually about one hour.

Besides layered-structured materials such as CoO_2, with the end-of-charge voltage of 4.2 V, there are other cathode materials, and they have different end-of-charge voltage. As an example, for olivine-group materials (iron phosphate) cathode, paired with graphite anode, the end-of-charge voltage is 3.65 V; and when paired with another anode, titanium oxide, the end-of-charge voltage is 2.85 V. This further accentuates the importance of correctly identifying the lithium battery anode material (or so-called chemistry) and pairing battery with the correct charger. Use of incompatible chargers would lead either to undercharge or to overcharge, with potentially catastrophic consequences.

Safety

It has been previously described that the energy of a battery during normal operation consists of electrical energy (i.e., chemical energy convertible into electricity) and thermal energy (i.e., chemical energy only convertible into heat). In case of an electrical short, which can occur through several different unwanted circumstances, the electrical energy can also be converted into heat (Figure 92).

A lithium-ion 18650 cell has a specific capacity of around 1000 kJ $kg^{-1}K^{-1}$ and a specific cell energy of around 200 Wh kg^{-1}; and if this combined energy would be released in a short circuit event, under adiabatic conditions, the result would be an enormous increase in temperature of 720 K. In practical, nonadiabatic conditions and for incomplete cell discharge during short circuit occurrence, the temperature increase is significantly lower.

Examination of the thermal energy shows that the largest part is due to the combustion energy of the electrolyte, and expectedly significant effort has been directed toward developing nonflammable electrolytes and solid electrolytes to minimize the amount of energy released from the battery in the form of explosion or fire.

Further analysis shows that there are four outcomes or hazards that can occur in lithium batteries outside of normal operation:

- *Thermal hazard* occurs when the cell temperature increases to about 100°C, triggering exothermal chemical reaction of electrolyte and release of flammable gases. At this temperature, the flashpoint of hydrocarbon gases is exceeded, but there is no fire because there is no free oxygen inside the cell. The heat evolution, however, becomes uncontrollable and causes decomposition of metal oxide in the cathode-active mass, accompanied by oxygen release and subsequent

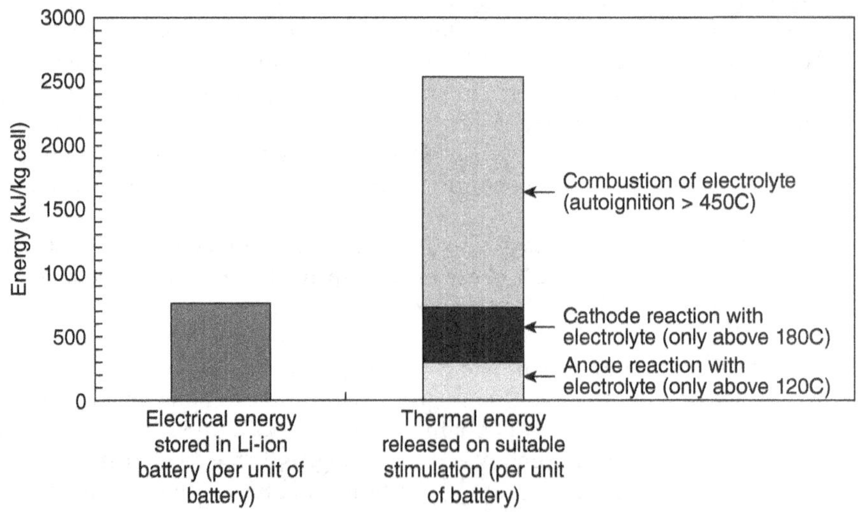

Figure 92 Electrical and thermal energy for a 18650 lithium cobalt oxide cell. (Reproduced from R. Stringfellow, D. Ofer, S. Sriramulu, B. Barnett, 218th Meeting of the Electrochemical Society, Las Vegas, NV, Oct. 12, 2010.)

burning of the gases from the electrolyte. This leads to massive heat release, temperature increase, and subsequent fire and explosion.

- *Kinetic hazard* is created by the temperature increase and burning of decomposition products of the electrolyte. Next, pressure in the cell rapidly increases and can lead to blast waves and material expulsion from the cell. The explosion risk is usually assessed for the lithium cells using the Berthelot-Roth product (BRP) that takes into account gas volume and reaction enthalpy (BRP ~ $V_{gas} \cdot \Delta_r H$). The higher BRP value means higher risk of explosion. As an illustration, a BRP value of 1200 GJ/m^3 (GJ = 10^9 J) indicates the explosion limit, the value for "black powder" (a mixture of sulfur, charcoal, and potassium nitrate) is 3000 GJ/m^3, a primary Li/SiO_2 battery has the value 4000 GJ/m^3, and a rechargeable Li/LICO (cobalt oxide) cell with inorganic electrolyte $LiAlCl_4/SO_2$ has a lower BRP value of 200 GJ/m^3. However, a graphite/LCO cell with organic electrolyte has a higher BRP value of 3200 GJ/m^3 because of the gaseous products of electrolyte decomposition.
- *Chemical hazard* is measured by the possible health effects and impact on the environment. The biggest chemical hazard factor in lithium batteries is the release of hydrofluoric acid (HF) and carbon monoxide (CO).
- *Electrical hazard* is created in applications where batteries serially connected in stacks have voltages of up to 800 V as for some electrical vehicles.

Practical safety considerations evaluate the risks not only on the cell level, but more importantly for a battery. In case of fire or explosion in one of the cells, the propagation takes place and a hazardous condition spreads throughout the stack. Protection schemes have been developed to minimize the risk through a number of sensing and safety devices in the system (Figure 93).

Safety is, therefore, a system-level approach and can be accomplished even for batteries with very high energy content. The first step is to determine the hazard risk through mechanical, electrical, thermal, and environmental testing. These tests are covered mainly in the following standards: IEC 62660-2, ISO 12405-3, ISO 6469-1 and 3, QC/T 743, SAE J2464, J2929, or SAND 2005-3123. A battery is subjected to these tests by applying different excitation parameters and the responses are measured. The safety test results are classified on a scale of 0 to 7, where 0 means that the cell has no safety risk. A cell passes the test with the hazard level of less than or equal to 4 (Figure 94).

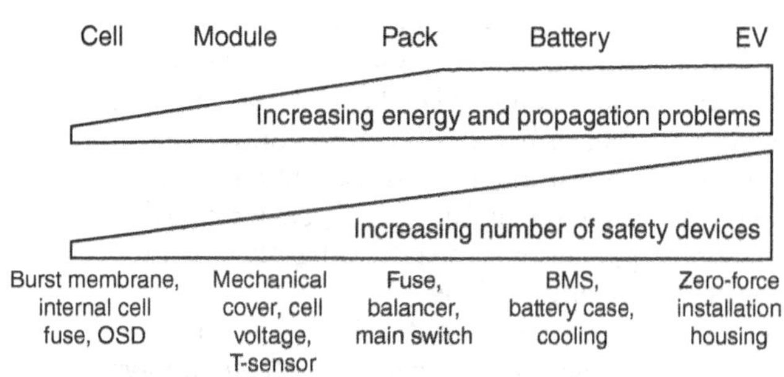

Figure 93 A simplified representation of the increased safety risk from cell to battery and implementation in an electric vehicle, juxtaposed against increased controls and safety measures.

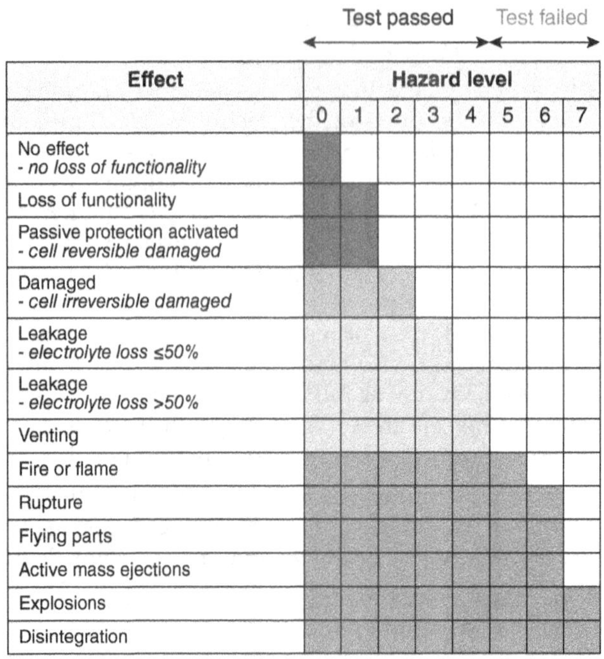

Figure 94 Effect-based criteria for determination of hazard level for lithium batteries. Developed by European Council for Automotive R&D (EUCAR).

Positive Electrode Materials

The positive electrode behaves as a cathode during discharging and undergoes electrochemical reduction by accepting electrons. During discharge, Li^+ ions are extracted from the negative electrode and inserted into the positive electrode.

Electrode properties, such as specific energy, energy density, and specific capacity, are related to the intrinsic properties of the material. Cycle life, lifetime, and safety depend on the nature and stability of the electrode–electrolyte interfaces as well as the structural stability of the electrode materials (Table 18).

The majority of commercially important cathode materials are olivine, layered or spinel structure (Figure 95). The olivine lattice allows a linear Li^+ ion mobility and provides high safety and cycle life at moderate cost. However, it is characterized by low specific capacity and low voltage. The best-known olivine $LiFePO_4$ is used for applications where high cycle life is required, such as storage applications or remote PV systems.

Layered structures exhibit Li^+ ion mobility in two directions and high specific capacitance but are relatively unsafe and expensive caused by Ni and Co. $LiCoO_2$ is still used in consumer applications; mixed oxides such as $LiNi_{0.33}Mn_{0.33}Co_{0.(NMC111)}$ are suitable for electric vehicle applications.

A spinel lattice (Figure 95—right) allows Li^+ ion conductance in the three spatial dimensions and provides high specific power at moderate cost, but with stability problems in the electrolyte solution. $LiMn_2O_4$ is typically used in a blend with other cathode materials.

		$LiCoO_2$	$LiMn_2O_4$ Spinel	$LiNi_{0.8}Co_{0.15}Al_{0.05}O_2$	$LiNi_{0.33}Mn_{0.33}Co_{0.33}O_2$	$LiNiO_2$	$LiFePO_4$
		(LCO)	(LMO)	(NCA)	(NMC111, NCM111)	(LNO)	(LFP)
Cell voltage	V	3.7	3.8	3.7	3.7	3.2–4.2	~3.5
Potential vs. Li/Li⁺:	V	3.0–4.4	3.0–4.5	—	—	—	—
Specific capacity: - theoretical	Ah kg^{-1}	274	296	—	—	192	168
	Ah L^{-1}	706	634	—	—	919	—
- practical capacity:	Ah kg^{-1}	~ 140	~ 120	~ 190	~ 160	—	—
Specific energy:	Wh kg^{-1}	90–180	160	140	< 180	high	80–120
Energy density:	Wh L^{-1}	220-350	270	—	—		—
Specific power:	W kg^{-1}	760	1800	High	Fair	Fair	High
Stability (cycle life)		~ 1000	> 1000	Fair	Fair	Fair	> 4000
Safety		Fair	Good	Fair	Fair	Fair	Safe
Cost (cathode material)		High	Low	High	High	Fair	Fair
	1 Ah kg^{-1} = 1 mAh g^{-1}; 1 Wh kg^{-1} = 3.6 J g^{-1}. $E(Li/Li^+) \approx -3$ V SHE						

TABLE 18 Properties of Rechargeable Lithium-Ion Battery Positive Electrode Materials (Cathodes)

Lithium Cobalt Oxide

Lithium cobalt oxide is used in most commercial consumer lithium-ion batteries. It is typically working at voltages in the range of 3.5–4.3 V (versus Li/Li^+). During charging, about $x = 0.5$ Li^+ ions per mole Li_xCoO_2 (or more depending on the end-of-charge voltage) are extracted from the positive electrode and inserted into the negative

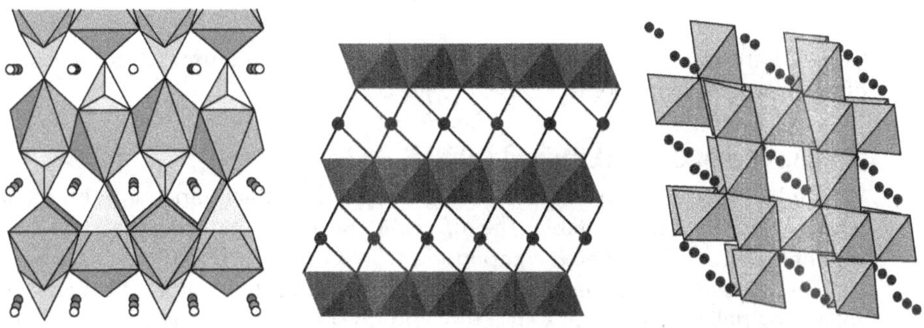

FIGURE 95 Olivine (left), $LiFePO_4$ (• Li), in projection along [001], with PO_4 tetrahedra and FeO_6 octahedra (linked together through corner sharing) show one-dimensional mobility of Li^+. In the middle, layered structure of $LiCoO_2$ with cobalt-centered octahedra shows two-dimensional mobility of Li^+. On the right, $LiMn_2O_4$ spinel with Mn-centered octahedra shows three-dimensional mobility of Li^+.

electrode. Half of the cobalt is oxidized from Co(III) to Co(IV), increasing the voltage of the cell from about 3.0 V to 4.2 V.

Cobalt oxide undergoes several phase changes during charge and discharge. The main phase transition point is located near 75% of the full charge and there is also a minor one near 93%. During charging, the positive electrode material acts as an electron donor and undergoes electrochemical oxidation, while the negative electrode material works as an electron acceptor and is electrochemically reduced. During charging, deintercalation reaction takes place at the positive electrode and intercalation on the negative electrode (Equations 95 and 96).

$$LiCo^{III}{}_{x}Co^{IV}{}_{1-x}O_2 \rightleftharpoons Li_{1-x}Co^{IV}O_2 + x\,Li^+ + x\,e^- \quad (95)$$

$$x\,Li^+ + x\,e^- + 6\,C \rightleftharpoons Li_xC_6 \quad (96)$$

$LiCoO_2$ provides long cycle life and a moderate gravimetric capacity, but also good volumetric capacity due to its high crystallographic and packing density. It is considered to be mature and safe for consumer applications, but not for large batteries such as those used in electric vehicles. Cobalt is associated with environmental and toxic hazards and large price fluctuation due to its relative scarcity.

The reversible delithiation (i.e., removing of lithium) from lithium cobalt oxide beyond 0.5 or 50% lithium is possible and sometimes used to maximize energy density by charging up to 4.4 V versus Li/Li^+. However, this leads to a decrease in cycle life and safety of the battery. The rate at which cobalt oxide can be lithiated and delithiated depends on the mobility of lithium ions in the crystal lattice. The diffusion coefficient is roughly $D(Li^+) = 5 \cdot 10^{-11}\ m^2\ s^{-1}$ in lithium cobalt oxide at room temperature. The low mobility of Li ions in the lattice is compensated by using positive electrode material powders with small particle size. Power density at the end of discharge is limited by the rapidly decreasing lithium-ion mobility in $Li_{>0.96}CoO_2$ at cell voltages < 3.8 V versus Li/Li^+.

Overdischarge causes decomposition of lithium cobalt oxide in cells with excess lithium, which is not the case in lithium-ion batteries (Equation 97).

$$LiCoO_2 + Li^+ + e^- \rightarrow Li_2O + CoO \quad (97)$$

Overcharge up to 5.2 V results in generation of cobalt (IV) oxide (Equation 98).

$$LiCoO_2 \rightarrow Li^+ + e^- + CoO_2 \quad (98)$$

Cobalt (IV) oxide is unstable and decomposes at elevated temperatures to lower oxides of cobalt and in the process releases oxygen.

Lithium Nickel Oxide (LNO and NCA)

Layered $LiNiO_2$ (LNO) has up to 30% higher specific capacity than lithium cobalt oxide, which is more thermally stable. The diffusion coefficient that describes the rate of diffusion of lithium in the layered compound $Li_{0.95}NiO_2$ is $D(Li^+) = 2 \cdot 10^{-11}\ m^2\ s^{-1}$.

There are safety concerns with $LiNiO_2$ because of the exothermic oxidation of the collapsing delithiated Li_xNiO_2 structure with the organic electrolyte. Because of the distorted structure of the M(III) cation, lithium nickel oxide ($LiNiO_2$) decomposes during delithiation to a three-dimensional spinel structure.

Partial substitution of nickel with cobalt and further stabilization of the lattice with Al^{3+} has made the material $LiNi_{1-x-y}Co_xAl_yO_2$ (NCA), with typical values of $x = 0.15$ and

$y = 0.05$, stable enough for commercial applications even at high currents (~3.6 V vs. Li/Li^+, ~190 Ah kg^{-1}). Unfortunately, the synthesis is complicated. The critical stability at high temperatures can be improved by an AlF_3 coating. The NCA material is commonly used in combination with different components such as LFP or LMO for the purpose of increasing the cell safety.

Layered Oxide Solid Solutions: $Li(Ni,Mn,Co)O_2$ (NMC, NCM)

The layered structural $LiMO_2$ (M = Co, Ni) solutions are popularly known as NMCs (sometimes also called NCM). Their framework can be stabilized by multivalent cations aluminum, gallium, magnesium, or titanium.

Lithium nickel cobalt manganese oxide (NCM) is described with the chemical formula $Li_{1-x}(Ni_{0.33}Co_{0.33}Mn_{0.33})O_2$ ($0 < x < 0.7$). This popular and broadly used technology is a solid, 1:1:1 solution of $LiCoO_2$, $LiNiO_2$, and $LiMnO_2$. The NCM cells combine improved safety and low cost, without significantly compromising performance (3.8 V vs. Li/Li+, 160 Ah kg^{-1}) and undergoing minimal volume change of 1–2%. Because of the stable binding between manganese and oxygen, the NCM structure is more stable than LCO or NCA. In combination with a graphite negative electrode, the NCM positive electrode can be charged to 4.4V, which is higher than in the case of conventional lithium-ion batteries (4.2 V).

Besides NMC111 (150 mAh g^{-1}), the NMC versions NMC532 (160 mAh g^{-1}), NMC622 (175 mAh g^{-1}), and NMC811 (200 mAh g^{-1}) are also developed. Besides the increasing capacity with increasing Ni content, unfortunately also a decrease of thermal stability and lifetime is observed.

Lithium-Rich Manganese-Based Layered Materials

$LiMO_2$ (M = Co, Mn) cannot be fully loaded with Li^+ ion because it leads to the structure collapse. Inert components such as Li_2MnO_3 can be used between the layers to stabilize the structure. "Over-lithiated" material such as $x\ Li_2MnO_3 \cdot (1\text{-}x)\ LiMO_2$ provides high specific capacities at charging voltage of 5 V. For example, $0.3\ Li_2MnO_3 \cdot 0.7\ LiMn_{0.5}Ni_{0.5}O_2$ has specific capacity of 250 Ah kg^{-1}. However, upon continuous cycling the material slowly undergoes phase transitions that change its voltage characteristics. These materials are not in commercial use at this time.

Lithium Manganese Spinel (LMO)

The mixed oxides $LiMn_2O_4$ or $Li_{1+x}Mn_{2-x}O_4$ ($x \leq 0.8$) have potential of about 4.0 V vs. Li/Li^+ and a gravimetric capacity of about 120 Ah kg^{-1}, which is 10–20% below that of lithium cobalt oxide. Manganese exhibits a mixed valence of +3.5, because Mn^{III} and Mn^{IV} coexist, but during charging more Mn^{IV} is formed (Equation 99).

$$LiMn_2O_4 \rightarrow Li_{1-x}Mn_2O_4 + x\ Li^+ + x\ e^- \qquad (99)$$

Even at potentials up to 5 V versus Li/Li^+, it is not possible to extract all the lithium from lithium manganese oxide and form manganese dioxide. The cross-linked channels in the three-dimensional spinel framework enable fast lithium-ion insertion and high power. The volume expansion upon lithium intercalation is less than that for the layered compounds.

Good high temperature performance, chemical stability, and lower cost compared to lithium cobalt oxide have made lithium manganese oxide an inherently safe, non-toxic, and environmentally benign positive electrode material.

Lithium-ion cells with manganese spinel positive electrodes show poor characteristics during cycling and reduced storage stability at elevated temperatures because of structural changes. Furthermore, manganese dissolves in the electrolyte causing poisoning of the negative electrode. In acidic solutions above 60°C, $LiPF_6$ hydrolyses generating HF and LMO disproportionates (Equation 100).

$$2\ Mn^{3+}(s) \rightarrow Mn^{2+}(l) + Mn^{4+}(s) \tag{100}$$

The cycling instability of $Li/LiMn_2O_4$ cells near 3 V is attributed to an asymmetric lattice expansion/contraction during discharge/charge. The phase transition of the cubic lithium manganese oxide to the tetragonal structure leads to a marked loss of capacity. LMO electrodes should not be stored below 30% state of charge. LMO is also used in combination with different cathode materials, for example NCA. These blends reduce manganese dissolution and increase safety compared to the layered oxide when used alone.

Lithium Iron Phosphate (LFP)

Lithiated transition metal phosphates are low-cost and environmentally benign positive electrode materials. Lithium iron phosphate ($LiFePO_4$) is a stable three-dimensional phospho-olivine, which occurs as the natural mineral triphylite. In a battery it has potential of 3.3–3.6 V and more than 90% of its theoretical capacity of 168 Ah kg^{-1} can be recovered. The material offers long cycle life and superior thermal and chemical stability. During charge, lithium iron phosphate is converted into iron phosphate ($FePO_4$).

Lithium iron phosphate withstands high temperatures without decomposition; it is incombustible and rather stable under overcharge and short circuit conditions. In the event of mishandling, the phosphate-based material will not burn or release oxygen and is not prone to thermal runaway.

Energy density is more than 14% lower than that of lithium cobalt oxide. The volume of the three-dimensional olivine structure changes upon intercalation and deintercalation of lithium ions. Phosphate anion makes olivine a poor electrical conductor in comparison with other electrode materials ($<10^{-9}$ S cm^{-1}). Small lithium iron phosphate particles, in intimate contact with conductive carbon, must be used to avoid inactive areas in the bulk electrode and to reduce the distance for Li^+ transport in the solid. $LiFePO_4$ can be synthesized, for example, from LiOH, H_3PO_4, and $Fe(NO_3)_3$ at 700°C. The conductivity of the olivine is improved by surface coating with nano-sized carbon or by slightly reducing the surface to FeP_x or magnetite (Fe_3O_4).

Transition Metal Sulfides

The disulfides MS_2 (M = Ti, Nb, Ta, Mo, W) and the oxides MO_2 (M = V, Cr, Fe, Co, Ni) have been investigated since the 1970s as positive electrode materials.

Titanium disulfide (TiS_2) can store lithium in the intercalation range up to $LiTiS_2$ with only 10% lattice expansion (Equation 101). The diffusion coefficient for Li^+ in TiS_2 is $D(Li^+) = 10^{-12}$ m^2s^{-1}. The average discharge voltage is 2.1 V. A theoretical specific energy for a rechargeable cell with a lithium metal negative electrode is 450 Wh kg^{-1}. The cathodes last more than 400 cycles with 20% loss in the electrode utilization.

$$TiS_2 + x\ Li + x\ e^- \rightarrow Li_xTiS_2 \tag{101}$$

The open-circuit voltage of about 2.5 V follows Nernst's equation:

$$E = 2.2\ \text{V} + \ln\frac{x}{1-x} \qquad (0 < x < 1) \tag{102}$$

Crystalline molybdenum disulfide (MoS_2) is not stable at discharge depth larger than 10% because lithium intercalation is accompanied by a structural change and the formation of lithium sulfide (Li_2S) and molybdenum metal. Amorphous molybdenum disulfide (MoS_2) behaves better.

Negative Electrode Materials

The negative electrode undergoes electrochemical oxidation during discharge and is an anode. Average charge and discharge potentials should be well above the lithium electrode potential (~0.5 to 1 V vs. Li) to avoid deposition of lithium dendrites during prolonged cycling (Table 19). At the same time, a high positive electrode potential (~5 V vs. lithium) is required for high-energy batteries.

Different anode materials include graphite, amorphous hard carbon, titanium alloy, metal oxides, tin composite, silicon alloy, and Li metal (Table 19).

Graphite and Amorphous Carbon

Commercial cells use carbon for negative electrodes because of the dimensional stability, high specific charge for lithium intercalation, and the more negative redox potentials than most metal oxides, chalcogenides, and polymers. There are many types of carbon, graphitizing-soft and nongraphitizing-hard carbons.

Natural graphite is a low-cost material which, however, limits the choice of electrolyte as some organic solvents, for example, propylene carbonate (PC), cointercalate with the solvated lithium ion, leading to exfoliation and electrolyte decomposition at fresh plane edges. The graphite structure is built up of many hexagonal mesh layers of carbon atoms. During charging, lithium ions are inserted between these layers up to a theoretical composition of LiC_6, resulting in capacity of 372 Ah kg^{-1} (Equation 103).

$$C_6 + Li^+ + e^- \rightarrow LiC_6 \tag{103}$$

In practice, 0.9 Li^+ per six carbon atoms and about 335 Ah kg^{-1} (graphite) or 0.5 Li^+ per six C (petrol coke) are achieved. The intercalation occurs in a small voltage range between 200 and 50 mV versus Li/Li^+. The potential–capacity curve (Figure 96, upper left) shows stages at 210 mV (random $1 \rightarrow 4$), 120 mV ($2L \rightarrow 2$), and 90 mV versus Li/Li^+ ($2 \rightarrow 1$).

Stage 4 means that every fourth graphite layer is filled with Li^+ ions; 2L denotes a liquid-like second stage that has no in-plane ordering; first stage is LiC_6. In these two-phase coexistence regions, the electrode potential is fixed to a constant value for thermodynamic reasons (Gibb's phase rule). During charge (intercalation), the distance between the graphite layer planes expands by about 10% to accommodate the Li^+ ions. The diffusion coefficient of lithium is largest in the dilute stage 1 (10^{-6}–10^{-5} cm^2s^{-1}), and smallest in stage 2 (10^{-8}–10^{-7} cm^2s^{-1}).

During the first intercalation, the electrolyte is decomposed and the Solid Electrolyte Interphase (SEI) is formed, whereby the cell undergoes an irreversible loss of capacity between 30 and 80 Ah kg^{-1}. The SEI consists of lithium salts (Li_2O, LiF, Li_2CO_3, etc.) and

136

	Graphite (Graphitized Soft Carbon)	Amorphous (Hard Carbon)	Alloy $Li_4Ti_5O_{12}$	Metal Oxides (SnO_2, SiO_2)	Tin Composite (TCO)	Silicon Alloy $Li_{4.4}Si$	Li Metal
Potential vs. Li/Li^+ (V)	0.05–0.2	0.1–0.7	1.56	0.05–0.6	0.05–0.6	0.05–0.6	0
Theoretical specific capacity (Ah kg^{-1})	372	—	175	—	~1000	4000	3830
Pract. specific capacity (Ah kg^{-1})	~350	~200	~170	~750	~600	—	—
Pract. specific power (W kg^{-1})	Medium	Good	Very good	Good	Good	Good	Bad
Stability (cycle life)	Good	Medium	Very good	Bad	Bad	Bad	Bad
Safety	Good	Good	Very good	Bad	Bad	Bad	Bad
State of the art	Commercial	Commercial	Commercial	Midterm	Midterm	Long-term	—
Price	Good	Medium	Medium	Medium	Very good	Very good	Good

TABLE 19 Overview of Rechargeable Lithium-Ion Battery Negative Electrode Materials (Anodes)

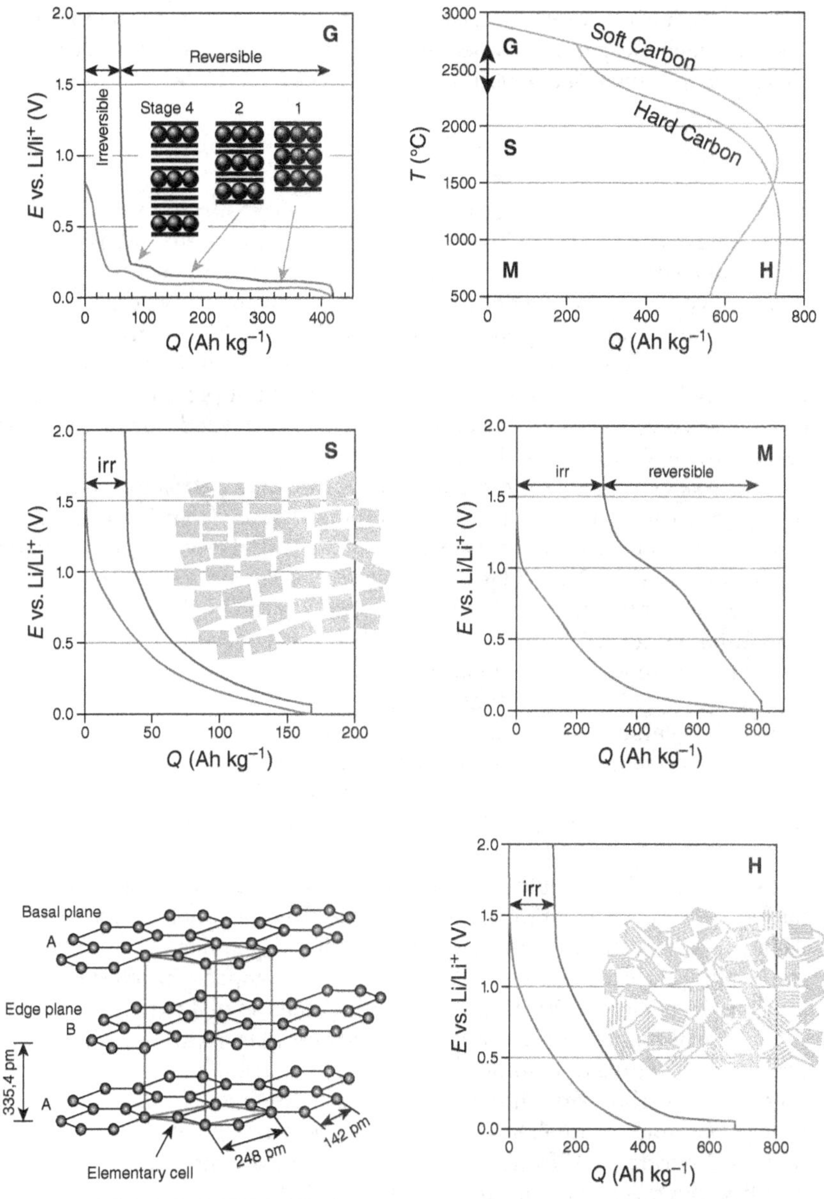

Figure 96 Charge-discharge characteristics of soft carbons and hard carbons (1-molar $LiClO_4$ in EC/DEC 1:1) and specific capacity depending on heat treatment temperature: **G** = graphite, **S** = soft carbon produced at 1800°C, **M** = soft carbon (700°C), **H** = hard carbon, nongraphitizable. (Adapted from M. Inaba, *Encyclopedia of Electrochemical Power Sources*, Vol. 5, Elsevier: Amsterdam, 2009, pp. 198–208.)

polyolefins; it is a poor electronic conductor but conducts ionically through Li^+ mobility. The irreversible capacity loss between lithium storage capacity (intercalation) and discharge (ex = deintercalation) is associated with electrolyte decomposition (Equation 104).

$$Q_{irr} = Q_{in} - Q_{ex} \text{ (in Ah kg}^{-1}\text{)} \qquad (104)$$

The capacity loss Q_{irr} increases linearly with rising surface area and depends on the electrolyte chosen. In graphitized carbon, the oxidation rate and oxygen chemisorption at the edge sites are much higher than that in the basal plane, whereby edge sites are more active for electrolyte decomposition. In subsequent cycles, the reversible lithium storage charge capacity approaches $Q_{rev} = Q_{in}$, which is strongly dependent on the carbon material, electrolyte composition, current density, and potential range. In the manufacture of lithium-ion batteries, an excess amount of positive electrode material must be used to compensate for the additional capacity needed during the initial formation cycles (irreversible capacity loss).

Artificial graphite is fabricated by heat treatment of petroleum coke, coal-tar pitch, or oil. Specific capacity is reduced at high temperatures from 1000 Ah kg^{-1} (at 500°C) to 150 Ah kg^{-1} (1800°C). Above 2000°C, an increasing degree of graphitization improves capacity again (200–300 Ah kg^{-1}) and minimizes irreversible charge.

Graphitizable *soft carbons* are composed of graphene layers. These anodes work in a potential range of 50–200 mV versus Li/Li^+ with a specific capacity of about 350 Ah kg^{-1}. They have good stability and safety, and are moderately priced, but offer lower power. During charging of highly graphitized carbon, the potential drops rapidly to near 0.8 V versus Li/Li^+ and then remains nearly constant. Following electrolyte decomposition and formation of the SEI layer, the potential declines again, while the majority of Li^+-ion intercalation occurs below 0.25 V. The potential–capacity curve of heat-treated carbons loses the inflections and plateau related to the staging phenomena of natural graphite.

An intermediate liquid-like mesophase facilitates the three-dimensional ordering. Such more expensive artificial *spherical graphite* in meso-carbon microbeads (MCMB) exhibits higher energy density than natural and artificial graphite (300–900 Ah kg^{-1}). However, reversibility, cycle life, and hysteresis in its potential–capacity profiles are poor. MCMB that are graphitized above 2700 K show a voltage profile similar to natural graphite.

Artificial amorphous carbons (*hard carbons*, glassy carbon, and nongraphitizable) have a high degree of disorder and cannot be graphitized. At about 1000°C, they are obtained by carbonizing thermosetting polymers (phenol-formaldehyde resins, furfuryl alcohol, and divinylbenzene–styrene copolymers), cellulose, charcoal, petroleum pitch, saccharides, and fruit shells. Some hard carbons are known for their good cycling stability; however, they are hygroscopic, and their capacity, which is below that of graphite, varies widely, depending on the degree and type of disorder. Anodes with amorphous carbons show a potential of 100–700 mV versus Li/Li^+ and ~200 Ah kg^{-1}, good power and safety, moderate price, and mechanical stability. Their potential is more positive than that of graphite and allows higher charge rates, as lithium is not easily plated on the negative electrode. The potential–capacity profile does not show any stages, but a plateau at ~0.05 V. In the unlikely event when no lithium intercalation takes place, electrolyte decomposition and gas evolution can occur.

Multiwalled *carbon nanotubes* (prepared by pyrolysis of acetylene and subsequent heat treatment, show less capacity than graphite and low coulombic efficiency (<50%).

Vapor-grown carbon nanofibers (VGCF) show 360 Ah kg^{-1} at 80% efficiency. However, due to their high electronic conductivity, these materials are being used as well as conductive additives in electrode formulations, replacing carbon black at a much lower concentration.

Lithium Titanate (LTO)

The defect spinel lithium titanate, $Li_4Ti_5O_{12}$ or $Li[Li_{0.33}Ti_{1.67}]O_4$ or $2\ Li_2O \cdot 5\ TiO_2$, combines high power and thermal stability with moderate cost. The specific capacity of about 170 Ah kg^{-1} (theoretically 175 Ah kg^{-1}) have been achieved. In contrast to the 2D-structure of graphite layers, the 3D-structure of LTO is considered a zero-strain material that allows Li^+ intercalation without volume change (Equation 105).

$$Li_{1.33}Ti_{1.67}O_4 + Li^+ + e^- \rightarrow Li_{2.33}Ti_{1.67}O_4 \tag{105}$$

The potential of +1.562 V versus Li/Li^+ remains nearly constant between 0 and 170 Ah kg^{-1} capacity and remains below +1 V versus Li/Li^+ and within the electrochemical stability window of the electrolyte, so that SEI does not form. The energy capacity is roughly half that of graphite-based lithium-ion cells because of the lower cell voltage compared with a graphite negative electrode. Low resistance, high power, and high and extended cycle life also characterize this material. Reversible capacity during cycling of Li-Ti-O materials can be further improved through doping with aluminum, magnesium, or vanadium.

Lithium Alloys

Lithium alloy anodes work in the potential range of 50–600 mV versus Li/Li^+ and offer promise of specific capacities of 1000 Ah kg^{-1} for Sn alloy and 4000 Ah kg^{-1} for Si alloy. This comes at the increased risk of compromised stability and safety. The main challenge with lithium alloys are volume swings (up to 400%) during charging and discharging, which cause disintegration and loss of electrical contact between the particles.

The specific capacity of negative electrode materials declines from 3861 Ah kg^{-1} for lithium metal to 339 Ah kg^{-1} for lithium graphite, in the following order:

$$\text{Li metal} > Li_2Si_5 > Li_3As > Li_2Sn_5 \approx LiAl > Li_3Sb > Li_3Bi,\ LiC_6$$

At the same time, the undesirable variations in volume change from 412% for $Li_{22}Si_5$ to 10% for lithium graphite anode:

$$Li_2Si_5 > Li_2Sn_5 > Li_3Bi,\ Li_3As > Li_3Sb > LiAl,\ LiZn >> LiC_6$$

Lithium-aluminum alloys (LiAl) expand by about 94% during the phase transition from aluminum to LiAl. Despite good practical capacities of 250–300 Ah kg^{-1}, intermetallic alloys have poor cyclability. Sufficient cycle life requires a large excess of lithium, because the cyclic efficiency of lithium metal is about 99%. The higher negative electrode potential (~0.3 to 1.0 V vs. Li/Li^+) compared with graphite (~0.1 V) means a lower cell voltage.

Alloy electrodes are mainly used in rechargeable coin cells because thick electrodes with inflexible, large alloy particles degrade mechanically during cycling, so that cycling depth is better when limited to a very thin reaction layer.

Tin Composite Oxide (TCO)

In 1997, Fuji introduced lithium-ion technology (STALION) using an amorphous *tin composite oxide* (TCO, $SnB_{0.56}P_{0.4}Al_{0.42}O_{3.6}$) negative electrode. This electrode material reacts reversibly with lithium at about 0.5 V (Equation 106) and has twice the capacity of graphite (2200 Ah L^{-1} vs. 600 Ah kg^{-1}). Metallic tin, formed in the first step, creates $Li_{4.4}Sn$ nanodomains embedded within the lithium oxide (Li_2O) matrix, formed in the chemical reaction between metal oxide and lithium (Equation 107), and results in poor long-term cyclability and large irreversible capacity loss during the first cycle.

$$M + x\,Li^+ + x\,e^- \rightarrow Li_xM \tag{106}$$

$$MO_2 + 4\,Li \rightarrow 2\,Li_2O + M \tag{107}$$

Sn–Fe–C composites are characterized with low-voltage reversible reactivity, in spite of the initial irreversibility and temporary poor capacities. Active Sn_2Fe (with strong volume effect) is finely dispersed in a mixed-conducting $SnFe_3C$ matrix (with small volume change). Metallic tin, when precipitated at the grain boundaries, improves the cycling performance but lowers capacity from 285 to 170 Ah kg^{-1} after 80 cycles.

Electrolytes for Lithium Cells

The electrolyte provides ionic conductivity between the electrodes. It must be chemically and electrochemically stable in a wide range of temperatures. It should be nonpoisonous, have a high flashpoint, and a low combustion heat. Electrolytes may be liquids and polymers with solvated ions, or solids with Li^+ intrinsic conductivity of disordered inorganic salts.

Solvents for Liquid Electrolytes

Liquid electrolytes consist of a lithium salt in an organic solvent. Nonaqueous electrolytes can temporarily operate outside their window of thermodynamic stability both on the positive electrode side (3.5–5.5 V vs. Li/Li^+) and on the negative electrode side (<1 V vs. Li/Li^+), because electrolyte decomposition is kinetically controlled. Especially on the negative electrodes, the organic solvents are decomposed during charging, and the SEI is formed at first charge. Highly viscous solvent mixtures decompose slower but perform worse. High temperatures and high terminal voltages will accelerate the decomposition and oxidation processes.

Most solvents are mixtures of cyclic and linear carbonates. Ethylene carbonate (EC)—but not propylene carbonate (PC)—provides a protective layer (probably lithium ethylene dicarbonate) on the surface of graphite that prevents further continuous electrolyte decomposition and self-discharge. EC melts at +36°C and boils at 248°C under decomposition, and exhibits a dielectric constant, ε, of 90 (at 25°C), a flash point of 160°C, reduction potentials of 0.9 V versus Li/Li^+ (at glassy carbon) and 0.8 V (at graphite), and is oxidized above +6.2 V versus Li/Li^+ (at glassy carbon in lithium-free solution). Lithium-ion intercalation into graphite proceeds unhindered especially in lithium hexafluorophosphate in EC + diethyl carbonate (DEC). Dimethyl carbonate (DMC) is also frequently used.

PC, with a melting point of −49°C, boiling point of 242°C, dielectric constant of 65, flash point 132°C, and oxidation at +6.6 V versus Li/Li^+, promotes electrolyte decomposition and exfoliation of the graphite structure by solvent cointercalation. Although it is not forming a protective interphase, PC-based electrolytes are acceptable for use with amorphous carbons such as petroleum coke. Decomposition products at the negative electrode are Li_2CO_3 and propene, as well as linear dicarbonates and polymeric moieties. Linear carbonates, CO_2, alkenes, and polyether carbonates are formed in oxidation processes on the positive electrode and by thermal decomposition.

Conducting Salts

Conductive salts, such as $LiPF_6$, $LiBF_4$, and $LiClO_4$, contain a noncoordinating anion with extensive charge delocalization. The Li^+ ions act as current carriers between positive electrode and negative electrode when the battery is charged or discharged through an external circuit. During charging, the lithium ions move through the electrolyte from the positive to the negative electrode and, during discharging, they move in the opposite direction.

The conductivity σ (in $S\,cm^{-1}$) depends on the charge number z, the number N or the number concentration n (in m^{-3}), and the mobility u (in $m^2V^{-1}s^{-1}$) of the charge carriers in a given electrolyte volume V (Equation 108).

$$\sigma = e\,(n_\ominus\, u_\ominus + n_\oplus\, u_\oplus), \qquad \text{and} \qquad n_i = z_i\, N_i / V \tag{108}$$

It is also important to measure and understand the diffusion coefficient of the Li^+ ions in electrode materials as the basis for determination of the rates of intercalation and de-intercalation (Equation 109):

$$D = \frac{u\,R\,T}{zF} \tag{109}$$

where $R = 8.3144$ J mol^{-1}K^{-1}, $F = 96{,}496$ C mol^{-1}, T is the absolute temperature (in K), and $z = 1$ is the charge number of lithium.

The transference number (Equation 110) is the fraction of the total electrical current carried in an electrolyte by ionic species (i = cation or anion):

$$t_i = \frac{I_i}{I} \tag{110}$$

The practical transference number of the lithium ion, $t(Li^+)$, is well below 1 in liquid electrolytes and close to 1 in solid electrolytes, that is, the lithium ion transports almost 100% of the total ionic current through the solid matrix.

Lithium hexafluorophosphate ($LiPF_6$) is most commonly used in commercial lithium-ion cells. One of the major concerns with this electrolyte is the presence of trace amounts of water, which causes generation of hydrofluoric acid (Equation 111).

$$LiPF_6 + 4\,H_2O \rightarrow 5\,HF + LiF + H_3PO_4 \tag{111}$$

The reductive and thermal chemical decomposition of PF_6^- to PF_3 and PF_5, and the presence of LiF at the aluminum current collector is accompanied by the formation of protective aluminum fluoride (AlF_3). State-of-the-art electrolytes consist of $LiPF_6$ in a mixture of EC, DEC, and DMC. The electrolyte formulation of 0.75 mol L^{-1} $LiPF_6$ in EC/EMC (1:1 ratio) has the conductivity of 9.7 mS cm^{-1} at 25°C and can be used between –20°C and 60°C. The electrolyte formulation of 1 mol L^{-1} $LiPF_6$ has the conductivity of 5.8 mS cm^{-1} in PC, and 10.7 mS cm^{-1} in EC/DMC and is oxidized above 4.65 V versus Li/Li^+.

Lithium perfluorosulfonimide (LiTFSI) with the chemical formula $Li[(CF_3SO_2)_2N]$ has good conductivity and is nontoxic, safe, and extremely resistant to oxidation.

The conductivity of this electrolyte in 1-molar solution at 25°C is 5.1 mS cm^{-1} in PC and 9.0 mS cm^{-1} in EC/DMC. The same 1-molar solution at 25°C is resistant to oxidation up to 4.0 V vs. Li/Li^+. Comparatively, the conventional triflate $LiCF_3SO_3$ electrolyte has the conductivity of 1.7 mS cm^{-1} in PC and the decomposition voltage of 3.9 V. However, the aluminum collector of the positive electrode corrodes under those conditions and the passivating salt must be added.

Lithium bis(oxalato)borate (LiBOB) with the chemical formula $Li[B(C_2O_4)_2]$ is considered a promising alternative electrolyte. This electrolyte has the conductivity of 7.5 mS cm^{-1} in EC/DMC in 1-molar solution at 25°C. It is sometimes used in combination with lithium hexafluorophosphate ($LiPF_6$). It has improved cycle stability but lower high-current performance.

Large anions improve the conductivity of organic solutions, because the wide distance between the charge centers of an anion and Li^+ cation in the salt weakens the Coulomb forces, $F \sim [\varepsilon_0 r^2]^{-1}$. However, the mobility of big anions is low. As a result, conductivity, which requires both high dissociation and high mobility, follows the pattern:

$LiPF_6$, $LiAsF_6$ > LiTFSI (lithium perfluorosulfonimide, which has the best dissociation), $LiClO_4$ > $LiBF_4$ (best mobility) > $LiCF_3SO_3$ > $LiC_4F_9SO_3$.

Ionic Liquids

Ionic liquids are salts that are in a molten state below 100°C. The best conducting ionic liquids are based on 1-ethyl-3-methylimidazolium cation (EMI) and sulfonylimide anions (15 mS cm^{-1}, 25°C). However, ionic liquids are rather viscous and if a lithium salt is added, viscosity increases, causing conductivity decrease because new mixed-salt crystalline phases are formed.

Additives

Additives can be divided into the following categories based on their function:

1. SEI forming improving agent
2. Positive electrode protection agent
3. Lithium hexafluorophosphate salt stabilizer
4. Safety protection agent
5. Lithium deposition improving agent
6. Other agents such as solvation enhancer, aluminum corrosion inhibitor, and wetting agents

Surface-Film-Forming Additives

Vinylene carbonate and vinylethylene carbonate passivate the negative anode by forming polyvinylidene carbonate, which restricts the inorganic components.

Protection Agents

At the positive electrode (cathode), tributylamine, silicones, and dimethyl acetamide scavenge water and acidic impurities and prevent carbonate-based solvents from irreversible oxidation and CO_2 gasing. Ethers such as 12-crown-4, which preferentially solvate lithium ions, prevent PC from cointercalation into graphite.

Stabilizers

These impede the thermal decomposition of the conductive salt, $LiPF_6$ to PF_5 and LiF, and weaken the reactivity of phosphorous pentafluoride (PF_5) with organic solvents.

Overcharge Protectors

These are expected to function as highly reversible internal dischargers with an oxidation potential slightly below the electrolyte decomposition potential and above the end-of-charge voltage of the cathode. However, they mostly have negative impact on the cell performance. Anisole derivatives, for example, 2,5-di-*tert*-butyl-1,4-dimethoxybenzene, with oxidation potential of over 3.9 V versus Li/Li^+ and 1,4-difluoro-2.5-dimethoxybenzene, with oxidation potential of 4.2 V, function as redox shuttle compounds, but show poor long-term stability.

Shutdown Additives

Additives, such as biphenyl or cyclohexylbenzene, insulate the positive electrode by forming a polymer layer, growing at unacceptably high voltages, while hydrogen is released at the negative electrode.

Fire-Retardant Additives

Fire-retardant additives generate insulating layers to prevent radical reactions necessary for the combustion reaction.

Lithium Deposition Improving Agents

These agents are expected to prevent lithium plating on the graphite negative electrode by reducing the undesirable growth of the electrode–electrolyte interface, especially the formation of needle-like dendrites and sponge-like lithium. Fluoroethylene carbonate, which decomposes to vinylene carbonate and hydrogen fluoride, seems to be a promising additive for efficient lithium deposition. However, excess vinyl carbonate results in low cycling efficiency and a high self-discharge rate. LiBOB is a conducting salt and a promising additive for a multifunctional improvement of lithium-ion batteries when used together with lithium hexafluorophosphate, as it stabilizes the electrode–electrolyte interface, prevents dissolution of the electrode material, provides overcharge tolerance, and prevents aluminum corrosion.

Inorganic Solid Electrolytes

In a true solid electrolyte, only Li^+ ions are mobile, and the transference number is close to 1. The ions move through vacant or interstitial sites in a crystalline or glassy matrix.

Solid electrolytes are not flammable, but their conductivity is typically low. Furthermore, some are unstable at either anode or cathode potential and problems have also been reported at the electrode–electrolyte interface.

Crystalline solid electrolytes are used in all-solid-state Li | LiI | I_2 cells for cardiac pacemakers. Their self-discharge is very low as only Li^+ ions are mobile in the electrolyte.

Lithium lanthanum zirconate, $Li_7La_3Zr_2O_{12}$ (LLZO), achieves a conductivity of 10^{-4} S cm^{-1}, as a result of disorder of delocalized Li^+ in the cubic garnet crystal lattice and is stabilized by a small amount of aluminum. Garnets $Li_5La_3M_2O_{12}$ (M = Nb, Ta) show high stability in the presence of lithium metal, at voltages of up to 6 V.

Lithium phosphorous oxynitride (LiPON), $Li_{2.9}PO_{3.3}N_{0.46}$, made by sputtering $LiPO_4$ in N_2 atmosphere, is an amorphous, electronic insulator, which exhibits lithium-ionic conductivity of 3.3·× 10^{-6} S cm^{-1} at 25°C and good electrochemical stability at cell potentials up to 5.5 V versus lithium.

Sulfidic glasses (Li_2S–SiS_2–LiI, Li_2S–SiS_2–Li_3PO_4) have higher conductivity than oxide-based solid electrolytes. However, they form poorly conducting layers at the positive electrode. The sulfide $Li_{3.25}Ge_{0.25}P_{0.75}S_4$, which is the most stable member of the thio-LISICON (Lithium super ionic conductor) family, has ionic conductivity of 0.0022 S cm^{-1} at 25°C. Another formulation, $Li_{10}GeP_2S_{12}$, provides conductivity of at least 0.02 S cm^{-1} at 27°C.

Polymer Electrolytes

Polymer electrolytes are largely based on solvent-free *polyethylene oxide* (PEO) and a highly conductive salt, such as $LiN(CF_3SO_2)_2$ [$P(EO)_n$LiTFSI]. They can be fabricated into thin films in a simple fashion by casting or spin-coating. These compounds have a large electrochemical window and form a eutectic composition with PEO, that is, the composition at the lowest possible melting point of the mixture. The ionic conductivity of the "high-molecular-weight dry polymer" (10^{-6}–10^{-4} S cm^{-1} at 25°C) is directly linked to the degree of amorphousness and the lowest possible glass transition temperature controlling the ion mobility. Side chains with solvating groups increase the degrees of freedom and improve ionic conductivity but compromise the mechanical properties. Aligning or organizing the polymer chains improves ionic conductivity whereby liquid crystalline chain polymers reach liquid-like conductivity values at higher temperatures or when polarized. In some cases, it has been reported that the high conductivity values were retained upon cooling to room temperature.

The poor ionic conductivity of crystalline complexes PEO:$LiXF_6$ (X = P, As, Sb) can be raised by a factor of 100 by partial replacement of the XF_6 ion with other mono- or divalent anions.

Hybrid polymer electrolytes contain plasticizers, used to create lubrication between the polymer chains. These lubricants include polar solvents such as PC, γ-butyrolactone, or polyethylene glycol ethers or short-chain PEO with 4–25 monomer units. Conductive additives improve conductivity by a full order of magnitude when added as 10–25% wt.

Gels contain 60–95% liquid electrolyte and are 2–5 times less conductive than liquid solutions. PEO-based gels achieve ionic conductivities above 0.001 S cm^{-1}. For polyethers that contain electron pair donors as gelling agents, most of the solvation takes place through the polymer chains rather than carbonate solvents. Thus, lightly plasticized systems can be used in lithium metal batteries, while the soft gels work well only in lithium-ion configurations.

Nanoparticle fillers such as aluminum oxide (Al_2O_3), titanium dioxide, or silica (SiO_2) significantly increase the conductivity, prevent crystallization for several weeks, and increase apparent lithium transport number for $PEO_{20}LiClO_4$/8% Al_2O_3 from 0.31 to 0.77, and for $PEO_{20}LiBF_4$/ZrO_2 film from 0.32 to 0.81. Nanoscale ceramic additives are covered with Lewis acidic and basic surface groups, which interact with the ionic species of the PEO:$LiXF_6$ electrolyte.

Separators

The separator in liquid electrolyte batteries has a role to insulate the electrodes and prevent electronic contact between the electrodes and enable free ionic transport. The separator materials must be chemically stable, be roughly 8–25 μm thick, and should exhibit a porosity of about 50% and a pore size less than 1 μm, in order to hold sufficient liquid electrolyte and to prevent, at the same time, penetration of particles from the electrodes. To ensure a uniform current distribution and to prevent the growth of dendrites on the negative electrode, the permeability of the separator must be uniform. The resistance of a separator to ion transport depends on the thickness, porosity, and tortuosity. As a measure of its transport properties, the air permeability (Gurley value) is determined as the time required for a specific amount of air to pass at a specific pressure and cross-sectional area through the separator material of given porosity and thickness.

The separator mechanical strength is described by the Young's modulus. For example, a 25 μm-thick separator should have tensile strength of 1000 kg cm^{-2} in the machine direction and pass a puncture stress with a maximum load of 300 g on a needle. The separator should wet easily, lay flat, and be dimensionally stable.

Shrinkage and softening of a separator at elevated temperatures can cause an internal electrical short and subsequently a thermal runaway. The thermal shrinkage should be less than 5% after 60 min at 90°C, for example, polyethylene (PE) membranes shrink 10% during 10 min at 120°C. Common PE–polypropylene (PP) bilayer separators in lithium-ion batteries shut down at about 130°C without loss of mechanical integrity (melting temperature of PP is ~165°C).

Polyolefin separators made by dry process (e.g., Celgard, PP–PE–PP) have an open and straight pore structure, whereas separators from wet process (e.g., Exxon Mobile "Tonen," PE) show a tortuous pore structure.

Finally, addition of inorganic fillers improves wettability and liquid retention.

Outlook for Lithium Batteries

Numerous research activities and challenges exist for further development of lithium batteries. Future, near-term lithium batteries are expected to meet the following requirements:

1. Compact and light, without heavy metal housing
2. Free of poisonous metals
3. Reliability and inherent safety
4. Availability in a variety of design and sizes
5. High-energy density values of at least 250 Wh kg^{-1} and 600 Wh L^{-1}, and very high cycle life
6. High power density and high discharge voltage
7. Energy loss per month less than 1% when the battery is not in use
8. Applicable for wide range of temperatures without performance loss
9. Low cost

Commercially available 3.6 V lithium-ion technology used in consumer devices has specific energy of maximum 240 Wh kg^{-1} and energy density of up to 670 Wh L^{-1} (e.g., Panasonic 18650 cells). Lithium polymer delivers 260 Wh kg^{-1} (600 Wh L^{-1}), and thin-film lithium-ion battery provides 250 Wh kg^{-1}. Different batteries must meet specific application requirements, such as higher energy, higher power, longer life, or lower cost.

Most cells currently made worldwide are graphite/NMC cells (roughly 50% in 2020) and graphite LFP cells (25% in 2020), followed by NCA and LCO cells at 10% each, and LMO (5%).

The main research direction when searching for new active materials are high-voltage positive electrodes. Interesting materials are spinels such as $LiCoMnO_4$ (5.0 V), olivines such as $LiNiPO_4$ (5.2 V), and vanadates such as $LiNiVO_4$ (4.8 V). The challenge in using 5 V materials to improve the stored energy lies in the development of a stable electrolyte.

Lithium-Sulfur Batteries

Lithium sulfur is a promising version of lithium batteries with much higher specific energy than lithium-ion metal oxide chemistries. Theoretical specific energy is 2600 Wh kg^{-1} and it is estimated that about 378 Wh kg^{-1} can be achieved in practice. Lithium-sulfur prototypes have been demonstrated with the specific energies above 350 Wh kg^{-1} (350 Wh L^{-1}). No extensive reports are available on the cycle life and self-discharge.

The electrolyte used in lithium-sulfur batteries is usually a mixture of tetraethylene glycol dimethylether (TEGDME) and 1,3-dioxolane, which forms a stable solid–electrolyte interface. During cycling the electrolyte is decomposed reducing the lifetime to less than 300 cycles. More electrolyte increases the lifetime but reduces the specific energy.

The lithium-sulfur battery consists of a lithium anode (negative electrode) and sulfur cathode (positive electrode). During discharge, lithium is oxidized on the anode, while sulfur is reduced, and successive lithium sulfides are formed on the cathode (Equations 90, 112, and 113). The product Li_2S is deposited on the carbon matrix. Self-discharge proceeds in the similar mechanistic way.

$$2\,Li \rightarrow 2\,Li^+ + 2\,e^- \quad (90)$$

$$S + 2\,Li^+ \rightarrow Li_2S + 2\,e^- \quad (112)$$

$$S_8 \rightarrow Li_2S_8 \rightarrow Li_2S_6 \rightarrow Li_2S_4 \rightarrow Li_2S_3 \rightarrow Li_2S_2 \rightarrow Li_2S \quad (113)$$

During charge, Li_2S is not converted back into sulfur; rather, polysulfide anions $[S_x]^{2-}$ are formed, which diffuse through the electrolyte as a shuttle. The cell voltage is only 2.1 V, but lithium-sulfur cells tolerate overvoltage.

Sulfur is an insulator (5×10^{-30} S cm^{-1} at 25°C) and must be incorporated into an electronically conducting structure such as carbon and held together using a polymer binder. At the sulfur cathode, between S_8 (fully charged) and the formation of Li_2S, different reduced species are formed depending on the depth-of-discharge (DoD): Li_2S_8 at 12.5% DoD (2.4 V), Li_2S_4 at 25% DoD (2.2 V), Li_2S_2 at 50% DoD, and finally Li_2S at 100% discharge (2.05 V).

The sulfides Li_2S_2 and Li_2S are insoluble in the electrolyte and cause a passivation layer on the electrode surface. Therefore, the DoD must be limited in the practical batteries. The lithium polysulfides Li_2S_8 to Li_2S_3 are soluble in the electrolyte and can react in a chemical self-discharge reaction without generating electrical current. Commercialization of the Li-S system is expected after 2030.

Lithium-Air Batteries

Lithium-air batteries have a high theoretical specific energy of 3,505 Wh kg^{-1} and energy density of 3,436 Wh L^{-1}. Practical values at the time of writing this text are in the 400–500 Wh/kg^{-1} and Wh/L^{-1} range. The battery reactions are described in Equations 114 to 117.

Negative electrode: $$Li \rightleftharpoons Li^+ + e^- \quad (114)$$

Positive electrode: $$Li^+ + \frac{1}{2}O_2 + e^- \rightleftharpoons \frac{1}{2}Li_2O_2 \quad (115)$$

$$Li^+ + \frac{1}{4}O_2 + e^- \rightleftharpoons \frac{1}{2}Li_2O \quad (116)$$

Overall: $$2\,Li + O_2 \rightleftharpoons Li_2O_2 \quad (117)$$

Both oxygen electrode reactions (Equations 115 and 116) are kinetically limited, which means that the charging voltage is high and discharging voltage low, resulting in poor efficiency and poor specific energy and power. For example, at commonly used porous carbon electrodes, the charging voltage is 4.5 V and discharging voltage is 2.5 V, even for low charging and discharging current density. This results in voltage efficiency of 55%, which can be increased to about 73% (charging voltage 3.7 V and discharging voltage 2.7 V) when a carbon-supported Pt/Au catalyst is used. However, the use of expensive noble metal catalysts is not practically feasible.

Most commonly used electrolytes for Li-air systems are organic and aprotic. The solubility of reaction products (Li_2O_2, Li_2O) is low in aprotic solvents, resulting in precipitation in the pores of the oxygen electrode and capacity reduction.

When aqueous electrolytes are used, the reaction product (LiOH) has higher solubility, preventing pore blockage, and the capacity is increased. However, aqueous electrolytes react with metallic lithium, so direct contact must be prevented by separating anode and cathode compartment electrolytes with a water-stable lithium-conducting solid electrolyte such as NASICON-type LATP [$Li_{1+x}Al_xTi_{2-x}(PO_4)_3$]. To complicate the matter further, this solid electrolyte is also not stable in contact with Li metal and an additional buffer layer is necessary to protect metallic lithium. A frequently used material for this layer is polyethylene oxide [Li_3N, lithium phosphorous oxynitride (LiPON) glass film]. The theoretical specific energy of aqueous Li-air systems is 1,900 Wh/kg^{-1} and energy density is 2,004 Wh/L^{-1}, calculated based on the reaction product being LiOH (Equation 118).

$$4Li + O_2 + 6H_2O \Rightarrow 4(LiOH \cdot H_2O) \quad (118)$$

The development of aqueous Li-air systems is at the moment limited to primary batteries. Commercial viability of rechargeable Li-air batteries depends on improvements in reaction reversibility and reduced dendrite formation. The poor reaction reversibility leads to low cycle life of about 100 cycles and severe capacity fade. Additional problems include the need for water management due to humid oxygen, irreversible formation of lithium carbonate, slow oxygen transport to the electrode, and clogging of the electrode pores by lithium peroxide.

Further Reading and References

- P. Balbuena, Y. X. Wang (eds.), *Lithium Ion Batteries: Solid Electrolyte Interphase,* London: Imperial College Press, 2004.
- C. Daniel, J. O. Besenhard (eds.), *Handbook of Battery Materials,* Weinheim: Wiley-VCH, 2011.
- S. Dou, Review and prospect of layered lithium nickel manganese oxide as cathode materials for Li-ion batteries, *Journal of Solid State Electrochemistry* 17 (2013) 911–926.
- P. Kurzweil, K. Brandt, Overview: Lithium Rechargeable Systems, in J. Garche, C. Dyer, P. Moseley, Z. Ogumi, D. Rand, B. Scrosati (eds.), *Encyclopedia of Electrochemical Power Sources*, Vol. 5, Amsterdam: Elsevier, 2009, pp. 1–26.
- W. van Schalkwijk, B. Scrosati (eds.), *Advances in Lithium-Ion Batteries,* New York: Kluwer Academic/Plenum Publishers, 2002.

PART 2

Hydrogen and Fuel Cells

CHAPTER 7
Hydrogen Storage

Can you imagine why hydrogen is considered as the promising energy carrier in the future? Because it can be converted into different forms of energy by environmentally clean methods with reasonable efficiency. However, there is a long way to go before liquid hydrogen compounds are able to replace fossil fuels and the existing filling station infrastructure.

The term *storage* denotes the retention of produced hydrogen until its consumption by physical and chemical methods. Usually, hydrogen is stored in pressurized gas cylinders. Chemical compounds (hydrides) are suitable for vehicle applications with respect to safety and mass reasons.

Hydrogen Economy

In industry, hydrogen is used as a chemical for a number of processes:

- Synthesis of ammonia (Equation 119):

$$3\,H_2 + N_2 \rightarrow 2\,NH_3 \tag{119}$$

- Petroleum refining
- Hydrogenation of fats and unsaturated compounds (Equation 120):

$$\text{RCH=CH-COOH} + H_2 \rightarrow \text{RCH}_2\text{CH}_2\text{COOH} \tag{120}$$

- Production of methanol and hydrocarbons from synthesis gas (Equation 121):

$$2\,H_2 + CO \rightarrow CH_3OH \tag{121}$$

- Reduction processes in powder metallurgy (Equation 122):

$$CuO + H_2 \rightarrow Cu + H_2O \tag{122}$$

Moreover, hydrogen is used in microelectronics; in metal-working; in the production of glass, synthetic ruby, and metal hydrides; and in hydrogen peroxide H_2O_2 synthesis.

The pros and cons of an economical concept based on hydrogen as the major source of energy—from central or local production sites, storage, distribution, and consumption by the final users—have been discussed since the oil crisis in the 1970s. The term *hydrogen economy* was coined by the electrochemist John Bockris. Visionary ideas foresee gigantic solar plants in the world's deserts, which generate hydrogen by the electrolysis of sea water. The gas is distributed by tank ships and pipelines, supplying fuel for automotive applications, on-board auxiliary power in vehicles, stationary power generation in buildings, and as a long-term storage medium for excess electric power generated off-peak.

With poor sustainability, hydrogen is produced by the steam reforming of natural gas, by partial oxidation of oil residues, or by the gasification of coal or biomass, which generate CO, CO_2, and a variety of harmful byproducts. Hydrogen usage in combustion engines emits nitrogen oxides (NO_x), the concentration of which depends on the combustion temperature. In a fuel cell, however, hydrogen and air are converted to electric power, heat, and pure water, without any exhaust gases harmful to the environment.

Cryogenic hydrogen in tank trucks, hydrogen cars, and fuel cells are state-of-the-art, which strongly differ in their degree of maturity. Private transportation using hydrogen-driven vehicles requires an extensive infrastructure of filling stations, which has to be established in the near future.

Plants manage the photolytic splitting of water, which does not satisfactorily work in the present technology, and therefore it is the subject of current research.

Properties of Hydrogen

Hydrogen, H_2, the lightest of all elements, is a molecule H–H consisting of two hydrogen atoms ^1H (besides traces of "heavy" hydrogen), each having the average molar mass of 1.008 g mol^{-1}. Single H atoms behave extremely reactive and are therefore rare. *Antoine Lavoisier*, who discovered the elemental nature of this gas in 1783, gave hydrogen its name from Greek *hydor* = water and *gignomai* = to give birth. Hydrogen is the most abundant element in the universe and in tenth position on earth: about 1400 mg/kg in the earth's crust, 10.8% in sea water, 5 10^{-7} m^3/m^3 in air. Small amounts of deuterium, the heavy hydrogen isotope ^2H (<0.0003 mol/mol), and traces of radioactive tritium ^3H play a minor role in natural samples of hydrogen and water. The major part of hydrogen exists in the bound state, in the form of water, natural gas, petroleum, biomass, and organic compounds.

Mass Density Due to its very low boiling point (–252.98°C ≈ 20 K) and the negligible attracting forces between the molecules, hydrogen may be calculated as an ideal gas. Only helium melts and boils at lower temperatures. The density of hydrogen is 0.090 g/L at 0°C (Equation 123):

$$\varrho = \frac{m}{V} = \frac{M(H_2)}{V_m} = \frac{2 \cdot 1.008 \text{ g mol}^{-1}}{22.414 \text{ L mol}^{-1}} \approx 0.090 \text{ g L}^{-1} \text{ (at 0°C)} \quad (123)$$

where ϱ is the density, m is the mass, V is the volume, M is the molar mass, and V_m is the molar standard volume.

Amount Concentration The amount concentration of hydrogen in ideal gas mixtures is calculated by the help of the volume fraction or partial pressure (Equations 124 and 125):

$$w_i = \frac{\varphi_i M_i}{\bar{M}} = \frac{\varphi_i \varrho_i}{\bar{\varrho}} = \frac{m_i RT}{M_i pV} = \frac{c_i RT}{p} \quad (124)$$

$$x_i = \frac{p_i}{p} \approx \frac{V_i}{V} = \varphi_i \quad (125)$$

where w is the mass fraction (1% = 0.01 kg kg^{-1}), ϕ is the volume fraction (1 vol% = 0.01 m^3 m^{-3}), x is the mole fraction (1 mol% = 0.01 mol mol^{-1}), $\bar{M}$ is the average molar mass of the mixture, $\bar{\varrho}$ is the average density of the mixture, i is the component in the mixture, ϱ is the density, c is the molar concentration, m is the mass, p is the pressure, R is the molar gas constant 8.3144 J mol^{-1}K^{-1}, T is the temperature (K), and V is the volume.

Storage Density Liquid hydrogen is the lightest of all liquids (70.8 kg m^{-3} at the boiling point). The useable combustion heat (heating value) H_u (Equation 126) is the reaction enthalpy of the oxyhydrogen reaction $2H_2 + O_2 \rightarrow H_2O$ that equals 286.2 kJ mol^{-1} (25°C), minus the evaporation heat of water (44.016 kJ mol^{-1}).

$$W_m = \frac{W}{m} = \frac{H_u}{M} = \frac{241.1 \text{ kJ mol}^{-1}}{2 \cdot 1.008 \text{ g mol}^{-1}} = 119.6 \frac{\text{MJ}}{\text{kg}} = 33.22 \frac{\text{kWh}}{\text{kg}} \tag{126}$$

This specific energy of hydrogen is at least double the value of natural gas (Table 20). However, the storage of hydrogen requires roughly the threefold volume of a natural gas tank (Equation 127).

$$W_v = \frac{W}{V} = \frac{H_u}{V_m} = \varrho \cdot W_m = \frac{241.1 \text{ kJ mol}^{-1}}{22.414 \text{ L mol}^{-1}} = 10.76 \frac{\text{MJ}}{\text{m}^3} \approx 3 \frac{\text{kWh}}{\text{L}} \tag{127}$$

The evaporation heat of liquid hydrogen is 0.46 kJ mol^{-1}, making the storage of gaseous hydrogen more efficient. For comparison, diesel yields 42.5 MJ kg^{-1} and 33 MJ L^{-1}.

Transport Properties Among all gases, hydrogen exhibits the lowest viscosity and the highest heat conductivity. Small hydrogen molecules diffuse quickly in the air and through various materials; therefore, hydrogen-air mixtures are rapidly diluted and within short time reach safe ambient conditions. According to the kinetic theory of gases, two hydrogen molecules collide after a mean free path of $\bar{s} \approx 70$ nm. Their average speed is 56 m s^{-1}, and the diffusion coefficient is $D(H_2$ in air$) = 1.3 \cdot 10^{-6}$ m^{2}s^{-1} at 25°C (Equation 128):

$$D = \frac{1}{3}\bar{v}\bar{s}, \quad \text{wherein } \bar{v} = \sqrt{\frac{8RT}{\pi M}} \quad \text{and } \bar{s} = \frac{M}{\sqrt{2}\pi d^2 N_A \varrho} \tag{128}$$

and where D is the diffusion coefficient, d is the effective distance between two colliding molecules, M is the molar mass, N_A is the Avogadro constant, R is the molar gas constant, s is the mean free path, v is the average molecule speed, and ϱ is the density of the gas.

	Liquid Hydrogen (LH_2)	Compressed Hydrogen (GH_2), 700 bar	Natural Gas, (CNG) 200 bar	Gasoline	Diesel	Carbon (Coal)	Methanol	Lithium Battery
kWh kg^{-1}	33.3	33.3	12.2–13.9	11.1–12.1	11.8–11.9	7.8	5.6–6.2	0.1–0.2
kWh L^{-1}	2.1–2.4	1.86	≈2.58	8.2–8.8	≈9.8	11.7	4.4	0.25–0.67

TABLE 20 Specific Energy and Energy Density of Fuels: 1 kWh = 3600 kJ = 3.6 MJ

Chemical Processes Involving Hydrogen

Hydrogen is an odorless, colorless, and tasteless gas; it is not toxic but does not sustain life. Hydrogen is poorly soluble in water (1.78 $cm^3\,g^{-1}$ at 20°C) and somewhat better in methanol (11 $cm^3\,g^{-1}$). Hydrogen is released when strong acids act on ignoble materials such as zinc, iron, and aluminum. Hydrogen acts as a reducing agent and reacts with certain metals and alloys to produce hydrides, in which hydrogen is physically dissolved or present as a negatively charged ion. Adsorbed hydrogen promotes the autoignition of fine-grain metals in air. On catalytically active surfaces, hydrogen and oxygen recombine to water.

Molecular hydrogen is able to absorb spontaneously on most solid surfaces and establishes strong interaction of the hydrogen atoms with the surface atoms: $H_2 + 2M \rightarrow 2M\text{–}H$. On hydrogen-saturated surfaces, almost every metal atom is occupied by a hydrogen atom. Furthermore, the hydrogen atoms can penetrate into deeper layers to form subsurface hydrogen. The heat of hydrogen adsorption drops with rising surface coverage, often by a linear law for intermediate coverages (Temkin's surfaces).

Dissolution of hydrogen in metals and intermetallic compounds involves diffusion, adsorption under dissociation to atoms, movement into the subsurface state and substrate bulk, and finally the formation of solid solutions (α-phase) and hydrides (β-phase). Hydrides are stoichiometric chemical compounds, whereas solid solutions are composed arbitrarily. Sieverts' law (Adolf Sieverts 1929) predicts the *solubility* c (in mol/L) of diatomic gases (H_2, O_2, N_2) in metals, based on the equilibrium constant K of the reaction $H_2 \rightarrow 2H$ and the partial pressure p (Equation 129).

$$c(\mathrm{H}) = \sqrt{K \cdot \frac{p(\mathrm{H_2})}{p^0}} \tag{129}$$

According to the Gibbs phase rule, the hydride phase is formed at a constant hydrogen pressure, that is, the ideal sorption isotherm has a horizontal plateau (in practice a hysteresis). Above a certain critical temperature, the coexistence of two phases disappears (Figure 97).

Compressed Hydrogen

Think about whether hydrogen is better stored as a gas or liquid. Compressed hydrogen in gas holders (gasometers) and gas cylinders is commercially developed, although little economically reasonable. At pressures above 10 bar, the ideal gas law, $pV = nRT$, must be replaced by the Beattie-Bridgeman equation for real gases (Equation 130 and Figure 98):

$$p(T,\varrho) = \varrho^2 RT\left(1 - \frac{\varrho c}{T^3}\right)\left[\frac{1}{\varrho} + B(1 - bp)\right] - A\varrho^2(1 - ap) \tag{130}$$

where p is the pressure (Pa), ϱ is the density (kg m^{-3}), T is the temperature (K), R is the molar gas constant: 8.3144 $mol^{-1}K^{-1}$, constants: $a = -5.06 \cdot 10^{-6}$ m^3mol^{-1}, $b = -43.56 \cdot 10^{-6}$ m^3mol^{-1}, $c = 5.04$ $m^3K^3mol^{-1}$, $A = 0.02$ Pa m^3 mol^{-2}, $B = 20.96 \cdot 10^{-6}$ m^3mol^{-1}.

Gaseous hydrogen or *compressed hydrogen* (G-H_2) is an attractive choice based on its low weight for vehicle tanks: 23 kg m^{-3} at 350 bar and 39 kg m^{-3} at 700 bar (300 K). Below 700 bar, ideal gases reach the theoretical density of 57 kg m^{-3}. Hydrogen exhibits high

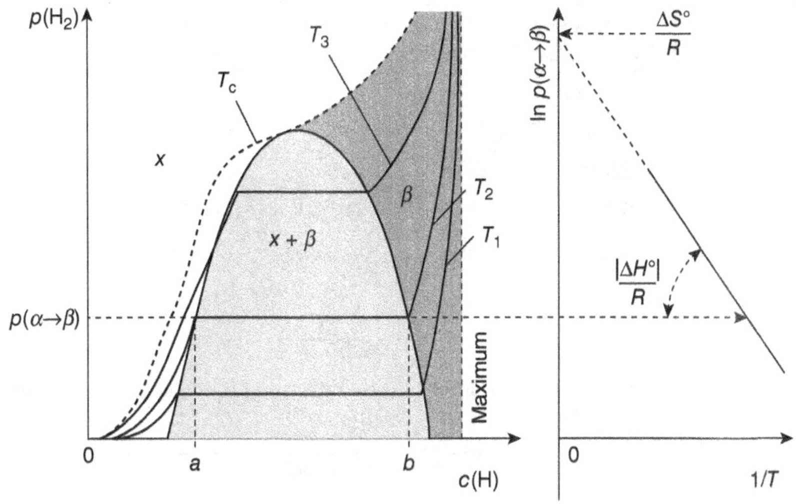

FIGURE 97 General phase diagram of metal-hydrogen (left); c hydrogen concentration in metal. The phase transition pressure $p(\alpha \rightarrow \beta)$ depends on temperature T (right), hydride formation enthalpy ΔH^0, and entropy change $\Delta S^0 \approx 120$ J $mol^{-1}K^{-1}$ during the transition from gaseous H_2 to the solid state.

specific energy (33.3 kWh kg^{-1}) but low energy density at standard conditions (0.003 kWh L^{-1}). The maximum energy density of G-H_2 (1.3 kWh L^{-1}) at 700 bar approaches the storage density of liquid hydrogen. When compared with fossil fuels, it is evident that hydrogen in the liquid or compressed gas form has the highest energy content per weight of all substances (Table 21).

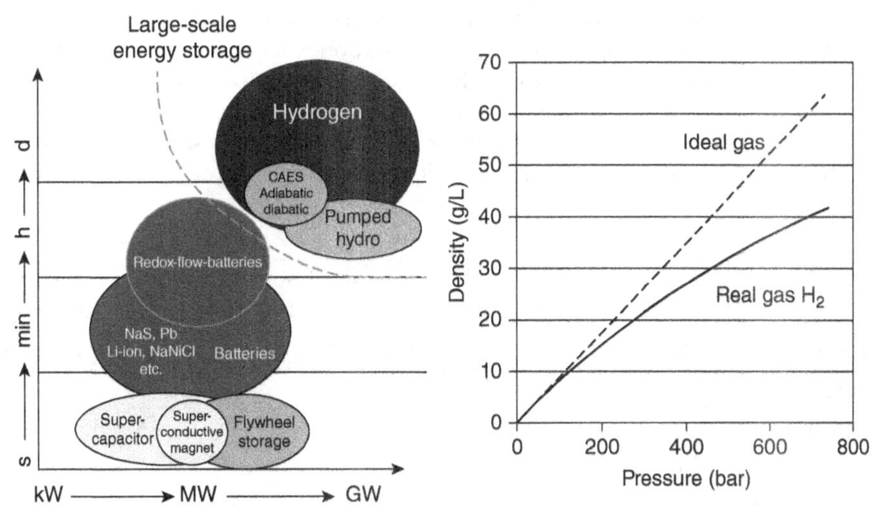

FIGURE 98 Power and storage time of different storage systems (left) and deviation from the ideal gas law at high pressures (right).

	Hydrogen Liquid (20 K)	Hydrogen 700 bar	Hydrogen 25°C	Methane 25°C	Natural Gas: 95% Methane	Gasoline
H_o: MJ kg^{-1}	142.0			55.6	54.0	46.5
H_u: MJ kg^{-1}	119.6			49.9	—	42.5
H_u: kWh kg^{-1}	33.3			13.9	—	11.8
H_u: kWh L^{-1}	2.36	1.86	0.0030	0.010	—	
ϱ: kg m^{-3}	70.83	39	0.0824	0.6556	—	700
Volume (L) of 1 kg	14.1	26	12140	1525	—	1.4

Note: $H_o = -\Delta H_v$ gross calorific value, combustion enthalpy; H_u lower heating value; 1 MJ m^{-3} = 2.778 · 10^{-4} kWh L^{-1}; ϱ density; 1 kg m^{-3} = 0.001 g cm^{-3}.

Table 21 Energy Density of Hydrogen and Fossil Fuels

No cooling units are required, but the high expenditure of energy for compression of about 12% of the energy content of hydrogen may be uneconomical. Current materials are sufficiently safe and impermeable. G-H_2 is transported in pipelines (1–2 MPa) or tank trucks (20–35 MPa); 2000–6200 m^3 G-H_2 weigh 150–500 kg.

Liquid Hydrogen

What an extremely cold liquid! Liquefied gas is stored in cryogenic refrigerators and special Dewar vessels by repeated cooling and compressing. Liquid hydrogen L-H_2, below the boiling temperature of −252.8°C = 20.4 K, provides a density of 71 kg m^{-3}, which is the 800-fold of the density of hydrogen gas at room temperature. Moreover, it is hardly reactive. The energy density equals 2.4 kWh L^{-1}. Above the critical point (−240°C, 1.3 MPa), hydrogen can no longer be liquefied. Although the energy expenditure for condensation, cooling, and evaporation is less efficient than with compressed gas, liquid hydrogen is useful for the storage of large quantities (Table 22).

Liquid hydrogen can easily be transported in tank trucks; 50 m^3 L-H_2 weigh 3700 kg. The liquefaction consumes roughly 28–46% of the energy stored in the hydrogen and additionally up to 6% for transportation between producer and hydrogen gas station

	Large Storage (Above 100 m^3)		Small Storage (Up to 100 m^3)		Vehicle Tank (Up to 0.5 m^3)	
Storage	kg H_2 kg^{-1}	kg H_2 m^{-3}	kg H_2 kg^{-1}	kg H_2 m^{-3}	kg H_2 kg^{-1}	kg H_2 m^{-3}
Underground	—	5–10	—	—	—	—
Compressed gas	0.01–0.014	2–16	0.012	~15	0.05	15
Metal hydride	0.013–0.015	50–55	0.012–0.014	50–53	0.02	55
Liquid hydrogen	≈1	65–69	0.15–0.50	~65	0.09–0.13	50–60

Table 22 Storage Density of Different Hydrogen Containers

(gasoline and diesel: only 0.2%). Evaporation losses occur during filling, and the L-H_2 tank suffers from inevitable boil-off losses of up to 3% per day, whereby the technical weaknesses are largely solved. For comparison, the efficiency of a coal-fired power station yields roughly 40%.

Adsorption on Surfaces

Adsorption means the adhesion of a substance on an arbitrary surface by chemical bounds (chemisorption) or interaction forces (physisorption). So far, the storage of hydrogen on highly porous surfaces has not been realized satisfactorily. Both reversibility and storage density are poor (Table 23). Some materials require inefficient cooling with liquid nitrogen (–196°C).

Activated carbon is widely used as the most important adsorbent in chemical industry. Despite the high specific surface area of 1000–2000 $m^2 g^{-1}$, only 1–2% hydrogen is stored at 77 K and 1 bar, whereby the smallest pores contribute most. The theoretical hydrogen sorption on microporous carbon can reach 9.6%. However, experimental data are hardly reproducible, depending on synthesis and purification methods, materials quality, and specific features of the measuring instruments. Clean and well-standardized carbons do not feature high adsorption; incorporated residues of catalysts from the synthesis stage or purposeful doping, for example, alkali ions, cause *spillover*, that is the migration of atomic hydrogen from a catalytically active metal to the carbon material. Hence, hydrogen atoms chemisorb, although the dissociation $H_2 \rightarrow 2H$ cannot proceed on a

Class	Material	Theoretical Specific Energy ($Wh\ kg^{-1}$)	H_2 Loading (% = 0.01 kg kg^{-1})	Sorption Temperature (°C)	Desorption Temperature (°C)
Sorbents	MOF-74		2.3	–200	—
	Carbon aerogel		4.6	–200	—
	Polyaniline		1–3	—	≈200
Metal hydrides	$Mg(BH_4)_2$		11	—	≈350
	AlH_3		9.3	—	≈100
	MgH_2		7.8	—	≈400
	$LiNH_2$, $MgNH_2$		4.5	—	≈240
	$NaAlH_4$		3.5	—	≈120
	$LiMn(BH_4)_3$		3	—	≈100
Chemical reactions	LiH + water	2540	7.8	Irreversible	Room temp.
	CaH_2 + water	1700	5.2	Irreversible	Room temp.
	$NaBH_4$ + water	3590	10.9	Irreversible	Room temp.
Nonmetal hydrides	$NH_3{\cdot}BH_3$		16	—	≈150
	1,6-Diazanaphthalene		7	—	≈280

Table 23 Examples of Hydrogen Storage Materials

pure carbon surface. Carbon powders in aqueous solutions show a pseudocapacitance of 10–100 F g^{-1} and redox transitions involving the electroactive groups of the material (see Supercapacitors, Part 3).

Carbon nanotubes (CNT) are able to incorporate 5–10% hydrogen in graphene-type structures of 1 nm thickness and up to 100 μm length. Storage density is poor, and desorption succeeds incompletely: only 2% at 25°C, and 100% above 530°C.

Graphite nanofibers (GNF) store about 14% H_2 at 120 bar, which is theoretically 75% of the total fiber mass. At medium pressures (<40 bar), hydrogen is released partly. Nanotubes of titanium dioxide and silicon carbide absorb roughly 2% hydrogen at 60 bar.

Clathrates. The water-hydrogen system shows cage structures at low temperatures (below 250 K) and high pressures (100–360 MPa). Hydrogen guest molecules are embedded into the labile frame of host water molecules. The maximum theoretical stoichiometry corresponds to the formula $H_2 \cdot 2.125\ H_2O$ (5.2% hydrogen). Double clathrates of various compositions can be formed with organic additives, water-soluble hydrating agents, or gaseous methane.

Metal-organic frameworks (MOF) are a group of promising crystalline materials with exactly defined pore sizes of 1–2 nm and specific surfaces up to 5000 m^2g^{-1}. Their heat stability is low. Regularly composed transition metal complexes (of copper, zinc, nickel, cobalt, manganese, aluminum, chromium, lanthanides) build up highly symmetric structures by the help of linking organic molecules. MOF-5, $Zn_4O[OOC–C_6H_4–COO]$, exhibits a specific surface of 2900 $m^2\ g^{-1}$, which is based on a cubic lattice of six $[ZnO_4]_4$ clusters in the corners of a cube and terephthalic acid ligands $[–CO\text{-}C_6H_4\text{-}CO–]$ at the edges. The oxocations play the role of "host" lattice sites and polycarbonate ions serve as linkers. The free space in the frame is occupied by hydrogen.

Zeolites are regularly formed alumosilicates with well-defined pores and a large internal surface. They are used as molecular sieves for drying organic solvents. The hydrogen absorption capability is poor: up to 0.3% hydrogen at room temperature, and up to 1.8% at liquid nitrogen temperature at a pressure of 1.5 MPa.

Glass microspheres (<0.1 mm, wall thickness 1 μm) incorporate hydrogen at 400°C and extreme pressures (450–1000 bar) because glass is hydrogen permeable. Incomplete desorption from the cavities occurs with heating and expansion to atmospheric pressure. Storage density (2–14%; ≈ 12 g L^{-1}) and energy density are poor compared with the high risk of breakage with charging and decanting.

Conducting polymers such as polyaniline and polypyrole are able to store up to 8% H_2. Because of the relatively low storage capacity compared with gas cylinders, hydrogen storage in polymers has not yet been used in practical application.

Metal Hydrides and Storage Alloys

We know that gaseous and liquid hydrogen can be stored in a tank. Now imagine hydrogen that is captured in a chemical substance.

Hydrides, chemical compounds of metal ions and hydride ions, are considered to be a long-term alternative to hydrogen tanks. The chemical bonds in typical nonmetal hydrides (B_2H_6, CH_4, NH_3, and PH_3) are quite stable, so that any highly reversible storage system cannot be realized. Hydrogen storage in metal has attracted attention since the discovery of hydrogen dissolution in palladium (T. Graham) in 1866. In the

1960s, intermetallic compounds were discovered. Metal structures are prone to hydrogen embrittlement (hydrogen corrosion), that is, they are gradually destroyed under the action of hydrogen that incorporates into the metal lattice. Expensive high-alloyed materials are promising in combination with correct welding, protection from outdoor corrosion, and the use of special inhibitors and coatings.

Binary Metal Hydrides

Metal hydrides are chemical compounds or alloys that are able to incorporate more or less stoichiometric quantities of hydrogen in the metal lattice. With heating, hydrogen is released again. Unfortunately, reversibility and storage density are limited, so that the storage system is heavy and does not completely return the stored hydrogen (Equation 131).

$$\text{hydrogen} + \text{metal} \xrightarrow{\text{charging}} \text{hydride} + \text{heat} \tag{131}$$

The maximum *hydrogen content* (mass fraction in %) of the metal hydride can be calculated by the help of the molar mass, that is, the sum of the atomic masses listed in the periodic table of the elements.

$$w(\mathrm{H_2}) = \frac{M(n\mathrm{H})}{M(\mathrm{MH}_n)} = \frac{n \cdot 1.008\ \mathrm{g\ mol^{-1}}}{M(\text{metal hydride})} \tag{132}$$

Theoretical *specific energy* depends on the heating values of hydrogen (Equation 133).

$$W_m = w(\mathrm{H_2}) \cdot 119.6\ \mathrm{MJ\ kg^{-1}} = w(\mathrm{H_2}) \cdot 33.2\ \mathrm{kWh\ kg^{-1}} \tag{133}$$

Example: The metal hydride (MH, 2% H_2) should store 655 Wh kg^{-1}. In practice, only about 160 Wh kg^{-1} (420 Wh L^{-1}) were achieved.

Magnesium hydride works as a reversible hydrogen storage and latent heat accumulator at high temperatures (Equation 134). The lattice stores 7.66% H_2, which is denser than liquid hydrogen. The energy density is 2500 kWh kg^{-1} (for magnesium); experimentally about 90% of the theoretical value is reached. At 280–300°C, hydrogen is released, and charging takes place at 500°C (200 bar) or catalytically at 20–65°C (1–80 bar; catalyst anthracene, $TiCl_4$, $CrCl_3$, $FeCl_2$).

$$74\ \mathrm{kJ} + \mathrm{MgH_2} \rightleftharpoons \mathrm{Mg} + \mathrm{H_2} \qquad (p(\mathrm{H_2}) = 1\ \text{bar at } 284°\mathrm{C}) \tag{134}$$

Water irreversibly generates magnesium hydroxide, although traces of steam and atmospheric contaminations are tolerated. In contact with diborane, a mixed hydride (magnesium boranate) is formed, which releases hydrogen at lower temperatures (Equations 135 and 136).

$$\mathrm{MgH_2} + 2\,\mathrm{H_2O} \rightarrow \mathrm{Mg(OH)_2} + 2\,\mathrm{H_2} \tag{135}$$

$$\mathrm{MgH_2} + \mathrm{B_2H_6} \rightarrow \mathrm{MgH_2} \cdot 2\,\mathrm{BH_3} \equiv \mathrm{Mg(BH_4)_2} \tag{136}$$

Calcium hydride was used earlier for filling weather balloons. It is synthesized by the reaction of hydrogen with calcium metal at 400°C (Equation 137). The violent reaction with water is irreversible.

$$\mathrm{CaH_2} + 2\,\mathrm{H_2O} \rightarrow \mathrm{Ca(OH)_2} + 2\,\mathrm{H_2} \tag{137}$$

Lithium nitride (Li_3N), from lithium amide ($LiNH_2$) and lithium hydride (LiH), is able to store up to 7% hydrogen (theoretically 11.3% H_2) at 255–285°C (1 bar) in a two-step process. However, ammonia is released in the process (Equations 138 and 139).

$$3\,LiNH_2 \rightarrow NH_3 + 3\,LiH + N_2 \tag{138}$$

$$LiNH_2 + 2\,LiH \rightarrow Li_2NH + LiH + H_2 \rightarrow Li_3N + 2\,H_2 \tag{139}$$

Complex Hydrides (Ternary Hydrides)

Complex hydrides are hydrogen compounds of boron, aluminum, and transition metals, which provide a hydrogen-metal ratio of $H/M \geq 2$. As a rule, ionic hydrides are Lewis bases, which react with acidic hydrides to produce complex compounds, for example, lithium-aluminum hydride $Li[AlH_4]$ and sodium boronhydride $Na[BH_4]$.

Lithium-boron hydride ($LiBH_4$) yields such high theoretical storage density as 18% H_2 = 121 kg H_2/m^3, in practice 9% H_2. At the useful operating temperature of 200–400°C, recharge is difficult, and toxic borane is released (Equations 140 and 141). Doping with titanium allows lower temperatures (120–170°C).

$$\text{Thermal process:} \quad LiBH_4 \rightarrow LiH + B + \frac{3}{2}H_2 \tag{140}$$

$$\text{Hydrolysis:} \quad LiBH_4 + 2\,H_2O \rightarrow LiBO_2 + 4\,H_2 \tag{141}$$

Sodium-boron hydride ($NaBH_4$, sodium boranate) can incorporate hydrogen up to theoretically 18% of its mass. The water soluble, white, solid material can be used once only; above 400°C, it decomposes during contact with catalysts (e.g., Raney-Ni, Ru, Co). In alkaline solution, hydrogen and sodium borate are immediately formed (Equation 142).

$$NaBH_4 + 2H_2O \rightarrow 4H_2 + NaBO_2 \tag{142}$$

The aqueous waste sludge of $NaBO_2$ is recycled with hydrogen. Fresh $NaBH_4$ is synthesized (1) by the reaction of boric acid trimethylester and sodium hydride (Schlesinger process) or (2) from finely ground borosilicate glass and sodium and hydrogen.

Aluminum-boron hydride, $Al(BH_4)_3$, surpasses with 150 kg m^{-3} the storage density of liquid hydrogen.

Sodium-aluminum hydride ($NaAlH_4$, sodium alanate), doped with titanium, is able to store hydrogen up to theoretically 7.5% of its mass at 60–130°C and high pressures. About 5.5% are released reversibly at 100–200°C (Equations 143 and 144). However, the useful lifetime of sodium alanate is short.

$$3\,NaAlH_4 \rightarrow Na_3AlH_6 + 2\,Al + 3\,H_2 \qquad (3.7\%\ H_2) \tag{143}$$

$$Na_3AlH_6 \rightarrow 3\,NaH + Al + \frac{3}{2}H_2 \qquad (1.8\%\ H_2) \tag{144}$$

Magnesium-aluminum hydride, $Mg(AlH_4)_2$, stores theoretically 9.3% H_2 while practically 6.6% were reported at 200–250°C. The compound is thermodynamically stable.

Storage Alloys and Transition Metal Hydrides (Quaternary Hydrides)

Alloys may be composed stoichiometrically (*intermetallic compounds*) or non-stoichiometrically (*solid solutions*). Hydrogen molecules, adsorbed of the surface, are split into hydrogen atoms and penetrate into the metal lattice. Hydrogen occupies the interstitial sites, whereby the crystal lattice is strongly expanded. Charging takes place by direct action of hydrogen into the finely dispersed metal under pressure and elevated temperature. Discharge succeeds largely in a reversible way by decompression or heating.

Interstitial hydrides are used in commercial nickel-metal hydride batteries. The AB_5 type comprises $LaNi_5{\cdot}M_x$ (where M = Cr, Fe, Co, Cu, Mn, Al, Si, Sn, Ge, In, Tl, Pd, Ag) and $La_{1-x}R_xNi_5$ (where R = Ce, composition metal, Ca, Y, Gd, Zr, Pr, Nd, Sm); the AB_2 type consists of alloys and Laves structures (e.g., $ZrMn_2$, $ZrCr_2$, $TiMn_2$, $ZrCo_2$, $ZrMo_2$, $ZrNi_2$, $MgNi_2$, $HfCr_2$).

Low-temperature hydrides (−30°C to 150°C) partly react with hydrogen already at room temperature: $Ti(Cr_{1-x}Mn_x)_2$, $TiCr_{1,8}H_{1.7}$ (0°C), $CaNi_5$, $LaNi_5H_x$ ($x \leq 6.5$), $BaReH_9$, $TiFeH_x$ (5% Mn, $x \leq 2$, <60°C), $MgNiH_x$ ($x \leq 4$), Mg_2NiH_4, Mg_2FeH_6, $CaNi_5H_4$ (<100°C). Safe technical applications are known from lanthanum-nickel alloys in submarines with fuel cell propulsion systems.

An example of a medium-temperature hydride (100 to 200°C) is $LaNi_4AlH_5$.

Some high-temperature hydrides (above 200°C) are Mg_2NiH_4, $PdH_{0.6}$, $ZrV_2H_{5.5}$ (Laves phase), $CeNi_3H_4$, $Y_2Ni_7H_3$, and $Ho_6Fe_{23}H_{12}$. $NaReH_9$ provides the highest hydrogen-metal ratio of 4.5; unfortunately, hydrogen is slowly released even at high operating temperatures. The intermetallic compound $LaMg_2Ni$ accepts hydrogen above 200°C in H_2 atmosphere (0.8 MPa) until the composition $LaMg_2NiH_7$ is reached; it goes over from a metallic conductor to an insulator.

Inorganic Nonmetal Hydrides

What shall be done with high quantities of hydrogen from electrolyzer plants using renewable power? Compounds, easy to manage, are wanted, which can be used in the existing infrastructure of gas stations. Inorganic nonmetal hydrides are small molecules, which are generated from special precursors and hydrogen and are split vice versa when needed. However, the process is ineffective due to the strong chemical bonds.

Ammonia

The use of ammonia for storing industrial hydrogen from wind and solar power plants is controversial. NH_3 is a corrosive, strongly caustic and toxic gas. Worldwide, in 2012, 198 million tons were produced using the exothermic Haber-Bosch process (Equation 145) from air nitrogen and hydrogen in fixed bed reactors (Fe catalyst, 150–250 bar, 200–500°C).

$$2H_2 + N_2 \rightarrow 2NH_3 + 92\ \text{kJ mol}^{-1} \tag{145}$$

As ammonia boils at −33°C, it is stored in liquid form at 10 bar. 1 m^3 liquid NH_3 at 27°C contains 106 kg H_2, which is 36 kg more than can be stored in the same volume using liquid H_2, and 66 kg more than can be stored in compressed H_2 (700 bar, 20°C).

On the other hand, the recovery of hydrogen is quite inefficient and uneconomic. The technically difficult combustion delivers 17.8 MJ kg^{-1} (heating value), which is half of the energy of diesel. Moreover, toxic nitrogen oxides are generated with ammonia combustion. The catalytic decomposition (Ru, Pt, 600°C) and also the direct conversion in PEM fuel cells is possible but not yet generally feasible.

Boron Nitrogen Hydrides

Ammonium trihydroborate (borazane), the product from ammonia and borane (in tetrahydrofuran as a solvent), is a solid material, which melts at 104°C, polymerizes with heating, and splits off about 6.5–12% hydrogen, until boron nitride (BN) results above 1000°C (Equations 146 and 147). The storage density is greater than that of liquid H_2. Unfortunately, ammonia can be set free and reversible charging is difficult.

$$NH_3 + BH_3{\cdot}THF \rightarrow NH_3BH_3 + THF \tag{146}$$

$$n\,NH_3BH_3 \xrightarrow{<120^\circ C} (NH_2BH_2)_n + n\,H_2 \xrightarrow{160^\circ C} (NHBH)_n + 2n\,H_2 \tag{147}$$

Organic Molecules as Hydrogen Storage

Organic compounds consist of carbon, hydrogen, and hetero atoms such as oxygen, nitrogen, and sulfur. Hydrocarbon and alcohol molecules carry bound hydrogen, which can be released by chemical processes. So far, organic compounds have not been able to work as reversible hydrogen storages, but they can be fired in combustion engines.

Methane

Methane (CH_4) is the main component in natural gas. The global reserves of $210{\cdot}10^{12}$ m^3, including shale gas and methane hydrate, rest in the Near East, Russia, and China. Shale gas consists of 80–95% methane, 3–16% ethane, up to 4% propane, and up to 2% butane.

Methane also occurs as a *synthesis gas* at supported catalysts (Ni or Ni-Ru, Ru, Rh) at 300–600°C and atmospheric pressure—which is the exothermic reversal of the steam reforming process (Equations 148 and 149).

$$3\,H_2 + CO \rightarrow CH_4 + H_2O + 206 \text{ kJ mol}^{-1} \tag{148}$$

$$4\,H_2 + CO_2 \rightarrow CH_4 + 2\,H_2O + 165 \text{ kJ mol}^{-1} \tag{149}$$

The water gas shift reaction is an undesired side reaction (Equation 150).

$$H_2O + CO \rightarrow CO_2 + H_2 \tag{150}$$

Synthesis gas consists of 98.5 vol% CH_4, 0.5 vol% CO_2, 1.0 vol% H_2, and less than 0.1 vol% CO. Natural gas must be desulfurized before use in fuel cells. The product water must be condensed out of the synthesis gas stream because it inhibits the reaction (wet methanization); in a second step, dry methanization follows. Supported catalysts, such as CeO_2, ZrO_2, and TiO_2, favor partial reduction and CO_2 activation more actively than dispersed nickel. The high reaction heat requires steam dilution in a loop-reactor, which converts low H_2 syngas (e.g., from coal gasification) into two wet and a dry methanization without coking. Fluidized-bed reactors (without steam feed, ≤60 bar) have reached the pilot stage.

Methane is stored (1) as a gas (200–300 bar, 9.2 MJ L^{-1}), whereby the losses due to compression amount to 5% of the stored energy, or (2) as a liquid (–162°C, 22 MJ L^{-1} = 64% of gasoline). Prior to liquefaction in cooling cascades, suffering 8–10% loss (of the stored energy), dust, acid gases (H_2S, CO_2), helium, water, and heavy hydrocarbons must be separated. The leakage of L-CH_4 (0.15% per day) in tank ships is three times slower than that of liquid hydrogen.

Methane adsorbs reversibly both on microporous-spongy metal organic frameworks (MOF) and polymers. The material UTSA-20 stores 0.222 g cm^{-3} = 178 m^3 m^{-3} (27°C, 35 bar), which is nearly the density of methane at 350 bar. Water and CO_2 must be excluded because they bind more strongly to the MOF than methane.

Methane can be fired in gas turbines and combustion engines. The combustion of methane generates fewer emissions (kg CO_2 per MJ energy) than other fossil fuels. The heating value (50 MJ kg^{-1}) surpasses that of gasoline and diesel (43–44 MJ kg^{-1}).

Synfuels

Synthetic hydrocarbons (synfuels) have been produced since the 1920s by the exothermic Fischer-Tropsch process using synthesis gas (Equation 151). The water gas shift side reaction is undesirable (Equation 152).

$$(2n+1)\,H_2 + n\,CO \rightarrow [CH_2]_n + n\,H_2O + 165\text{ kJ mol}^{-1} \tag{151}$$

$$H_2O + CO \rightarrow CO_2 + H_2 \tag{152}$$

Syngas from natural gas reforming (H_2/CO = 1.8 to 2.5) or from coal gasification (H_2/CO = 0.5 to 1.7) is cleaned of impurities such as sulfur and nitrogen compounds first, and then converted into an alkane-alkene mixture, which is separated by distillation: liquefied petroleum gas (LPG, C_1 to C_5), synthetic gasoline (C_5–C_{11}) and diesel (C_9–C_{25}), and paraffin wax (C_{35}–C_{120}). At high temperatures (300–350°C, 10–40 bar, Fe), more gasoline and alkenes and undesired methane are generated. At lower temperatures (200–240°C, Co) and high pressures, middle distillate and waxes predominate. By after-treatment (hydroisomerization and hydrocracking), long hydrocarbon chains are cracked into short, branched compounds. In water-cooled fixed bed tube reactors and low-temperature slurry bubble columns, yields of 50% are reached using catalysts based on ruthenium-doped cobalt. Low-cost iron (Fe_2O_3/H_2) supports the unwanted water gas shift reaction.

Methanol

Methanol, CH_3OH, is produced in the 30 Megaton range per year from natural gas or coal with about 70% efficiency (Equations 153 and 154). The reactors are fixed bed or tube type.

$$2\,H_2 + CO \rightarrow CH_3OH + 91\text{ kJ mol}^{-1} \quad \text{(high pressure)} \tag{153}$$

$$3\,H_2 + CO_2 \rightarrow CH_3OH + H_2O + 50\text{ kJ mol}^{-1} \text{(high temperature)} \tag{154}$$

The original catalyst for the process, Zn/Cr_2O_3 (320–380°C, 300 bar), was replaced by the more stable but sulfur-sensitive Cu/ZnO/Al_2O_3 (≈250°C, 50 bar).

Methanol can easily be stored in the existing gas station infrastructure. It is a valuable starting material for the synthesis of formaldehyde, methyl-*tert*-butyl ether, and acetic acid. Methanol can be mixed up with gasoline or can be burned directly; the emissions (soot, NO_x, CO) are lower, but only half of the combustion heat of gasoline is achieved (20 kJ kg^{-1}), so that large tank volumes are required. A process of methanol synthesis from hydrogen generated in the steam methane reforming reaction is less efficient than the direct combustion.

Dimethyl Ether

Dimethyl ether (DME, CH_3–O–CH_3) is widely considered to be a promising substitute for liquefied petroleum gas (LPG) for automotive applications. DME is produced mainly in China from coal gas. Two processes are practically used as follows:

1. The two-step methanol synthesis is supplemented by the exothermic dehydrogenation over the acid catalyst, whereby coking takes place as an unwanted side reaction (Equations 155 and 156).

$$2H_2 + CO \rightarrow CH_3OH \tag{155}$$

$$2CH_3OH \rightarrow CH_3\text{–O–}CH_3 + H_2O + 23\ \text{kJ mol}^{-1} \tag{156}$$

2. Direct synthesis from synthesis gas in a bubble column (260°C, 50 bar) using a bifunctional catalyst, which affects both methanol synthesis and dehydrogenation. The consecutive reaction pulls methanol out of the chemical equilibrium, and the produced water drives the desired water-gas shift reaction (≈93% CO turnover at 260°C, 50 bar) (Equations 157 to 159).

$$4H_2 + 2\,CO \rightarrow (2CH_3OH) \rightarrow CH_3\text{–O–}CH_3 + H_2O \tag{157}$$

$$CO + H_2O \rightarrow CO_2 + H_2 \tag{158}$$

$$\text{Overall reaction: } 3H_2 + 3CO \rightarrow CH_3\text{–O–}CH_3 + CO_2 + 249\ \text{kJ mol}^{-1} \tag{159}$$

The ratio $H_2/CO \approx 1$ means that the syngas from biomass gasification has 96% turnover rate and 93% selectivity for DME (JFE, Japan).

Liquid DME (6 bar, 25°C) provides a combustion heat of 20.5 MJ L^{-1}, similar to LPG (24.7 MJ L^{-1}) and liquid methane (–162°C; 22.2 MJ L^{-1}). The high cetane number is good (55–60), as well as the "clean" combustion (generating low levels of soot, NO_x, and CO) and the low global warming potential ($^1/_{50}$ of methane). Unfortunately, the low viscosity of DME requires modified diesel injection systems. On the other hand, the well-to-wheel efficiency is better by a quarter compared with Fischer-Tropsch diesel.

Higher Alcohols and Ethers

A 2–10% content of oxygen-rich molecules in the fuel improves combustion temperature in engines and reduces the emissions of exhaust gases (e.g., soot, NO_x, CO). The addition of ethanol is common in Brazil and in the United States. The low solubility of the short-chain alcohols in diesel requires adapted fuel feed, otherwise phase separation, vapor barrier, engine misfire, and performance losses occur.

Butanol as a fuel additive provides high heating value of 33 MJ kg^{-1} (gasoline: 44 MJ kg^{-1}). The worldwide production of 3 Mt per year is based on the exothermic low-pressure hydroformylation of propene, on homogeneous rhodium phosphine catalysts (<100°C, 20 bar). The produced *n*- and *iso*-butanal is hydrogenated in the exothermic second stage (Equations 160 and 161). Propene is gained by catalytic cracking of naphtha (crude petroleum).

$$CH_3CH{=}CH_2 + H_2 + CO \rightarrow CH_3CH_2CH_2CHO \quad +130\ \text{kJ mol}^{-1} \tag{160}$$

$$CH_3CH_2CH_2CHO + H_2 \rightarrow CH_3CH_2CH_2CH_2OH \quad +81\ \text{kJ mol}^{-1} \tag{161}$$

The direct synthesis of higher alcohols from biomass (syngas) requires bifunctional catalysts, which are able to activate CO and manage hydrogenation and chain propagation. The yields of alcohols above C_3 have been poor so far compared with methanol synthesis.

Polyoxymethylene dimethyl ethers, $CH_3O\text{–}[CH_2O]_n\text{–}CH_3$ (POM-DME or OME), are used as diesel additives to gain improved exhaust gas emissions. They are synthesized fairly inefficiently from methanol and formaldehyde on acid zeolite catalysts of sulfonated polystyrene (Amberlyst) at 50–100°C and ambient pressure (Equation 162). Short-chain molecules are generated at higher temperatures.

$$2\,CH_3OH + n\,HCHO \rightarrow CH_3\text{–}(O\text{–}CH_2)_n\text{–}OCH_3 + H_2O \qquad (n \geq 3) \qquad (162)$$

The side products, polyoxymethylene glycols $H(O\text{–}CH_2)_n\text{–}OH$ and hemiformals, can be separated by distillation.

Formaldehyde and Formic Acid

Formaldehyde (methanol) and formic acid have been proposed as industrial storage media for hydrogen. Formaldehyde and its aqueous solution "formalin" are technically produced by methanol oxidation on Ag catalyst and through dehydrogenation over Ru catalyst (Equations 163 and 164). In aqueous solutions, mostly the hydrate is present.

$$\text{Synthesis:} \qquad CH_3OH + \frac{1}{2}O_2 \rightarrow HCHO + H_2O \qquad (\text{Ag, }600°\text{C}) \qquad (163)$$

$$\text{Dehydrogenation:} \qquad HCHO + H_2O \rightarrow HO\text{–}CH_2\text{–}OH \rightarrow CO_2 + 2H_2 \quad (\text{Ru}) \qquad (164)$$

Formic acid is the oxidation product of formaldehyde. However, it is technically synthesized from solid sodium hydroxide and carbon monoxide (Equations 165 and 166). At platinum catalysts, hydrogen is produced even at room temperature.

$$\text{Synthesis:} \qquad CO + NaOH \rightarrow HCOONa \qquad (6\text{–}8\text{ bar, }120\text{–}130°\text{C}) \qquad (165)$$

$$\text{Dehydrogenation:} \qquad HCOOH \rightarrow CO_2 + H_2 \qquad (166)$$

Liquid Organic Hydrogen Carriers (LOHCs)

LOHCs are a pair of molecules that can be reversibly hydrogenated and dehydrogenated based on its aromatic rings (Ar) (Equation 167). Different systems are currently under research, and the most important ones are toluene and dibenzyl toluene (Table 24). Toluene can be easily dehydrogenated to methyl cyclohexane.

$$\text{Ar} + n\,H_2 \rightarrow \text{ArH}_{2n} \qquad (167)$$

Perhydro-N-ethyl carbazole + N-ethyl carbazole (H_{12}-NEtC/NEtC) have a ring system that incorporates hydrogen in an exothermic reaction in the presence of a Ru/Al_2O_3 catalyst at 20–50 bar and 150–200°C. At low pressure of 1–5 bar, temperature of 160–260°C, and Pt/Al_2O_3-catalyst, H_{12}-NEtC is reversibly dehydrogenated (endothermic 55 kJ mol^{-1} H_2). Volatile side products and catalyst poisons such as CO do not form. However, the catalytic loss of the ethyl groups and the thermal decomposition of carbazole are unwanted.

Perhydrodibenzyl toluene + dibenzyl toluene is a model system based on an industrial heat transfer oil. Dibenzyl toluene maintains liquid state down to –30°C. The hydrogenation enthalpy of H_2 is 70 kJ mol^{-1} and dehydrogenation requires high temperatures (280–360°C).

	Benzene Toluene (Marlotherm® LH) MLH/H_{12}-MLH	Dibenzyl Toluene (Marlotherm SH) MSH/H_{18}-MSH	N-Ethylcarbazole NEtC/H_{12}-NEtC
Hydrogenated form	$CH_3(C_6H_4)–CH_2–(C_6H_5)$	$(C_6H_5)CH_2–(C_6H_3)(CH_3)–CH_2(C_6H_5)$	$[C_6H_4]N(C_2H_5)[C_6H_4]$
Melting point (°C)	–30	–39 to –34	+68
Boiling point (°C)	280	390	270
Dehydrogenated form	$CH_3(C_6H_{10})–CH_2–(C_6H_{11})$	$(C_6H_{11})CH_2–(C_6H_8)(CH_3)–CH_2(C_6H_{11})$	$[C_6H_8]N(C_2H_5)[C_6H_8]$
Storage capacity	6.2% H_2	6.2% H_2	5.8% H_2
Hydrogenation heat (kJ mol^{-1})	71	71	55

TABLE 24 Properties of Liquid Organic Hydrogen Carriers

Safety Issues

The safety of hydrogen has always been a controversial issue. The storage of hydrogen in tanks requires material resistant to stress cracking and gas warning devices. In general, using hydrogen is as dangerous as using gasoline and natural gas. A *leakage of hydrogen* is less alarming than a leakage of natural gas at the same pressure and size of the environment because of the smaller energy loss, although more H_2 molecules escape per second. Mixtures of hydrogen in air are flammable in the weight ratio range of 4–75% hydrogen, and explosive above 18% H_2. Wet hydrogen does not detonate above 40% humidity and is not inflammable above 60% humidity. The ignition temperature is high (585°C), but ignition energy (0.02 mJ) is 14.5 times lower than that of methane. The flame propagation rate (3.5 m/s in air) is 8 times higher than that of methane. The velocity of detonation (1.5–2.2 km/s) and specific explosion energy (1 g H_2 corresponds to 24 g TNT) surpass natural gas and gasoline. The flame temperature (2045°C) is close to that of propane. This makes the requirements of gas equipment more stringent with respect to hydrogen leakage.

Because of the safety concerns, *detection of hydrogen* is extremely important. Thermocatalytic sensors measure the heat efficiency of hydrogen combustion in the presence of a catalyst in the range of 0.1–10 vol% hydrogen in air. Semiconducting sensors record the physical changes during the chemisorption of hydrogen on their surface, usually in the range above 1 ppm (mL/m^3). Fiber-optic detectors use the optical properties of films of rare earth element alloys.

Electrochemical sensors are two-electrode cells with a solid electrolyte and are able to detect less than 10^{-3} vol% (µL m^{-3}) hydrogen. Potentiometric sensors measure the working electrode potential against a reference electrode. Amperometric sensors record the current, which is proportional to the hydrogen concentration consumed in a redox process, at a fixed potential difference in a fixed time interval. Only one electrode is in contact with the hydrogen.

CHAPTER 8

Hydrogen Production and Purification

Where does the hydrogen come from that we want to store in technical quantities? Global large-scale production of hydrogen ($500 \cdot 10^9$ m^3 = 45 million tons) is mostly based on natural gas (40%), oil (30%), coal (18%), and electrolysis (4%). Hydrogen is essential for the synthesis of ammonia, fertilizers, and the catalytic hydrogenation of hydrocarbons. Synthetic fuels are commonly grouped under the headings gas-to-liquid (GTL), coal-to-liquid (CTL), and biomass-to-liquid (BTL). This chapter outlines technical processes except water electrolysis.

Hydrogen from Fossil Fuels

The most important technical processes that provide large quantities of hydrogen also generate harmful carbon dioxide emissions, which must be separated from the hydrogen stream. There are six important reactions for production of hydrogen.

1. Steam reforming: $CH_4 + H_2O + 209\text{ kJ} \rightarrow 3H_2 + CO$
   Water-gas shift reaction: $CO + H_2O \rightarrow H_2 + CO_2 + 90\text{ kJ}$
2. Partial oxidation (POX): $2CH_4 + O_2 \rightarrow 2CO + 4H_2$
3. Autothermal reforming (ATR): $2CH_4 + \frac{1}{2}O_2 + H_2O \rightarrow 2CO + 5H_2$
4. Carbon dioxide reforming: $CH_4 + CO_2 \rightarrow 2H_2 + 2CO$
5. Pyrolysis: $CH_4 \rightarrow 2H_2 + C$
6. Steam decoking: $C + H_2O \rightarrow H_2 + CO$

Steam Reforming Natural gas reacts with an excess of steam at supported nickel catalysts (700–850°C, 25–30 bar), followed by the exothermic water-gas shift reaction in two steps at lower temperature. The produced gas is purified by CO_2-absorption in ethanolamine and conversion of CO into methane. Steam reforming generally provides the highest hydrogen yield and efficiency of all processes; its thermal management is complex and bulky.

Long-chain hydrocarbons, biomass, and methanol are useable after converting them into methane in a prereformer (450–500°C, 25–30 bar), and subsequent purification from sulfur, chlorine, and halogen compounds.

Partial Oxidation Natural gas, coal, and higher hydrocarbons are oxidized under lack of oxygen (900–1000°C). This exothermic reaction may be used for the start-up of steam reforming. In China and South Africa, suspensions of ground coal and water are employed. Water acts as a coolant and limits soot generation. The POX process is the least efficient but has the simplest configurations.

Autothermal Reforming This combination of steam reforming and partial oxidation, preferably in small plants or mobile systems, converts methanol or hydrocarbons into hydrogen. Air and steam must be dosed exactly, and the catalyst must support reforming, water-gas shift, and oxidation.

Hydrogen purification is sometimes necessary for mostly low temperature fuel cells. For coarse cleaning of carbon monoxide in hydrogen-rich synthesis gas, chemical processes are used, while fine cleaning requires physical methods.

- Preferential oxidation (PROX): $CO + \frac{1}{2}O_2 \rightarrow CO_2$
- Methanation at nickel catalysts: $CO + 3H_2 \rightarrow CH_4 + H_2O$
- Gas absorption of CO_2 and H_2S in basic or alcoholic solutions (Rectisol process)
- Pressure swing adsorption (PSA) of impurities on solid-state surfaces
- Hydrogen diffusion through a palladium-silver membrane, which holds back other gases

Pyrolysis Hydrocarbons are cleaved in carbon and hydrogen under exclusion of air. The Kværner process uses a plasma burner at 1600°C (Equation 168).

$$C_nH_m \rightarrow n\,C + \frac{m}{2}H_2 \tag{168}$$

Hydrogen from Biomass

Making hydrogen from nonfossil fuel sources is necessary once coal, gas, and oil run out. Biomass comprises woodchips, straw, cereals, and other biogenic resources. Due to the oxygen content of about 40%, the steam reforming of biomass requires less oxygen feed from outside and yields higher efficiency than coal. Combustion needs stoichiometric oxygen supply while gasification is understoichiometric, and pyrolysis proceeds in the absence of air.

Coal gasification delivers hydrogen-rich gas. At 1000°C, pulverized coal reacts with small quantities of pure oxygen in order to produce heat ($C + \frac{1}{2}O_2 \rightarrow CO$) and is brought in contact with overheated steam as a hydrogen source. The process generates two times more CO_2 (per H_2) and is more expensive than steam reforming.

Biomass gasification involves pyrolysis and coking under exclusion of air to generate syngas ($H_2 + CO/CO_2$), coke, and methanol. The process is more expensive than coal gasification because of the lower energy density, water content, and heterogeneous composition of biomass. Fast-growing plants such as poplar trees and sugar beets, agricultural residual, and foodstuffs might also be economic. Moreover, biomass can be converted to methane, followed by steam reforming to produce hydrogen.

Fermentation involves anaerobic microorganisms capable of releasing hydrogen directly from biomass, as long as hydrogen concentration does not get too high.

Methane-producing bacteria and sulfate-reducing bacteria that consume hydrogen are disruptive to this process. Direct fermentation converts maximum 33% of the combustion heat of glucose into hydrogen generation. Methane fermentation is more effective with 85%.

Hydrogen from Ignoble Metals

The idea of using metals with negative standard potential to liberate hydrogen from diluted acids and bases has been known since 1783, when Charles filled gas balloons with hydrogen generated from iron chips and sulfuric acid. In Germany, prior to the First World War, Zeppelin's airships were filled with hydrogen generated from finely dispersed silicon or ferrosilicon and sodium hydroxide solution because silicon is resistant to all acids (except hydrofluoric acid) (Equations 169 and 170).

$$\mathrm{Si + 2\,NaOH + H_2O \rightarrow NaSiO_3 + 2\,H_2} \qquad \text{(60–150°C)} \tag{169}$$

$$\mathrm{NaSiO_3 + (1+\mathit{x})\,H_2O \rightarrow SiO_2 \cdot \mathit{x}\,H_2O + 2\,NaOH} \qquad \text{(regeneration: 250–300°C)} \tag{170}$$

1 mol (28 g) silicon gives theoretically 44.8 L hydrogen, which is 1.6 m^3 H_2 per kg Si. The overall efficiency is poor, and too much chemical waste is produced. Also, passivation by oxide cover layers must be avoided.

Solar Thermal and Thermochemical Water Splitting

Seawater is an inexhaustible source of hydrogen. The European Hydrosol Project, for example, aims at the CO_2-free hydrogen production by the help of the energy from the sun. At present, hydrogen generation from solar water splitting is very inefficient.

Thermolysis Chemical compounds such as water can theoretically be split by solar thermal heat from hundreds of mirrors directing sunrays onto a ceramic absorber positioned on a tower. The thermal energy is used to split water into hydrogen and oxygen (Equation 171).

$$\mathrm{2\,H_2O \rightarrow 2\,H_2 + O_2} \tag{171}$$

Even at extreme temperatures above 1700°C, the efficiency of thermal dissociation of water is unsatisfactorily low. Furthermore, hydrogen and oxygen must be separated by a heat-resistant membrane of high efficiency.

Iron-Steam Cycle Thermochemical cycle processes use reactions that run under temperature control. Water can be split in contact with metal oxides at 900°C in two steps, but the efficiency of that process is low (Equations 172 and 173).

$$\text{Solar thermolysis:} \quad \mathrm{M^{IV}O_2 \rightarrow M^{II}O + \frac{1}{2}O_2} \qquad \text{(1000–1300°C, endothermic)} \tag{172}$$

$$\text{Chemical hydrolysis:} \quad \mathrm{M^{II}O + H_2O \rightarrow M^{IV}O_2 + H_2} \qquad \text{(800–1200°C, exothermic)} \tag{173}$$

The metal M and its oxides may be zinc ($Zn \rightarrow ZnO$), iron ($FeO \rightarrow Fe_2O_3 \rightarrow Fe_3O_4$), cobalt ($CoO \rightarrow Co_2O_3 \rightarrow Co_3O_4$), manganese ($MnO/MnO_2 \rightarrow Mn_3O_4$), or cerium ($Ce_2O_3 \rightarrow CeO_2$). The iron-steam process at spongy iron at red heat (800°C) has been known for a hundred years.

At around 560°C, iron (II)-oxide is generated from magnetite and further reacted to produce elemental iron used to generate hydrogen in a reaction with steam (Equations 174 to 176). Note that the overall reaction is simply decomposition of water (Equation 177). In practice, magnetite is converted into $FeCl_2$ by hydrochloric acid at 380°C.

$$Fe_3O_4 + H_2 \rightarrow 3\,FeO + H_2O \quad \text{and} \quad FeO + H_2 \rightarrow Fe + H_2O \tag{174}$$

$$Fe_3O_4 + CO \rightarrow 3\,FeO + CO_2 \quad \text{and} \quad FeO + CO \rightarrow Fe + CO_2 \tag{175}$$

$$3\,Fe + 4H_2O \rightarrow Fe_3O_4 + 4\,H_2 \tag{176}$$

$$\text{Sum:} \quad H_2O \rightarrow H_2 + \frac{1}{2}O_2 \tag{177}$$

The iron-steam power plant comprises three parts:

1. Syngas generation by coal gasification
2. Iron(II) regenerator and waste heat utilization in a gas and steam turbine
3. Iron-steam reactor

Photoelectrochemical Hydrogen Production

The photolysis of water is the process in green plants on the basis of photosynthesis.

Photosynthesis Oxygen, hydrogen (in the form of NADPH), and energy (in the form of ATP) all result from the light reaction. The dark reaction gives glucose through reduction of six carbon dioxide molecules (Equations 178 and 179).

$$H_2O + NADP^+ + n(ADP + P) \rightarrow \frac{1}{2}O_2 + NADPH + H^+ + n\,ATP \tag{178}$$

$$CO_2 + 2NADPH + 2H^+ + 3\,ATP \rightarrow \text{“}CH_2O\text{”} + H_2O + 3\,(ADP + P) + 2NADP^+ \tag{179}$$

The photooxidation of chlorophyll takes place at two pigment systems that absorb light at 682 nm and 700 nm (Equation 180). Antenna pigments collect wavelengths around 480 nm.

$$Chl + hf \rightarrow Chl^* \rightarrow Chl^+ + e^- \tag{180}$$

The photoelectron (e^-) jumps into an adjacent redox system within less than 10 ps, far too quickly for the exited chlorophyll Chl^* (singlet, spin combination ↑↓) to reach the low-energy triplet (↑↑). Pigment system P_{700} saves light energy in ATP molecules, reducing an iron-sulfur protein (ferredoxin) and starts further redox processes. The manganese-rich pigment system P_{682} gives back electrons to Chl^+ and procures the electrons by direct oxidation of water (Equation 181).

$$2H_2O \rightarrow O_2 + 4H^+ + 4e^- \tag{181}$$

Chemical technology uses photocatalysts instead of living biological systems.

Photocatalysis at Semiconductors Semiconductors that are hit by photons of light generate electron-hole pairs, which carry enough energy to split water in its components.

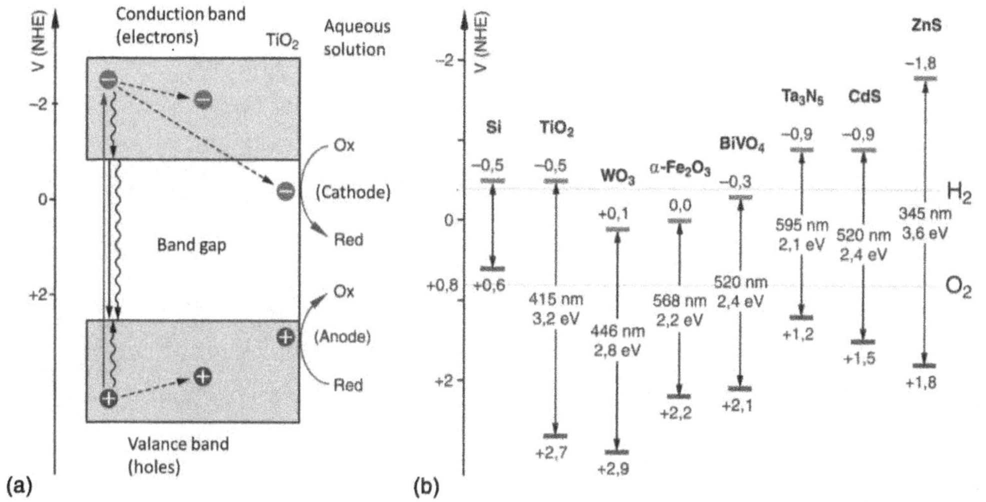

Figure 99 (a) Principle of photocatalysis between a semiconductor and an aqueous solution: generation of electron-hole pairs and charge carrier diffusion in the material. (b) Band gaps of semiconducting powders in contact with pure water (pH 7). (P. Kurzweil, *Angewandte Elektrochemie*, Springer Vieweg, Wiesbaden, 2020.)

In contrast to electrolysis which requires two solid-liquid interfaces, a single surface is sufficient for photocatalysis (Equations 182 and 183).

$$(-)\ \text{Cathodic reduction:}\quad 4H_2O + 4e^- \rightarrow 2H_2 + 4OH^- \qquad (E° = -0.41\ \text{V, pH 7}) \qquad (182)$$

$$(+)\ \text{Anodic oxidation:}\quad 4OH^- \rightarrow O_2 + 4e^- + 2H_2O \qquad (E° = +0.80\ \text{V, pH 7}) \qquad (183)$$

The band gap must be lower than 3 eV in order for a semiconductor to absorb visible light (Figure 99). The catalyst must be effective for both oxidation and reduction half-cell reactions. Some oxides (e.g., WO_3, Fe_2O_3, $BiVO_4$) are only useful for water oxidation. Oxynitrides (doped GaN-ZnO and $ZnGeN_2$-ZnO) tend to corrode and have low quantum yields of 0.1%, so that only small quantities of gases (a few mmol) are generated after serval months. In acid solutions, oxide defect materials (IrO_2, titanates, MnO_2) are used.

Photocatalysis with two active semiconductor powders, bismuth vanadate for H_2-evolution and Rh(III)-doped strontium titanate for O_2 generation, has been studied. The electron transfer takes place within the semiconductor particles.

Photoelectrochemical cells separate H_2 and O_2 by combining two semiconductor absorbers across the electrolyte. The cell looks like an electrolyzer without an external power source.

CHAPTER 9

Water Electrolysis

Electrolysis means the decomposition of an ionically conducting substance (electrolyte) by the action of the electric current. The electrolysis of aqueous solutions allows utilization of electricity generated from renewable energy technologies such as solar energy, wind power, hydropower, and tidal power.

Hydrogen from water electrolysis can be fed into existing natural gas networks or can be used in fuel cells at the time of demand. Because the thermolytic water splitting requires temperatures above 4700 K, a more efficient process is water electrolysis at room temperature (Equation 184).

$$H_2O_{(l)} + \underbrace{237\ \text{kJ mol}^{-1}}_{\text{Electricity}} + \underbrace{48.6\ \text{kJ mol}^{-1}}_{\text{Heat}} \rightarrow H_2 + \frac{1}{2}O_2 \tag{184}$$

Adriaan Paets van Troostwijk and Jan Rudolph Diemann discovered the electrolysis process in 1789. In 1800, William Nicholson and Anthony Carlisle decomposed water in two-volume-parts hydrogen and one-volume-part oxygen using a novel column, which was invented in 1792 by Alessandro Volta. The invention did not come without controversy. The paper by Nicholson and Carlisle was published 2 months before Volta's epochal work was released, because Sir Joseph Banks, president of the Royal Society of London, passed Volta's manuscript of March 1800 to his countrymen first. Later, William Cruikshank (London) electrolyzed salt water using his improved version of a Volta pile and discovered the anodic metal dissolution at high currents. Sir Humphry Davy in London in 1806 concluded that pure water is solely split in hydrogen and oxygen, and he proved the Grotthuss theory that H^+ and OH^- ions move between the electrodes during electrolysis.

From 1890, hydrogen was electrolytically generated for airships, while large-scale electrolysis was known since the 1920s. In 1948, the company Lonza introduced an alkaline electrolysis process under pressure and in the 1970s General Electric built solid-polymer electrolyte electrolyzers using DuPont's Nafion membranes.

Basic Electrode Processes

An electrolyzer is a simple device capable of applying electric voltage between two electrodes immersed in a liquid or molten electrolyte (Figure 100).

Water electrolysis is performed using highly conductive, porous electrodes that are inert in the electrolyte in the voltage range required for water decomposition. Pure water does not provide sufficient electrolytic conductivity for practical processes and for that reason diluted acids, bases, or salt solutions are used. It is important to note that

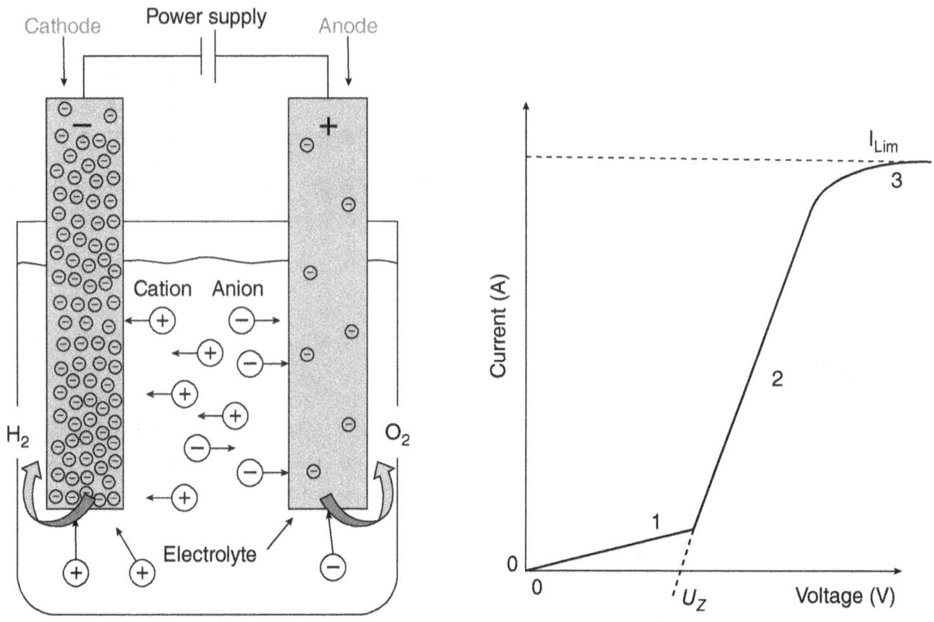

FIGURE 100 (Left) Schematic representation of a water electrolysis process. (Right) Current-voltage characteristics: 1, activation overvoltage; 2, decomposition voltage U_z and ohmic voltage drop in the electrolyte; 3, current limited by diffusion.

regardless of the composition of aqueous solution, only water is decomposed in the process. The overall reaction describes the decomposition of water into hydrogen and oxygen (Equation 185), with the input of electricity or heat.

$$2\,H_2O \rightarrow 2\,H_2 + O_2 \tag{185}$$

At the positive electrode (*anode*), electrochemical oxidation takes place resulting in a release of electrons and evolution of oxygen. At the negative electrode (*cathode*), electrochemical reduction takes place with electrons as reactants and hydrogen is evolved.

The electrode reactions in diluted acids show evolution of hydrogen at the cathode and oxygen on the anode, as well as net reaction (Equations 186 to 188).

(−) Cathode: $2\,H_3O^+ + 2\,e^- \rightarrow H_2 + 2\,H_2O \quad |\cdot 2 \qquad E^0 = 0\ V$ (186)

or simplified: $2\,H^+ + 2\,e^- \rightarrow H_2$

(+) Anode: $6\,H_2O \rightarrow O_2 + 4\,e^- + 4\,H_3O^+ \qquad E^0 = +1.23\ V$ (187)

or simplified: $2\,H_2O \rightarrow O_2 + 4\,e^- + 4\,H^+$

Net reaction: $2\,H_2O \rightarrow 2\,H_2 + O_2 \qquad U_0 = -\Delta E^0 = 1.23\ V$ (188)

When balancing the redox reactions, the cathodic reaction must be multiplied by 2 to achieve the same number of electrons in both anodic and cathodic half-cell reactions. The thermodynamically calculated reversible cell voltage is negative, because electrolysis is a nonspontaneous process. The standard cell voltage is given.

The electrode reactions in alkaline medium are given (Equations 189 to 191).

(–) Cathode: $2\,H_2O + 2\,e^- \rightarrow H_2 + 2\,OH^- \quad |\cdot 2$ (189)

(+) Anode: $4OH^- \rightarrow O_2 + 2H_2O + 4e^-$ (190)

Net reaction: $H_2O \rightarrow H_2 + ½O_2$ (191)

Thermodynamics of Water Electrolysis

Decomposition Voltage The theoretical terminal voltage in the cell with no current flow is based on the Gibbs' free enthalpy of the cell reaction ΔG, which can be calculated from tabulated values of the enthalpy of formation H and the entropy S of the products minus that of the educts (symbol Δ). The formation of a liquid water molecule from the elements at 25°C (T = 298.15 K) gives a reaction heat of –285.83 kJ and entropy change of –163.34 J mol^{-1}K^{-1}; the signs are reversed for the decomposition of water (Equations 192 to 195).

Reaction: $H_2O(l) \rightarrow H_2(g) + ½O_2(g)$ (192)

$$\Delta G^0 = \Delta H^0 - T\Delta S^0 \qquad (193)$$

$$\Delta G = \left[\Delta H^0(H_2)+\frac{1}{2}\Delta H^0(O_2)-\Delta H^0(H_2O)\right]-T\left[S^0(H_2)+\frac{1}{2}S^0(O_2)-S^0(H_2O)\right] \qquad (194)$$

$$\Delta G=\left[0+\frac{1}{2}\cdot 0-285.83\right]\cdot \text{kJ mol}^{-1}-298.15\text{ K}\cdot\left[\frac{1}{2}\cdot 205.14+130.68-69.91\right]$$

$$\text{J mol}^{-1}\text{K}^{-1}=+237.13\text{ kJ mol}^{-1} \qquad (195)$$

The positive sign of ΔG means that the electrolysis reaction is an endothermic process ($\Delta H > 0$) under increase of entropy ($\Delta S > 0$) and nonspontaneous process ($\Delta G > 0$). The reversible cell voltage ΔE has a negative value (Equation 196). Each water molecule consumes two electrons ($z = 2$) for the cell reaction $H_2O \rightarrow H_2 + ½O_2$.

$$U_0 = -\Delta E^0 = \frac{\Delta G^0}{zF} = \frac{237.13\text{ kJ mol}^{-1}}{2\cdot 96485\text{ C mol}^{-1}} \approx 1.23\text{ V} \qquad (196)$$

This result means that the cell voltage for water electrolysis at 25°C has a theoretical value of 1.23 V.

Thermoneutral Voltage During the electrolysis, the cell temperature decreases because the reaction is endothermic, and a lower electrical input is sufficient for water decomposition if heat is taken from the surroundings.

Electrical energy input: $\Delta G^0 \approx 237$ kJ (per 1 mol = 18 g of water)

Reaction enthalpy: $\Delta H^0 \approx 286$ kJ

Thermal energy from the environment: $T\Delta S^0 \approx 49$ kJ

A self-sustaining electrolysis reaction without thermal input from the outside requires voltage that corresponds to the combustion heat of water (calorific value) (Equation 197). The entropy of the reaction increases ($\Delta S > 0$) because of the larger number of particles in the gas phase: three molecules, 2 H_2 and O_2, are generated from two liquid H_2O molecules. The entropic part $T\Delta S^0 = Q$, called "anergy," of the free reaction enthalpy $\Delta G^0 = \Delta H^0 - T\Delta S^0$ cannot be converted into useful work.

$$U_{\mathrm{H}} = -\Delta E^0_{\mathrm{H}} = \frac{\Delta H^0}{zF} = \frac{285.83 \text{ kJ mol}^{-1}}{2 \cdot 96485 \text{ C mol}^{-1}} \approx 1.48 \text{ V} \tag{197}$$

Consequently, electrolysis of liquid water at 25°C works above 1.48 V without the need for external heating, and the electrolyzer cell gets hot although the cell reaction is endothermic. If water vapor is electrolyzed, the evaporation enthalpy of water must be taken into account and the heating value of steam ($\Delta H = 241.8$ kJ mol^{-1}) should be used.

This thermodynamic consideration must be amended in practice because waste heat is always generated when current passes through the cell because of the internal resistance of the electrolyte and the electrodes. In alkaline electrolysis and SPE electrolysis no heat is supplied from the outside.

Hot-Water Electrolysis The temperature dependence of cell voltage can be estimated using a difference quotient (Equation 198).

$$\left(\frac{\partial \Delta E}{\partial T}\right)_p = -\frac{1}{zF}\left(\frac{\partial \Delta G}{\partial T}\right)_p = \frac{\Delta S}{zF} \quad \Rightarrow$$

$$U(T) = -\Delta E(T) = U_0 - \left(\frac{\partial \Delta E}{\partial T}\right)_p (T - T^0) \tag{198}$$

The reversible cell voltage $U_0 = -\Delta E^0 = 1.23$ V is applicable only for standard conditions of $T^0 = 298.15$ K (25°C). At higher temperatures, for example at 60°C in liquid water, the Gibbs' free energy change is lower, and the necessary cell voltage is only 1.20 V. Based on this small change in cell voltage, the advantage of hot water electrolysis is negligible in practice.

Steam Electrolysis When the cell temperature further increases and the liquid electrolysis becomes steam electrolysis, the cell voltage further decreases, while the process efficiency improves. The energy input is less and it is more economical to conduct water electrolysis at high temperatures. The decomposition voltage at 25°C is:

$$H_2O(g) \rightarrow H_2(g) + ½O_2(g)$$

1. $\Delta G^0 = \Delta H^0 - T\Delta S^0$

$$= [-(-241.82)]\frac{\text{kJ}}{\text{mol}} - 298.15 \text{ K} \cdot \left[130.68 + \frac{1}{2} \cdot 205.15 - 188.83\right] \frac{\text{J}}{\text{mol K}}$$

$$= 228.57 \ \frac{\text{kJ}}{\text{mol}}$$

2. $U_0 = -\Delta E^0 = \dfrac{\Delta G^0}{zF} = \dfrac{228.57 \text{ kJ mol}^{-1}}{2 \cdot 96485 \text{ C mol}^{-1}} \approx 1.18 \text{ V}$

	Reaction Enthalpy ΔH (kJ mol^{-1})	Thermo-neutral Voltage U_H (V)	Gibbs' Free Reaction Enthalpy ΔG (kJ mol^{-1})	Reversible Cell Voltage $U_0 = -\Delta E$ (V)	Contribution of Reaction Entropy $T\,\Delta S/(2F)$ (V)
Liquid water 25°C	285.8	1.48	237.2	1.23	–0.25
Steam 100°C	242.6	1.26	225.1	1.17	–0.09
1000°C	249.4	1.29	177.1	0.92	–0.37

Table 25 Thermodynamic Data of Water Electrolysis: $H_2O \rightarrow H_2 + ½O_2$ at 101,325 Pa

The theoretical decomposition voltage is further reduced from 1.17 V at 100°C to 0.91 V at 1000°C because of the increasing contribution of the entropy term, $T\,\Delta S$ (Table 25 and Figure 101).

High-Pressure Electrolysis The impact of pressure on cell voltage is fairly minor at pressures of less than 10 bar. Assuming ideal gases, $pV = nRT$, the difference quotient, $(\partial G/\partial p)_T = V$, shows that the theoretical decomposition voltage is determined by the volume of gas and high pressure lowers the cell voltage at 25°C (Equation 199).

$$\left(\frac{\partial \Delta E}{\partial p}\right)_T = -\frac{1}{zF}\left(\frac{\partial \Delta G}{\partial p}\right)_T = -\frac{\Delta V}{zF} \Rightarrow$$

$$U(p) = -\Delta E(p) = U_0 - \frac{RT}{zF}\ln K_p = 1.23\ V - \frac{RT}{4F}\ln\frac{p(H_2)^2\,p(O_2)}{p(H_2O)^2} \qquad (199)$$

K_p is the equilibrium constant on the basis of pressure. The partial pressure of water in the gas phase can be neglected, $p(H_2O) \approx 1$. If pressure is increased from 1 bar to

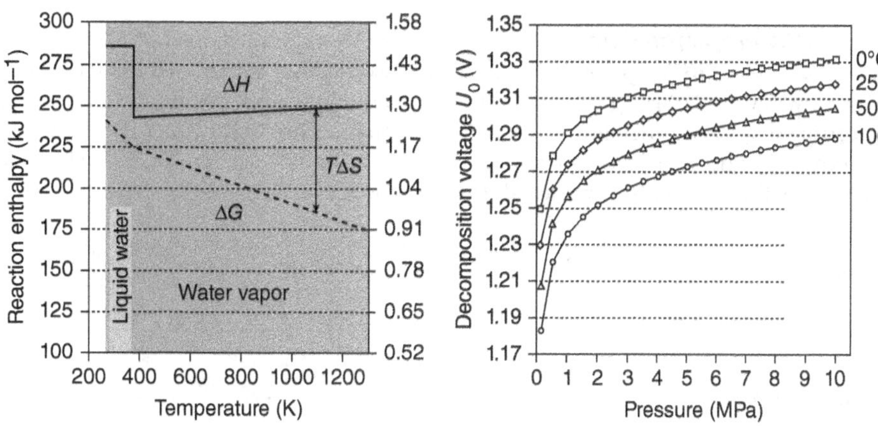

Figure 101 The decomposition voltage of water electrolysis depends on temperature and pressure.

30 bar, decomposition voltage drops by 0.065 V. In practice, the kinetic inhibitions of the electrode processes are rate-determining. In addition, the cost of pressure equipment must be considered when selecting the type of electrolysis process.

Efficiency

The overall efficiency of the water electrolysis process depends on the voltage efficiency and the current efficiency (Equations 200 to 202).

Voltage efficiency: $$\eta_U = \frac{\Delta G}{W} = \frac{1.23\ \text{V}}{U(I)} < 1 \quad \frac{\text{Free energy of water splitting}}{\text{Electrical input (voltage)}} \qquad (200)$$

Current efficiency: $$\eta_I = \frac{Q}{Q_F} = \frac{I}{I_F} = zF\frac{\dot{n}(\text{H}_2)}{I} \quad \frac{\text{Produced mass of hydrogen}}{\text{Electrical input (current)}} \qquad (201)$$

Overall cell efficiency: $$\eta = \eta_U \eta_I \qquad (202)$$

Thermodynamic Efficiencies The usefulness of thermodynamic considerations is rather limited because the endothermic cell reaction consumes heat from the environment, so thermal efficiencies may be greater than 100% (Equations 203 to 205).

Thermal efficiency: $$\eta_{th} = \frac{\Delta H}{Q} \approx \frac{1,48\ \text{V}}{U(I)} \quad \frac{\text{Enthalpy of cell reaction (combustion heat)}}{\text{Heat input}} \qquad (203)$$

Thermoelectrical efficiency: $$\eta_{the} = \frac{W}{Q} \quad \frac{\text{Produced electrical energy}}{\text{Heat input}} \qquad (204)$$

Electrochemical efficiency: $$\eta_{ech} = \frac{\Delta H}{W} \quad \frac{\text{Enthalpy of cell reaction}}{\text{Electrical input}} \qquad (205)$$

Just imagine, theoretically, efficiency, η_{ech}, of 120% is possible, if $W = \Delta G$ and $U = 1.23$ V.

Energy (Conversion) Efficiency With increasing electric power, the electrolyzer produces more and more hydrogen and oxygen, but the efficiency drops, and the increasing electric power does not proportionally generate more hydrogen because more waste heat is produced. Energy efficiency reflects the combustion heat ΔH or heating value H_u of the produced hydrogen per electric power input P (Equations 206 to 208). The heat flow dQ/dt is the externally supplied thermal input.

Liquid water electrolysis: $$\eta_H = \frac{\Delta H^0 \cdot \dot{n}(\text{H}_2)}{P} = \frac{285.8\ \text{kJ mol}^{-1}}{P} \frac{\dot{m}(\text{H}_2)}{2 \cdot 1,008\ \text{g mol}^{-1}} \qquad (206)$$

High-temperature electrolysis: $$\eta_H = \frac{\Delta H^0 \cdot \dot{n}(\text{H}_2)}{P + \dot{Q}} = \frac{285.8\ \text{kJ mol}^{-1}}{(P + \dot{Q})} \frac{\dot{m}(\text{H}_2)}{2 \cdot 1.008\ \text{g mol}^{-1}} \qquad (207)$$

Steam electrolysis: $$\eta_H = \frac{H_u \cdot \dot{n}(H_2)}{P+\dot{Q}} = \frac{241.8 \text{ kJ mol}^{-1}}{(P+\dot{Q})} \frac{\dot{m}(H_2)}{2 \cdot 1.008 \text{ g mol}^{-1}} \tag{208}$$

The practical energy expenditure of water splitting (in kWh per kg or m^3) is the ratio of electric power input to hydrogen flow (Equation 209).

$$\eta_P = \frac{P}{\dot{m}(H_2)} \quad \text{or} \quad \eta_P = \frac{P}{\dot{V}(H_2)} \tag{209}$$

Note that $1 \text{ m}^3\text{h}^{-1}\text{W} = 1/3600 \text{ m}^3\text{W}^{-1} = 0.2778 \text{ m}^3\text{kW}^{-1}$.

Faraday's Laws and Practical Calculations

Michael Faraday, Davy's assistant in London in 1832, worked on the dissociation in electrolyte solutions and electrochemical decomposition using electricity from induction machines and Volta batteries. Faraday found similarities among all sources of electricity. Although at that time, Faraday did not accept the newly introduced atomic theory and he did not understand the difference between voltage and current, he was able to find the connection between electric charge and molar mass. He also coined the terms anode (Greek ἄνω = upward), cathode (κατά = down), and anion and cation (όδός = the way up and down). Based on Faraday's measuring instrument for Volta electricity, which simply was a closed electrolysis cell, modern coulometers were later built. Then, in 1881, H. V. Helmholtz identified the elementary quantum of electricity, which was given the name "electron" by G. J. Stoney in 1891. Even today, the term "faradaic" current is used to describe the current that, for example, decomposes water or is generated by a battery and differs from a "capacitive" current for charging and discharging the electrolytic double-layer.

Faraday's First Law Suppose we want to know, for example, for how long a current must flow through an electrolysis cell to generate 1 kg hydrogen. The deposited mass m (in kg) is proportional to the flowing current I, in amperes, and time t, in seconds, which gives the electric charge $Q = I \cdot t$ in coulombs (C = As) (Equation 210).

$$Q = \int_0^{\Delta t} I(t)\,\mathrm{d}t = zFn = \frac{m}{k}, \quad \text{and} \quad k = \frac{M}{zF} \tag{210}$$

Applying the equation, a direct current of $I = 1$ A, flowing through a silver nitrate or copper sulfate solution for the time $t = 3600$ s produces a mass of 4.02 g of silver or 1.19 g of copper at the cathode. The equation is valid only if no side reactions or non-faradaic losses take place. The material-specific constant $k = 1.1179$ mg/C (silver) or $k = 0.3294$ mg/C (copper) is called the *electrochemical equivalent* and is derived from Faraday's second law.

Faraday's Second Law The deposition of 1 mol of a monovalent substance such as silver requires the electric charge of 1 mol of electrons, which is 96,485 C and is the Faraday constant F (Equation 211).

$$F = N_A e = 96{,}485 \text{ C mol}^{-1} \tag{211}$$

$N_A = 6.022 \cdot 10^{23}$ mol^{-1} is the Avogadro constant and $e = 1.602 \cdot 10^{-19}$ C is the elementary charge.

Equal quantities of electric charge deposit different masses from different solutions. The mass ratio is determined by the electrochemical equivalent, which is obtained by dividing the molar mass M by the charge number z of the ion (Equation 212).

$$\frac{n_1}{n_2} = \frac{z_1}{z_2} \quad \text{or} \quad \frac{m_1}{m_2} = \frac{M_1/z_1}{M_2/z_2} = \frac{k_1}{k_2} \tag{212}$$

Gas Production To yield 1 kg of hydrogen, the stoichiometric calculation shows that 9 kg of water must be decomposed, because H_2O consists of $2/18 \approx 11\%$ H_2. One liter of pure hydrogen has a volume of 22.4 L at 0°C and weighs 0.0899 g, because 1 mol of hydrogen is 2.016 g H_2. Hence, electrolysis requires 0.8 L of water and the electric charge of $1/k = 95{,}694$ As/kg = 8607 C/L = 2.39 Ah per liter hydrogen (Table 26).

Hydrogen and oxygen may be considered as *ideal gases* because their condensation temperatures are low, and the intermolecular forces are weak. The amount of substance of 1 mol ($n = 1$) of an ideal gas has a volume of $V_m = 22{,}414$ L mol^{-1} under standard conditions ($T^0 = 273$ K, $p^0 = 101{,}325$ Pa). The volume V at different temperatures T and pressures p is calculated using the "ideal gas law" equation where molar gas constant R is 8.3144 J mol^{-1}K^{-1} (Equation 213).

$$\frac{pV}{T} = \frac{p_0 V_0}{T_0} = nR \tag{213}$$

Coulombic Efficiency Technical electrolysis works at good coulombic efficiencies of up to 99%. The coulombic efficiency α, which includes current efficiency and electricity yield, is affected by the losses in electrode processes that run incompletely and by the side reactions (Equation 214).

$$\alpha = \frac{\text{Actually deposited mass (in mol or kg)}}{\text{Theoretical amount according to second Faraday law (in mol or kg)}} \tag{214}$$

The following formulas help calculating practical applications of direct current electrolysis:

- Deposited mass: $m = \frac{kQ}{\alpha} = \frac{M}{\alpha z F} It$

	$\frac{k}{\text{mg C}^{-1}}$	$\frac{V}{\text{mL C}^{-1}}$	$\frac{V}{\text{L Ah}^{-1}}$
Hydrogen (H_2)	0.01045	0.1162	0.4185
Oxygen (O_2)	0.0829	0.05802	0.2089
Oxyhydrogen ($2H_2 + O_2$)	0.09337	0.1743	0.6273
Conversion of units: 1 C = 1 As; 1 Ah = 3600 C			

TABLE 26 Electrochemical Equivalent and Generated Volume Flow for Water Electrolysis

- Deposited amount of substance: $n=\frac{m}{M}=\frac{Q}{\alpha zF}$
- Deposited mass of gas at a known volume V: $m=\frac{V_0 M}{V_m}=\frac{pV}{T}\cdot\frac{T_0}{p_0}\frac{M}{V_m}$
- Evolved gas volume at operating conditions: $V(T,p)=\frac{T}{p}\frac{p_0}{T_0}\frac{V_m}{zF}\,Q$

V_0 is the volume at $T_0 = 273.15$ K (0°C) and $p_0 = 101{,}325$ Pa. For dry and wet real gases, the molar volume V_m can be found in tables. For hydrogen, $M(H_2) = 2.026$ g mol^{-1}.

Calculation Examples

1. What is the amount of substance and the mass of 1.0 m³ hydrogen at standard conditions?

$$n=\frac{V}{V_m}=\frac{1\ \text{m}^3}{22.414\ \text{m}^3\ \text{kmol}^{-1}}\approx 0.045\ \text{kmol}$$

$$m=n\cdot M=\frac{V}{V_m}\cdot M=\frac{1\ \text{m}^3\cdot 2.016\ \text{kg kmol}^{-1}}{22.414\ \text{m}^3\ \text{kmol}^{-1}}\approx 0.090\ \text{kg}$$

2. What electric charge is required to generate this mass at 99% efficiency? (1 Ah = 3600 C)

$$Q=\alpha zFn=\frac{1}{0.99}\cdot 2\cdot 96485\frac{\text{C}}{\text{mol}}\cdot 44.6\ \text{mol}\approx 8.5\cdot 10^6\ \text{C}\approx 2400\ \text{Ah}$$

3. What electric energy is necessary at a cell voltage of 2.0 V?

$$W=QU=2400\ \text{Ah}\cdot 2.0\ \text{V}\approx 4800\ \text{Wh}$$

4. What hydrogen volume (at 35°C and 1039 kPa) is produced by the above electric charge? The molar volume of wet hydrogen equals 26.05 L/mol at these conditions.

$$V=\frac{T}{p}\frac{p_0}{T_0}\frac{V_m}{zF}\,Q=\frac{(273.15+35)\text{K}}{1039\ \text{hPa}}\cdot\frac{1013.25\ \text{hPa}}{273.15\ \text{K}}\cdot\frac{26.05\ \text{L mol}^{-1}}{2\cdot 96485\ \text{C mol}^{-1}}\cdot 8.5\cdot 10^6\text{C}\approx 1.3\ \text{m}^3$$

5. What is the energy efficiency of electrolysis? Compare the energy consumption to produce 1 m³ dry hydrogen at standard conditions in theory, and wet hydrogen at 2.0 V ($V_m = 22.55$ L/mol).

$$W_{\text{th}}=QU=U_0\frac{zFV}{V_m}=1.23\ \text{V}\cdot\frac{2\cdot 96485\ \text{C mol}^{-1}\cdot 1000\ \text{L}}{22.414\ \text{L mol}^{-1}}\approx 10.6\ \text{MJ}=2.94\ \text{kWh}$$

$$W=QU=U_0\frac{zFV}{V_m}=2.0\ \text{V}\cdot\frac{2\cdot 96485\ \text{C mol}^{-1}\cdot 1000\ \text{L}}{22.55\ \text{L mol}^{-1}}\approx 17.1\ \text{MJ}=4.75\ \text{kWh}$$

$$\eta=\frac{W}{W_{\text{th}}}\approx 62\ \%$$

Cell Voltage and Kinetic Inhibitions

It has been previously discussed that the theoretical cell voltage sufficient to electrolytically decompose water is 1.23 V. However, practical experiments show that more than 1.5 V is required. There are several reasons that explain the difference between the theory and practice.

Theoretical Cell Voltage At ambient conditions other than the temperature of 25°C, the pressure of 101,325 Pa, and the concentration of 1 mol L^{-1}, the open-circuit voltage between the two electrodes in the chemical equilibrium (at rest) can be calculated using the Nernst equation (Equation 215). Note that the obsolete term "electromotive force" (emf) is no longer used.

$$\Delta E = E_{\text{cathode}} - E_{\text{anode}} = \Delta E^0 - \frac{RT}{zF}\ln\frac{a(\text{products})}{a(\text{educts})} \tag{215}$$

When applied to the electrolysis of water at 25°C, $H_2O \rightarrow H_2 + \frac{1}{2}O_2$, the theoretical decomposition voltage (negative reversible cell voltage) depends on the partial pressures of the gases (Equation 216). The water activity is close to one.

$$U_0 = -\Delta E = -\left(\Delta E^0 - \frac{RT}{2F}\ln\frac{a(H_2)\cdot a(O_2)^{\frac{1}{2}}}{a(H_2O)}\right) \approx 1.23\ \text{V} + \frac{0.0592}{2}\log\frac{p(H_2)}{p^0}\cdot\sqrt{\frac{p(O_2)}{p^0}} \tag{216}$$

The calculated value of the terminal voltage U_0 is the minimum voltage necessary for the electrolysis to start.

Practical Cell Voltage The internal resistance of the cell R_i comprises all ohmic losses, which are caused by the electrolyte, the diaphragm, adsorption layers, gas bubbles, electrocatalysis (activation polarization), and mass transport (concentration polarization). The minimum voltage that has to be applied to the cell is mainly determined by the resistance of the electrolyte solution R_e and the resistance of the electrode processes (polarization resistance: R_p). With increasing current that flows through the cell, the voltage drop IR_e in the electrolyte increases as well and the solution temperature increases from the heat, which is the result of the inefficiencies. In addition, the so-called overvoltage, η, at the electrodes is increasing because of the losses from inefficient, non-faradaic electrode processes. The terminal voltage $U(I)$ drives a current across the cell and gas is produced. However, the losses in the electrolyte and at the electrodes increase as the current increases (Equation 217).

$$U(I) = U_0 + \eta_{\text{anode}} + |\eta_{\text{cathode}}| + IR_e = U_0 + IR_i \tag{217}$$

Kinetic Inhibitions Internal resistance and overvoltage originate from the kinetic inhibitions in the electrode reactions. The charge transfer across the electrode–electrolyte interface is not infinitely fast because the electrons have to overcome the activation energy (tunnel barrier). In addition, the diffusion of electroactive species and product gases in the electrolyte is a slow process, especially at high currents, leading to mass transport limitations. But the rate-determining reaction in electrolysis is the oxygen evolution. All kinetic inhibitions act like resistances (impedances) and combine to give the total polarization resistance.

The **internal resistance** of the cell is the sum of the electrolyte resistance between the electrodes and the polarization resistances of anode and cathode (Equation 218).

$$R_i = R_e + R_P = R_e + R_{\text{anode}} + R_{\text{cathode}} \tag{218}$$

The internal resistance due to the kinetic inhibitions requires that a higher terminal voltage than theoretical, the sum of the open-circuit voltage and the overvoltage, must be applied to start the reaction. The ohmic voltage drop in the electrolyte solution does not count to the overvoltage of the cell. The overpotential η comprises all deviations of the electrode potential under current flow conditions (Equations 219 and 220).

Electrode: $$\eta_i(I) = E(I) - E_0 \begin{cases} < 0 \text{ cathode} \\ > 0 \text{ anode} \end{cases} \tag{219}$$

Cell: $$\eta(I) = U(I) - U_0 > 0 \tag{220}$$

In practice, the overpotential of electrode or the overvoltage of a cell is measured as the difference between the open-circuit value and the terminal voltage under current flow. The overpotential depends on the electrode material and the type, composition, and flow properties of the electrolyte. The overpotential, η, increases with increasing temperature and current density, and decreases for electrodes with larger surface area.

In summary, the internal resistance comprises losses in the electrolyte, at the electrodes, in the membrane or the diaphragm, as well as those caused by the gas bubbles and the electric circuit.

Current-Voltage Characteristics of Electrolysis

Cell voltage is experimentally recorded at different applied currents (or vice versa) and the results are presented in form of current-voltage curves that are used to delineate the processes.

At the phase boundary between electrode and electrolyte, the reaction rate r (per unit area A) of a redox process A $\rightarrow$ B is mainly determined by the rate of charge transfer and diffusion processes.

$$r = \frac{1}{A}\frac{\mathrm{d}n_i}{\mathrm{d}t} = \underbrace{\frac{i}{zF}}_{\text{Faradaic reaction}} + \underbrace{\left(-D_i \frac{\mathrm{d}c_i}{\mathrm{d}x}\right)}_{\text{Diffusion}} + \underbrace{t_i \frac{i}{zF}}_{\text{Migration}} + \underbrace{c_i v_x}_{\text{Forced convection}} \tag{221}$$

Overall reaction rate and current are limited by the slowest process, the so-called rate-determining step. Migration plays no role in highly conductive electrolytes such as diluted acids and bases.

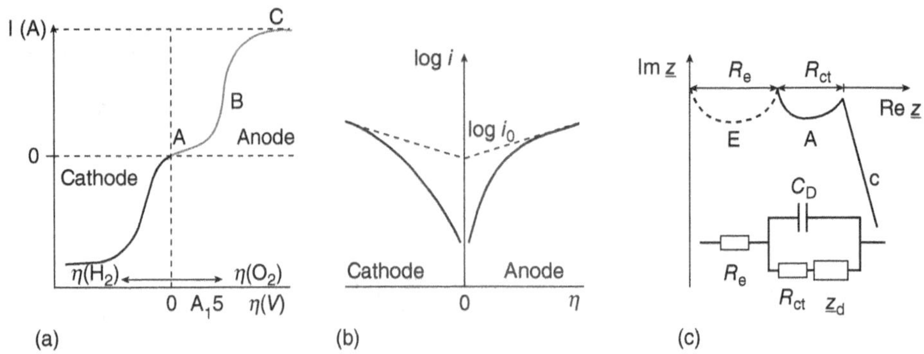

FIGURE 102 (a) Current-overpotential characteristics: 0, open-circuit potential; A, activation; B, ohmic drop in the electrolyte; C, diffusion-limited current. (b) Tafel plot. (c) Complex plane plot of impedance (in mathematical convention) and general equivalent circuit of an electrolyzer cell: E, electrolyte; A, charge transfer; C, diffusion impedance; R_e, electrolyte resistance; C_D, double-layer capacitance; R_{ct}, charge transfer resistance; $\underline{z}_d$, mass transfer impedance.

Electrode Processes and Impedance

The current–voltage characteristics (Figure 102) and the impedance spectrum show three regions:

1. The *charge transfer* reaction is the charge transition between electrode materials and adsorbed species. It controls the electrode potential at low current densities. The activation overpotential, η_{ct}, corresponds to the charge transfer resistance, R_{ct}, in the impedance spectrum. The anode (oxygen electrode) determines the overall overvoltage for most reactions, while the cathode (hydrogen electrode) shows small overpotential, also described as high reversibility. Especially low hydrogen overvoltage is observed on platinum electrodes. All irreversible electrode processes, that is, those with high overpotential, behave like resistances that produce heat during the current flow.

   Reversible processes, such as charging and discharging of the double layer, act as reactances, which do not generate waste heat. *Double-layer capacitance* can be estimated using the time constant, τ, of the charge transfer process, located at the semicircle minimum of the complex plane curve (Equation 215). Alternatively, the real part of complex capacitance, $C = \underline{Y}/\omega$, is extrapolated to high frequencies (Equation 222).

$$\tau = R_{ct} C_D = \frac{1}{2\pi f_m} \tag{222}$$

$$C_D = \lim_{\omega\to\infty} \frac{\text{Im } \underline{Y}(\omega)}{\omega} = \lim_{\omega\to\infty} \frac{-\text{Im } \underline{Z}(\omega)}{\omega \cdot [(\text{Re } \underline{Z}(\omega))^2 + (\text{Im } \underline{Z}(\omega))^2]} \tag{223}$$

2. At medium currents, the terminal voltage necessary for an electrolytic process increases by the value of the ohmic voltage drop in the electrolyte solution. Impedance in the complex plane shows the *electrolyte resistance* at high frequencies (Equation 217). In highly conducting solutions, the left arc at

frequencies above 1 kHz is often not completely visible, so that in a typical evaluation the first intersection with the real axis is evaluated.

$$R_e = \mathrm{Re}\ \underline{Z}(\omega \to \infty), \quad \text{at Im } \underline{Z} = 0 \tag{224}$$

3. *Mass transport* inhibition occurs at high current densities where the chemical conversion at the electrodes is much faster than the transport rate of H_3O^+ and OH^- to the electrodes, and the removal of the products H_2 and O_2 from the electrodes. As a result, the diffusion-limited current cannot be exceeded even at extremely high voltages. Diffusion impedance $\underline{Z}_d$ is often modeled by the so-called Warburg impedance, $\underline{Z}(\omega) = [A(\mathrm{j}\omega)]^{-0.5}$, or complex equivalent circuit networks.

The *total internal resistance of the cell* is obtained by adding all electric and kinetic resistances at both electrodes (Equation 225).

$$R_i = R_e + R_{ct} + R_d + \cdots \tag{225}$$

Charge Transfer Kinetics without Diffusion

At an ideal charge transfer electrode, adsorption processes play no role and the reactions obey *Faraday's law*, while the current density (Equation 226) reflects the reaction rate r of the redox reaction $2\ H^+ + 2\ e^- \to H_2$ ($z = 2$).

$$i = \frac{I}{A} = \frac{1}{A}\frac{\mathrm{d}Q}{\mathrm{d}t} = \frac{zF}{A}\frac{\mathrm{d}n}{\mathrm{d}t} = zFr \approx zFkc^b \tag{226}$$

The *Butler-Volmer equation* considers the reaction equilibrium between the cathodic reduction of hydrogen and the reverse reaction of hydrogen oxidation. In stirred solution, the surface concentrations, $c(0)$, equal the bulk concentrations c^b in the electrolyte. At low overpotential η the exponential function becomes approximately $e^x \approx 1+x$ (Equations 227 and 228).

$$i = i_\oplus - |i_\ominus| = i_0 \left[\frac{c_{\mathrm{ox}}(0,t)}{c_{\mathrm{ox}}^b} \cdot e^{\frac{\alpha zF\eta}{RT}} - \frac{c_{\mathrm{red}}(0,t)}{c_{\mathrm{red}}^b} \cdot e^{\frac{-(1-\alpha)zF\eta}{RT}}\right]$$

$$\approx i_0 \left[e^{\frac{\alpha zF\eta}{RT}} - e^{\frac{-(1-\alpha)zF\eta}{RT}}\right] = i_0 \left[e^{\ln 10 \cdot \frac{\eta}{b_\oplus}} - e^{-\ln 10 \cdot \frac{\eta}{b_\ominus}}\right] i \tag{227}$$

$$i \approx i_0 \frac{zF\eta}{RT} \quad (\text{at } \eta \to 0) \tag{228}$$

Constant b is the *Tafel slope* (in V dec^{-1}) for the $\oplus$ anode and $\ominus$ cathode. R is the molar gas constant. The factor F/RT is roughly 39 V^{-1} at 25°C.

The charge transfer resistance (in $\Omega\ \mathrm{m}^2$) at low overvoltage corresponds to the slope of the current-voltage curve near the equilibrium potential (Equation 229).

$$R_{ct} = \frac{\mathrm{d}\eta}{\mathrm{d}i} = \frac{RT}{zFi_0} \quad \text{at} \quad |\eta| \ll \frac{RT}{\alpha zF} \approx \frac{25\ \mathrm{mV}}{z} \tag{229}$$

The exchange current density, i_0, is the very small current at the equilibrium potential, which measures the catalytic activity of the electrode material and the reversibility of the electrode reaction (Equation 230).

$$i_0 = i_\oplus = |i_\ominus| \approx 0 \tag{230}$$

The transfer coefficient, α, reflects the activation barrier of the charge transfer reaction.

- $\alpha = 0.5$ means that both the anodic and the cathodic branch of the $i(\eta)$-curve are completely symmetrical.
- $\alpha < 0.5$ is typical for a flat curve, which shows small currents despite high overpotential. The reduced species are rate-determining, and the cathodic branch is steeper than the anodic branch.
- $\alpha > 0.5$ is characteristic for a steep anodic branch and high currents, where the oxidized species are controlling the rate.

The Tafel equation describes well the irreversible charge transfer at large overpotentials ($\eta >> RT/zF$), when the reverse reaction may be neglected. The current–voltage characteristics can be linearized on the logarithmic scale ($\eta = a + b \log I$) (Figure 102b).

The overpotential, η, is the measured voltage with respect to the open-circuit potential: $\eta = E - E_0$ (Equations 231 and 232). Because the measured potential is distorted by the ohmic voltage drop in the electrolyte, the electrolyte resistance (IR drop) should be corrected prior to the determination of Tafel slopes.

Anodic: $$|\eta| = \frac{RT \ln 10}{\alpha z F}(\log i - \log i_0) = b_\oplus(\log i - \log i_0) = a + b_\oplus \log i \tag{231}$$

Cathodic: $$|\eta| = \frac{RT \ln 10}{(1-\alpha) z F}(\log i - \log i_0) = b_\ominus(\log i - \log i_0) = a + b_\ominus \log i \tag{232}$$

The Tafel slope, $b = \mathrm{d}E/\mathrm{d}(\log I)$, is a measure of the catalytic activity of the electrode material (Equation 233). By using $b \approx (60\ \mathrm{mV})/z$, the electrochemical valency z of the reaction $A + z\,e^- \rightarrow B$ can be estimated. The value of b roughly 120 mV means that $z = 2$ and two electrons are exchanged in the redox reaction. A consecutive redox reaction $A \rightarrow B \rightarrow C$ causes two Tafel slopes, especially if the reaction of adsorbed intermediate takes place. The intercept a of the extrapolated Tafel line is used to determine the exchange current density.

$$\log i_0 = \frac{a}{b} \tag{233}$$

Charge Transfer Limited by Mass Transport

A more complicated case involves electrodes where reaction is controlled by diffusion processes. When a direct current is established from an external source across the electrolyzer cell, the diffusion boundary layers form in the time period from seconds to minutes until stable electrode potentials are reached. At a certain voltage, diffusion becomes rate-determining, where despite higher voltage the current is at a plateau and the rates of hydrogen and oxygen generation remain the same. This is the diffusion-limited current density, i_{lim}, governed by the limited concentration of electroactive species on the electrode surface ($c^s \to 0$), even in vigorously stirred solution (Equation 234).

$$i = i_{lim}\left[1 - e^{-\frac{zF\eta}{RT}}\right] = -zFr \tag{234}$$

The reaction rate r (per unit electrode area A) is described by Fick's diffusion law, whereby a diffusion boundary layer, δ, is 1 to 100 μm (Equation 235).

$$r = A\beta c^b = A\sqrt{\frac{D}{\delta}}c^b \approx -D\frac{(c^b - c^s)}{\delta} \tag{235}$$

The Tafel equation for the anode (oxygen evolution), at moderate overpotential (η >>70 mV, RT/F << 1) and neglected reverse reaction, shows the sum of charge transfer and diffusion inhibition, which is often called "concentration polarization" (Equation 236).

$$|\eta| = \frac{RT\ln 10}{\alpha zF}\left[\log\frac{i_0}{i_{lim}} + \log\frac{i_{lim} - i}{i}\right] = I \cdot (R_{ct} + R_d) \tag{236}$$

The *diffusion resistance*, R_d (in Ω m²), is determined using the Nernst equation and the concentration ratio $c_i^s / c_i^b = 1 - i / i_{lim}$, whereby the counter electrode factor is set as $c_j = 1$ (Equation 237).

$$R_d = \frac{\eta_d}{i} = \frac{RT}{zFi_{lim}} \tag{237}$$

The Hydrogen Electrode

At the negative terminal, the cathodic hydrogen evolution takes place. In general, the hydrogen electrode poses no challenge for technical electrolysis. Platinum is typically used as the electrode material and shows very good performance, but its high cost is always the reason for searching alternative, less expensive materials.

Open-Circuit Potential and pH Shift

The *standard hydrogen electrode* (SHE) consists of a platinum sheet in 1-molar hydrochloric acid (1.184 mol kg^{-1}) at 25°C, purged with hydrogen gas at 101,325 Pa. By definition, the electrode process $\frac{1}{2}H_2 \leftrightharpoons H^+ + e^-$ exhibits the potential $\varphi^0_{NHE} = 0$ at all temperatures; and the hydrated proton H^+(aq) in aqueous solution has zero entropy and enthalpy of formation: $S^0 = \Delta H_f^0 = \Delta G_f^0 = 0$.

The *reversible hydrogen electrode* (RHE) is a practical electrode made of a platinum group metal, which is purged with hydrogen gas bubbles and immersed directly in the working electrolyte. In aqueous solutions, on hydrogen-adsorbing materials, a potential difference occurs based on the equilibrium reaction $H_{2(g)} \to 2\,H_{(ad)} \to 2\,H^+ + 2\,e^-$ (in water: H_3O^+). According to the Nernst equation, the hydrogen ion activity, a, and the pressure, p, of hydrogen gas over the solution determine the cell voltage (Equation 238).

$$E_{RHE} = -\frac{RT}{F}\ln\frac{\sqrt{p(H_2)/p^0}}{a(H^+)} = -\frac{RT}{F}\left[pH + \frac{1}{2}\log\frac{p(H_2)}{p^0}\right] \approx -0.0592 \cdot pH \tag{238}$$

This means that the thermodynamic equilibrium potential of the hydrogen electrode depends on the pH value of the electrolyte solution. The electrode potential of the hydrogen electrode is 0 V in acid solution (pH 0), –0.414 V in neutral solution (pH 7), and –0.828 V in alkaline solution (pH 14) at 25°C.

pH Shift During electrolysis, the electrolyte-filled space around the cathode (catholyte) gradually becomes alkaline, and the anolyte becomes more acid, because water is decomposed.

Acid solution: $2H^+ + 2e^- \rightleftharpoons H_2$

Alkaline solution: $2H_2O + 2e^- \rightleftharpoons H_2 + 2OH^-$

Hydrogen and oxygen electrodes behave in the same way, so that the decomposition voltage does not change during electrolysis, although the pH values near the electrode change.

Dissolved Hydrogen Hydrogen partial pressure, $p(H_2)$, is 1 bar during electrolysis even in dilute aqueous solutions, for example, 0.0008 M. According to Henry's law, hydrogen dissolves in the electrolyte up to a certain mole fraction x (Equation 239).

$$x(H_2) = \frac{n(H_2)}{n(H_2)+n(H_2O)} \approx \frac{n(H_2)}{n(H_2O)} = \frac{p(H_2)}{H} = \frac{101325\ \text{Pa}}{7.12\cdot 10^9\ \text{Pa}} \approx 1.42\cdot 10^{-5}\,\frac{\text{mol}}{\text{mol}} \tag{239}$$

The solubility of hydrogen at 0°C is $c(H_2O) = x(H_2)\cdot c(H_2O) = 780\ \mu\text{mol L}^{-1}$, if the water concentration is assumed to be $c(H_2O) = 1000\ \text{g}/(1\ \text{L}\cdot 18\ \text{g mol}^{-1}) \approx 55.55\ \text{mol L}^{-1}$. Measured solubility of hydrogen in water at different temperatures are 86 μmol L^{-1} (20°C), 84.8 μmol L^{-1} (25°C), and 83.2 μmol L^{-1} (30°C).

Cyclic Voltammogram

The molecular processes at any electrode surface can be recorded and studied. The cyclic voltammogram (CV) (Figure 103a) represents the fingerprint of the surface and reactions. The differential adsorption isotherm (Figure 103b) reveals that hydrogen adsorption is structure-sensitive; on single-crystals, the surfaces Pt (111), Pt(100), and Pt(110) show different CV shapes. The hydrogen adsorption value is useful for estimating the true surface area of the electrocatalysts. The anodic potential range $E > 0$ V (from 0.05 to 0.4 V RHE) is called the region of hydrogen underpotential deposition (UPD). Evolution of H_2 occurs in the region of hydrogen overpotential deposition (OPD, $E \leq 0$ V RHE). Cathodic reduction and electrohydrogenation can often be carried out under milder conditions as compared with gas-phase hydrogenation.

The electrochemical isotherms, that is, the electrode potential versus electric charge $E(Q)$, in solutions look very similar to the sorption isotherms in the gas phase. Platinum is an excellent electrocatalyst, because hydrogen is both easily absorbed and desorbed, as a result of its moderate sorption enthalpy. In contrast to that, gold and lead inhibit hydrogen sorption layers. Titanium, on the other hand, experiences embrittlement due to the formation of hydrogen-rich phases.

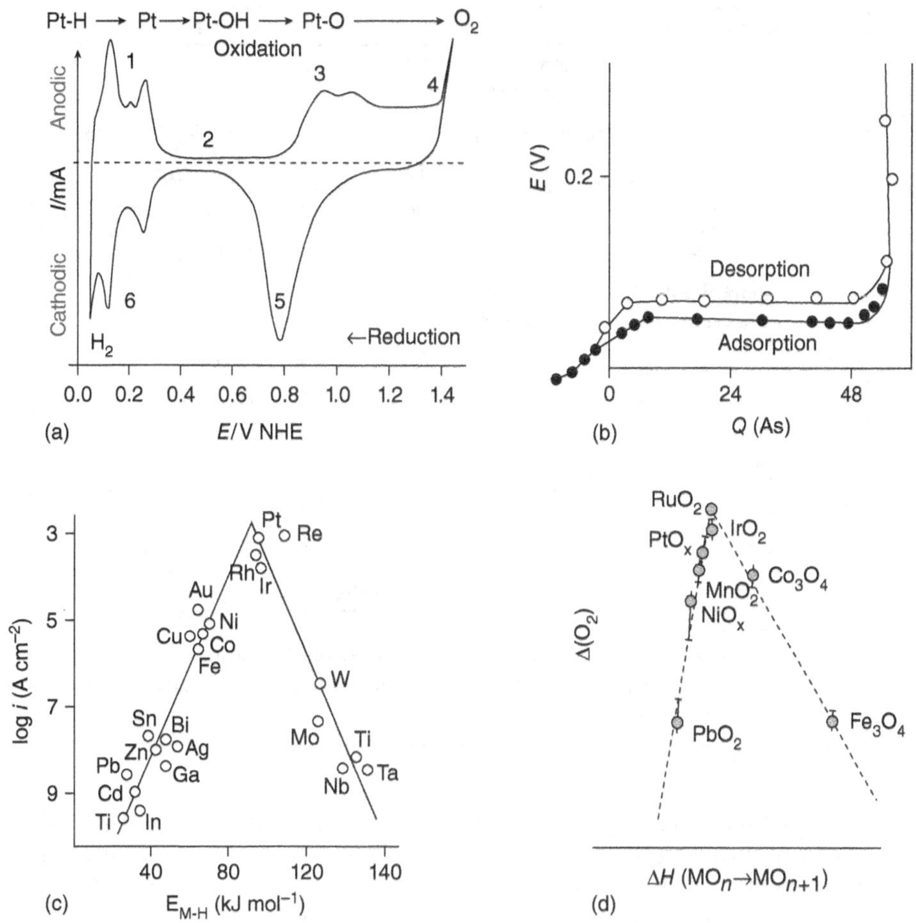

FIGURE 103 (a) Cyclic voltammogram of a smooth, polycrystalline platinum electrode in 0.5-molar sulfuric acid. (b) Electrochemical sorption during charge and discharge of a palladium electrode in 0.5-molar sulfuric acid. (c) Volcano curve: exchange current density of different metals, as a measure of catalytic activity, against the adsorption energy of hydrogen. (d) Electrochemical activity of metal oxides for oxygen deposition versus enthalpy of formation.

Elementary Reactions at the Hydrogen Electrode

At platinum metals in acid solution, the Volmer-Tafel mechanism takes place in two steps:

1. Proton diffusion from the bulk electrolyte to the electrode surface, followed by charge transfer reaction.
2. Recombination and formation of molecular hydrogen followed by desorption and mass transfer into the electrolyte (Equation 240 and Table 27).

$$2\,H^+ + 2\,e^- \rightarrow 2\,H_{ad} \rightarrow H_2 \qquad (240)$$

The type of this reaction is "charge transfer kinetics without diffusion" (see above), from two identical single-electron steps. The activation energy depends on the strength of the metal-hydrogen bond (Figure 103c).

	Acid Solution			Neutral and Alkaline Solution
	Volmer-Tafel mechanism at platinum and high current	*b* (mV)	*z*	
	$2H^+(bulk) \rightarrow 2H^+(double\ layer)$			$H^+(bulk) \rightarrow H^+(double\text{-}layer)\vert \cdot 2$
(1)	$2H^+ + 2e^- \rightarrow 2H_{ad}$	120	2	$H_2O + 2e^- \rightarrow H_{ad} + 2OH^- \vert \cdot 2$
(2)	$H_{ad} + H_{ad} \rightarrow H_{2,ad} \rightarrow H_2$	$30 \rightarrow \infty$	1	$H_{ad} + H_{ad} \rightarrow H_{2,ad} \rightarrow H_2$
	$2H^+ + 2e^- \rightarrow H_2$			$2H_2O + 2e^- \rightarrow H_2 + 2OH^-$
	Volmer-Heyrovsky mechanism at lead, mercury or low current			
(1′)	$H^+ + e^- \rightarrow H_{ad}$	120	1	$H_2O + e^- \rightarrow H_{ad} + OH^-$
(2′)	$H_{ad} + H^+ + e^- \rightarrow H_{2,ad}$ Intermediate $[H_{ad}\cdot H]^+$	$40 \rightarrow 120$	1	$H_{ad} + H_2O + e^- \rightarrow H_{2,ad} + OH^-$
	ECE mechanism (side reaction)			
(1a′)	$H^+ + e^- \rightarrow H_{ad}$	120	1	
(1b′)	$H^*_{ad} \rightarrow H_{ad}$	60	1	(surface diffusion)
(2′)	$H_{ad} + H^+ + e^- \rightarrow H_{2,ad}$	$40 \rightarrow 120$	1	

(1), Volmer reaction; (2), Tafel reaction; (2′), Heyrovsky reaction; ad, adsorbed; aq, solvated $[H(H_2O)_x]^+$; *b*, d*E*/d(log*I*) Tafel slope at $\theta_H = 0$ to 1 at diffusion-limiting current; *z*, electrochemical valency; C, chemical; E, electrochemical.

TABLE 27 Mechanism of Hydrogen Evolution

A platinum electrode acts as a catalyst and readily adsorbs and desorbs hydrogen. The medium adsorption heat of platinum (Pt-H) is strong enough to enable adsorption and not too strong to prevent desorption of the product. This creates best circumstances for fast H–H splitting. The adsorption of atomic hydrogen (so-called UPD) slightly above 0 V versus RHE resembles the replacement of water molecules.

$$Pt\text{–}H_2O + H^+ \rightarrow [Pt\text{-}H]^+ + H_2O \quad \text{or} \quad Pt\text{–}H_2O \rightarrow [Pt\text{-}H]^+ + OH^- \qquad (241)$$

Despite great catalytic properties of platinum, too much adsorbed hydrogen poisons the electrode surface. The cathodic Tafel slope is 120 mV per current decade.

As long as little hydrogen is adsorbed, the Tafel coefficient $b = 60$ mV is observed because adsorption is rate-determining (e.g., on silver at very low current densities). Exchange current density (catalytic activity) is the lowest at pH 0 and greatest at pH 12.5. In a strongly alkaline solution, surface roughness plays a role, while H_2 solubility is low and hydroxide ions compete for the surface sites. Passivation by an adsorbed oxygen layer occurs above 0.8 V versus RHE.

Cathode Materials

Noble Metals Hydrogen evolution requires that hydrogen is first adsorbed on the electrode surface. The best surfaces for this purpose are metals with free *d*-electrons, nickel, palladium, and platinum (atomic configurations: d^8s^2, $d^{10}s^0$, and d^9s^1) which exhibit low overpotential.

τ	Pd	Pt	Ir	Rh	Ni	Fe	Ag	Au	Cu	Zn	Ti	Graphite	Pb	Hg
H_2	0.05	0.07	0.1	0.1	0.12	0.2	0.35	0.4	0.35*	0.55	0.5	—	0.6	0.78
O_2	0.31*	0.3	0.3	0.3	0.23	0.3	0.32*	0.32	—	—	—	0.5*	0.89*	
i_0	—	$12 \cdot 10^{-6}$	$0.1 \cdot 10^{-6}$	$0{,}06 \cdot 10^{-6}$	0.042	0.17	—	—	—	—	—	—	—	—
b	—	46	40	42	95	191	—	—	—	—	—	—	—	—

TABLE 28 Hydrogen and Oxygen Overpotential of Different Metals in Potassium Hydroxide Solution at a Current Density of 1 mA cm^{-2} at 80°C (*25°C). Exchange Current Density i_0 (in A m^{-2}) and Tafel Slope b (in mV/decade) of Oxygen Deposition

Other metals that adsorb and desorb hydrogen readily are rhodium, palladium, platinum and its alloys (Pt_2Mo, TiPt, $HfPd_3$, $ZrPt_3$), ruthenium, and iridium. Sulfur and iron (II) poison the electrode surface and reduce activity. Gold and silver exhibit mid-range hydrogen overvoltage because of the poor H-adsorption (Table 28).

The cathode loading can be kept low (≤1 mg cm^{-2}) because the oxygen evolution reaction controls the electrolysis and hydrogen reduction will always be the faster of the two reactions. Carbon black is often used as support material, but may be replaced by carbon nanotubes, which have good electronic conductivity and corrosion stability. Platinum cathodes are also affected by the deposition of ions of Cu, Mo, Ag, Cd, Sn, and Pb present in the electrolyte; so pure water must be fed to the electrolyzer.

Ferrous Metals Oxygen adsorption affects cathodes made of iron, cobalt, and nickel, and disturbs the hydrogen evolution. In addition, hydrogen may penetrate the iron lattice and cause embrittlement. Nevertheless, steel cathodes are used in acid and weakly alkaline solutions.

Nickel and its alloys (Ni-S-Co, Ni-Fe-Mo-Zn, NiA150, NiMo25) are predominantly used in alkaline electrolysis. Additives such as titanium for stabilizing the nickel lattice, sulfur doping for improved hydrogen production, and MoS_2 for prolonged electrode life are often used.

Raney-nickel, a "spongy," highly porous material, fabricated by etching a nickel-aluminum alloy in hot sodium hydroxide solution, is also used. Intermetallic alloys (Ni_aM_b) of different metals often have long-term stability problems while amorphous metals ($Ni_{70}Mo_{29}Si_5B_5$, $Ni_{0,33}Zr_{0,66}$, $Fe_{40}Ni_{40}B_{20}$, $Fe_{60}Co_{20}Si_{10}B_{10}$) show high overvoltage.

Non-Precious Metals MoS_2, WO_3, and cobalt and nickel glyoximes show catalytic activity inferior to that of platinum. Other metals such as zinc, cadmium, mercury (atomic configuration: $d^{10}s^2$), tin, and lead are not considered for the cathode materials because of poor performance and inability to form stable M-H bonds.

The Oxygen Electrode

At the positive terminal, the anodic reaction of oxygen evolution takes place. The reaction is well studied, and various aspects of electrode material selection have been researched.

Open-Circuit Potential and pH Shift

The oxygen evolution reaction at platinum electrode is pH-dependent and has the equilibrium potential at 25°C change from 1.229 V at pH 0, to 0.185 V at pH 7, and +0.401 V at pH 14 (Equation 242).

$$E = 1.229 - 0.05916 \cdot \left[\mathrm{pH} - \frac{1}{4} \log \frac{p(\mathrm{O_2})}{p^0} \right] \tag{242}$$

Elementary Reactions at the Oxygen Electrode

In practice, the required cell voltage to generate oxygen increases with higher current due to overpotential and oxygen is evolved above 1.4 V. Oxygen overpotential is the greatest at pH 7, because of the low concentration of hydroxide ions in the solution next to the electrode.

The mechanism of oxygen evolution has not been completely understood (Table 29), but it appears to proceed through decomposition of instable oxides (Equation 243).

$$MO_x + H_2O \rightarrow MO_{x+1}\ (+\ 2\ H^+ + 2\ e^-) \rightarrow MO_x + ½O_2 \tag{243}$$

The electrochemical oxidation of water generates adsorbed hydroxyl radicals, which are further oxidized. At high temperatures, the recombination of oxygen atoms appears to be slow.

Anode Materials

Anodic oxygen deposition takes place in the voltage range from 800 mV to 1230 mV RHE (Figure 103a).

Metals Exchange current density is $i_0 \approx 10^{-10}$ to 10^{-11} A cm^{-2} and Tafel slope is 60–40 mV for low overpotential and 120 mV for high overpotential. Inner metal-metal oxide interfaces in the electrode create apparent Tafel slopes above 120 mV. Iridium, ruthenium, rhodium, and platinum are the most active anode catalysts. Lead and its alloys are electrochemically stable, but the overpotential for oxygen is high.

Nickel Raney nickel and its alloys (Ni-Co, Ni-Ir) are the most important, stable, and low-cost materials in alkaline solution, although a gradual degradation and metal dissolution occur during interruptions with intermittent operation.

Platinum Group Metal Oxides These materials are coated on etched nickel in alkaline solutions or on titanium supports in acid solutions. RuO_2 is more active than IrO_2,

	Acid Solution	Alkaline Solution
(1)	$H_2O \rightarrow OH_{ad} + H^+ + e^- \mid \cdot 2$ $OH_{ad} \rightarrow OH_{ad}$ (site change)	$OH^- \rightarrow OH_{ad} + e^- \mid \cdot 2$
(2)	$OH_{ad} \rightarrow O_{ad} + H^+ + e^- \mid \cdot 2$	$OH_{ad} + OH^- \rightarrow O_{ad} + H_2O + e^- \mid \cdot 2$
(3)	$2\ O_{ad} \rightarrow O_2$ $2\ H_2O \rightarrow O_2 + 4\ H^+ + 4\ e^-$	$2\ O_{ad} \rightarrow O_2$ $4\ OH^- \rightarrow 2\ H_2O + O_2 + 4\ e^-\ (z = 2)$

ad, adsorbed; aq, solvated $[H(H_2O)_x]^+$.

TABLE 29 Generalized Mechanism of Oxygen Evolution

but slow anodic dissolution occurs forming hydrated oxides of Ru(VI) and Ru(VII), as well as gaseous RuO_4.

Oxygen adsorbs readily on ruthenium at roughly 0.8 V versus RHE. The Ru-O bond is stronger than the Pt-O bond. The coating of OH groups on the electrode surface enables acid-base reactions. The RuO_2 surface is uncharged at about pH 5.5, and Co_3O_4 at pH 7 (Equation 244).

$$M\text{–}OH + H^+ \rightleftharpoons [M\text{–}OH_2]^+, \text{ and } M\text{–}OH + OH^- \rightleftharpoons [M\text{–}O]^- + H_2O \qquad (244)$$

Mixed oxides are less active than RuO_2 but the addition of IrO_2, Ta_2O_5, or SnO_2 improves stability. Mixed Ir-Ru oxide rutile structures (1:1 and 7:1) are the most active binary, and Ir-Ru-Sn (rutile, 2:1:1) are the most active ternary oxide catalysts. IrO_2/SnO_2 (2:1) provides lower cell voltage than pure IrO_2 because tin removes adsorbed hydroxyl species from the active iridium sites.

Tantalum improves the electronic conductivity of the $SnO_2/IrO_2/Ta_2O_5$ electrode and promotes the surface enrichment of iridium. $Ir_xRu_yTa_zO_2$ (2 mg cm^{-2}) achieves cell voltages of 1.57 V at 1 A cm^{-2} and 80°C, so that 3.75 kWh per m^3 of hydrogen are consumed with an efficiency of 94%. IrO_2 and RuO_2 anodes are not affected by metal ion contaminations in the electrolyte.

Transition Metal Oxides Spinel compounds of the type $AO_xB_2O_3$ (e.g., $NiCo_2O_4$, $Ni_{1+x}Fe_{2-x}O_4$, Co_3O_4) and perovskites ABO_3 based on lanthanum (e.g., $LaNiO_3$, $LaCoO_3$) have limited importance because of their moderate long-term stability. Doping with up to 3% lithium increases surface roughness and BET surface and porosity. NiO and Co_3O_4 are not stable in acid solution but can be used in alkaline solutions.

Ozone Formation Metal oxides such as SnO_2, TiO_2, PbO_2, and PtO_2 do not adsorb hydroxide and hence show high oxygen overpotential. Free hydroxyl radicals and ozone are generated instead. Lead dioxide PbO_2 can only be used as anode material, not as cathode.

Carbon Graphite is not considered a good oxygen electrode material because of its high oxygen overpotential of 0.5 V and formation of CO_2 at high current densities.

Electrolytes

Aqueous salt solutions, solid electrolytes, or molten salts are used for electrolysis. In 1827, Georg Simon Ohm discovered that his Ohm's law is valid for electrochemical cells and batteries. Since the 1840s, Kohlrausch had measured electrolyte resistances using alternating current in order to avoid electrode polarization. Daniell continued Faraday's work and, in 1843, he postulated the existence of free mobile ions in electrolyte. J. W. Hittorf discovered in 1853 that metal ions migrate to the electrode surface during electrolytic metal deposition and that some ions move faster than others, which led him to the concept of the transference number. In the 1890s, the theory of ions and dissociation got generally accepted thanks to Ostwald.

Aqueous Solutions

Electrolyte Conductivity The concept of molar conductivity proved Kohlrausch's square-root law in highly diluted electrolytes. The ion conductivities λ of the individual cations and anions contribute to the overall molar conductivity Λ (in S m² mol⁻¹) of the solution (Equations 245 to 247):

$$\Lambda_\infty = \lambda_{\infty\oplus} + \lambda_{\infty\ominus} \tag{245}$$

$$\Lambda \approx \Lambda_\infty - k\sqrt{c} \tag{246}$$

$$\kappa = c\Lambda = c(z_\oplus F u_\oplus + |z_\ominus| F u_\ominus) \tag{247}$$

where c is the molar concentration (in mol L^{-1}), F is the Faraday constant, k is the constant, κ is the conductivity (1 S cm^{-1} = 100 S m^{-1}), and u is the ion mobility (in $m^2V^{-1}s^{-1}$), ∞ in highly diluted solution.

Potassium Hydroxide Diluted sulfuric acid and potassium hydroxide show the best conductivity of all electrolytes (Table 30). Molar concentration, c (in mol L^{-1} of solution), mass concentration β (in g/L), mass fraction, w (1 % = 0.01 = 10 g/kg solution), and b molality in mol kg^{-1} of solvent can be converted using the molar mass M(KOH) = 56.11 g mol^{-1} and the density ϱ (1 g cm^{-3} = 1 kg L^{-1}) of the solution (Equation 248).

$$c = \frac{\beta}{M} = \frac{\varrho w}{M}, \quad \text{and} \quad b = \frac{1}{\frac{\varrho}{c} - M} \tag{248}$$

Conductivity (in S cm^{-1}) depends on molar concentration and temperature (Equation 249).

$$\kappa(T) = -2.041\ c - 0.0028\ c^2 + 0.005332\ c\ T + 207.2\ \frac{c}{T} + 0.001043\ c^3 - 0.0000003\ c^2T^2 \tag{249}$$

A solution of 30% KOH (7.61 mol/L) boils at 112°C and has the highest conductivity of 1.36 S cm^{-1} (= 0.733 Ω cm) at 80°C. However, KOH absorbs CO_2 from air and forms K_2CO_3, reducing the conductivity and clogging porous electrodes.

Electrolyte Additives Ionic additives improve the catalytic activity of the hydrogen electrode by electrodeposition of metal composites, for example, ethylenediamine complexes $(Co[en]_3)Cl_3$ or Na_2MoO_4 or Na_2WO_4 in KOH solution.

Sulfuric Acid Diluted acids are rarely used in technical water electrolysis because of corrosion problems and high volatility (vapor pressure). The conductivity of H_2SO_4 is the best in 3-molar solution (0.85 to 1 S cm^{-1}).

	25°C			50°C		55°C	75°C	80°C	
w (%)	25	26.83	30	25	30	35	40	26.83	32.6
ϱ (g cm^{-3})	1.236	1.255	1.288	—	—	1.341	1.396	1.255	1.315
κ (S cm^{-1})	0.618	—	0.624	0.910	0.960	1.010	1.250	1.310	1.364

Table 30 Density ϱ and Conductivity κ of Potassium Hydroxide Solutions at Different Temperatures

Sulfuric acid can be decomposed in cathodic reaction at 0.172 V to sulfurous acid H_2SO_3 and traces of H_2S and sulfur may be formed. Peroxydisulfuric acid, $H_2S_2O_8$, can be generated anodically at 2.01 V.

Cationic surfactants, such as hexadecyltrimethylammonium bromide (HTMAB), inhibit hydrogen evolution and promote oxygen evolution reaction through enhanced H_2O_2 production.

Hydrochloric Acid HCl is used for the anodic chlorine production ($E^0 = 1.40$ V).

Phosphoric Acid Commercial 85% H_3PO_4 (1.687 g cm^{-3}) melts at 21°C, boils at 158°C, and is electrochemically stable up to 225°C. However, temperature above 150°C is required for satisfactory conductivity.

A solution of 95% H_3PO_4 has conductivity of 0.6 S cm^{-1} at 200°C. Because of the evaporation of water, 98–100 % H_3PO_4 is present at 190°C and 1 bar; and 93% H_3PO_4 at 205°C and 8.2 bar. Below 95% H_3PO_4 the water vapor pressure is undesirably high (Table 31). When water evaporates in aqueous solutions, diphosphoric acid $H_4P_2O_7$ and higher phosphoric acids are formed above 150°C, as soon as the composition reaches $H_3PO_4 \cdot H_2O$. A solution of 92–95% phosphoric acid is prone to auto dehydration and autoprotolysis (Equation 250).

$$[H_4PO_4]^+ + [H_2PO_4]^- \rightleftharpoons 2\,H_3PO_4 \rightleftharpoons H_4P_2O_7 + H_2O \qquad (250)$$

Solid 100% phosphoric acid melts at 42°C and is used in phosphoric acid fuel cells (PAFC). The adsorption of oxygen is impeded by phosphate.

Molten Electrolytes These are used for electrowinning of magnesium and aluminum. Eutectic mixtures are used to lower the temperatures below that of the single pure components.

Inorganic Solid Electrolytes

Steam electrolysis requires inert, gas-tight, and mechanically stable solid electrolytes with high ionic conductivity (>10 S m^{-1}) and ideally no electronic conductivity.

Zirconia Ceramics based on yttrium-stabilized zirconium dioxide ($ZrO_2 \cdot Y_2O_3$, in short YSZ) are advanced oxide ion conductors at temperatures above 800°C (Figure 104). The annealed material tolerates acids, except hydrofluoric acid, and is attacked by molten alkali hydroxides. The ionic conductivity is based on the displacement of lattice voids V (Equation 251).

$$O_O^{\times} \rightarrow V_O^{\bullet\bullet} + 2\,e' + \frac{1}{2}\,O_2 \qquad (251)$$

w	85%	93%	97%	103%	108%
$p(H_2O)$ (mbar)	450	300	120	10	3

Table 31 Water Vapor Partial Pressure in Phosphoric Acid of Different Concentrations

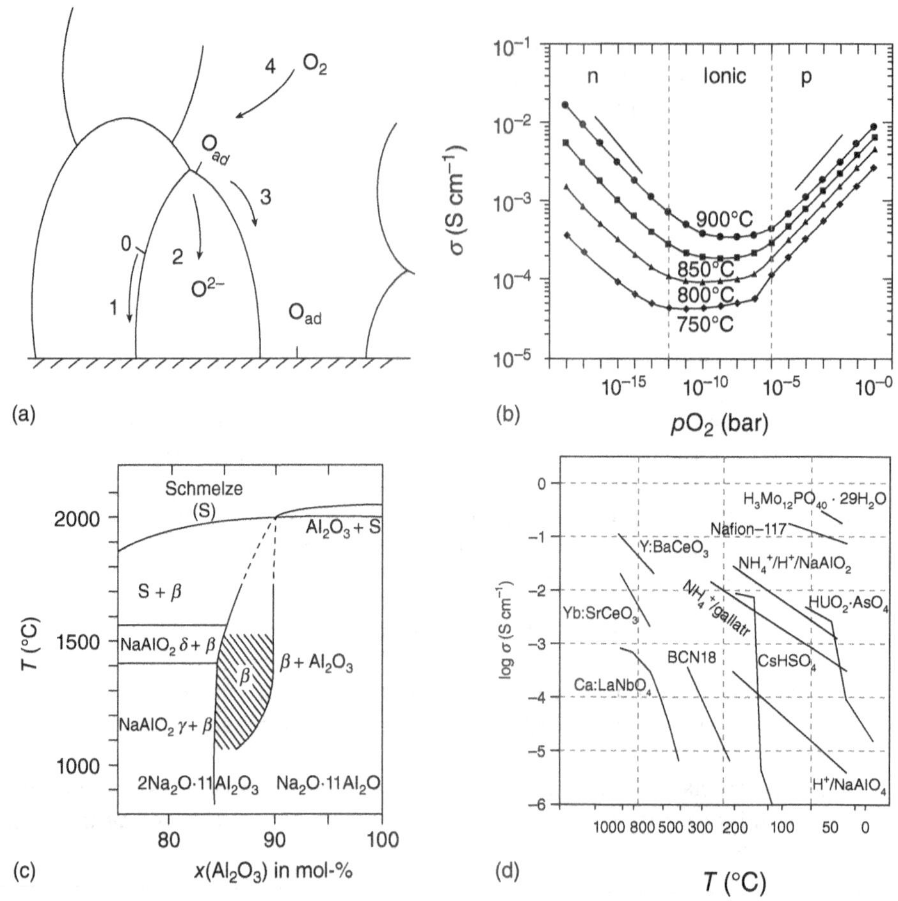

FIGURE 104 (a) Diffusion of oxygen in solid phases: 1, grain boundary diffusion; 2, bulk diffusion; 3, surface diffusion; 4, gas diffusion. (b) Conductivity of the mixed conductor SrTi0.99Fe0.0103 at different temperatures: 1, electronic n-conducting; 2, ionic; 3, holes p-conducting. (c) Arrhenius diagram of inorganic and organic proton conductors. (d) Phase diagram of aluminum oxide and β-alumina.

Y^{3+} ions cluster between the Zr^{4+} ions and stabilize the ZrO_2 lattice. Perovskites such as ABO_3 tolerate a certain oxygen deficit δ, so that more oxygen voids are generated by yttrium doping (Equations 252 and 253).

$$Y_2O_3 \rightarrow 2\ Y'_{Zr} + 3\ O_O^{\times} + V_O^{\bullet\bullet} \tag{252}$$

$$2\ Zr_{Zr}^{\times} + O_O^{\times} + Y_2O_3 \rightarrow 2\ Yb'_{Zr} + V_O^{\bullet\bullet} + 2\ ZrO_2 \tag{253}$$

The Kröger-Vink notation means: (atom, void V, electron e, hole h)$_{position}^{charge}$ at oxygen sites (O), without charge (×), positive (•) or negative (′). Here: O_O oxygen is at the correct position; the void V_O occupies an oxygen site. With increasing temperature T (in K), the intrinsic defects in the ZrO_2 lattice grow by oxygen vacancies and oxide

	σ (S m^{-1}) at 950°C	σ (S m^{-1}) at 650°C	σ (S m^{-1}) at 500°C
$La_{0,8}Sr_{0,2}Ga_{0,8}Mg_{0,2}O_3$	30	5	< 1
ZrO_2 + 10 % Sc_2O_3	30	3.5	0.3
$Ce_{0,9}Gd_{0,1}O_{1,95}$	20	4	1
ZrO_2 + 8 % Y_2O_3 (YSZ)	12	1.2	0.12
ZrO_2 + 6 % Sc_2O_3	12	1.5	0.,15
ZrO_2 + 4 % Sc_2O_3	8	0.9	0.1
ZrO_2 + 3 % Y_2O_3	4	0.4	< 0.05

Table 32 Approximate Ionic Conductivity of Inorganic Solid Electrolytes

ions on interstitial sites. Ionic conductivity, $\sigma \approx 10^{-10} \ldots 10^{-4}$ S m^{-1}, by displacement (hopping) of vacancies obeys the Arrhenius equation (Equation 254 and Table 32).

$$\sigma(T) = \sigma_0 \cdot e^{-\frac{E_A}{kT}} \rightleftharpoons \log\sigma = -\frac{E_A}{k \cdot \ln 10} \cdot \frac{1}{T} + \log\sigma_0 = b \cdot \frac{1}{T} + a \qquad (254)$$

The pre-exponential factor σ_0 is a function of the diffusion coefficient, activation energy is $E_A \approx 0.6$-1.2 eV (1 eV = $1.602 \cdot 10^{-19}$ J), k is the Boltzmann constant $1.381 \cdot 10^{-23}$ J/K; b is the slope and a the intercept of Arrhenius plot, ln 10 = (log e)$^{-1}$.

The impact of ambient air temperature on conductivity is negligible. At 1000°C, YSZ is an oxide ion conductor in a large range of oxygen partial pressures, $p(O_2) = 10^{-24} \ldots 10^{8}$ bar; while at highest pressures, YSZ becomes a p-conductor. Electronic conduction plays a role only in conditions of extremely low O_2 partial pressures and high temperature (2200°C, < 1 μbar).

Practical ceramic membranes contain greater than or equal to 16 mol% calcium oxide (CaO) or magnesium oxide (MgO) and 8–10 mol% yttrium oxide (called: 8 YSZ or 10 YSZ, *cubic-stabilized zirconia,* CSZ). These membranes have the conductivity of 10 S m^{-1} (100 mS cm^{-1}) at 1000°C. Microstructural changes and the formation of tetragonal phases cause a loss of conductivity in the first 1000 operating hours.

Partly stabilized zirconia (PSZ) contains less than 8 mol% Y_2O_3. The internal tension in such solid solutions provides good temperature change stability and low thermal expansion (≈ 10^{-5} K^{-1}). *Tetragonal zircona polycrystals* (TPC) with 3 mol% Y_2O_3 are more mechanically stable but have lower conductivity.

Doped Zirconia Small, rare earth metal ions significantly increase the ionic conductivity. Addition of 10% (mol) of scandium (ScSZ) to ZrO_2 fluorite lattice results in conductivity of 30 S m^{-1} (1000°C).

Gadolinium-Doped Ceria (GCO, $Ce_{1-x}M_xO_2$, where M is Sm, Gd) These are mechanically stable solid solutions with Gd^{3+} ions and oxygen voids in the CeO_2 fluorite lattice. Unfortunately, at low oxygen partial pressures, electronic conductivity proceeds through the $Ce^{4+} \rightarrow Ce^{3+}$ transition that requires operating temperatures below 550°C.

Lanthanum Gallate (La,Sr)(Ga,Mg)O_3 (LSGM) is an expensive, strontium-doped, ionically conducting perovskite. The mechanical and chemical stability are poor, because gallium evaporates at high temperature in reducing atmosphere, affecting the nickel oxide electrodes. LSGM tolerates mixed conducting electrodes and is used below 600°C.

Mixed Electron-Oxide Conductors

Metal oxides such as PbO_2 and ZnO become ion conductors with heating. Although unusable as electrolytes in high-temperature electrolysis, they can be used as electrode materials. Ionic conductivity (by anions and cations) and electronic conductivity, σ, (by electrons and holes) are combined in mixed conductors (Equation 255).

$$\sigma = \sigma_e + \sigma_i = \sum_i z_i e\ N_i u_i = e\ (n_e u_e + n_p u_p + n_\ominus u_\ominus + n_\oplus u_\oplus) \qquad (255)$$

In this equation, e is the elementary charge $1.602 \cdot 10^{-19}$ C; N is the number of particles, n (= N/V) is the number of particles per volume (m^{-3}), and u is the ion mobility t ($m^2V^{-1}s^{-1}$).

Strontium Titanate $SrTiO_3$ is a mixed conductor made by iron-doping of the perovskite lattice. Iron, substituted in a titanium site, accepts an electron to create an iron ion and a hole (Equation 256).

$$Fe_{Ti}^{\times} \rightarrow Fe'_{Ti} + h^{\bullet} \qquad (256)$$

In ambient air ($O_O^{\times} \rightarrow V_O^{\bullet\bullet}$ + 2 e′ + ½O_2), the oxide content changes slightly to $SrTi_{0.99}Fe_{0.01}O_{3-\delta}$ (Figure 104b). Such ceramics are used as oxide ion conductors, in amperometric oxygen sensors, and as filters for separation of oxygen from gas mixtures. In those applications, cathodic voltage is applied to force oxygen reduction (Equation 257).

$$O_2 + 4e^- \rightarrow 2\,O^{2-} \qquad (257)$$

Lanthanum Manganite This strontium- or calcium-doped electron conductor $La_{1-x}Sr_xMnO_3$ (LSN, $x \approx 0.15$ to 0.25) or $La_{1-x}Ca_xMnO_3$ (LCM) generates cation voids ($Mn^{3+} \rightarrow Mn^{4+}$) in contact with oxygen. Oxide ion conductivity in these materials is near zero. $La_{1-x}Sr_xMnO_3$ is thermos-mechanically stable ($\alpha = 12 \cdot 10^{-6}$ K^{-1}) and does not react with ZrO_{2-}solid electrolytes.

Lanthanum Strontium Cobalt Oxide (LSC) $(La,Sr)CoO_{3+}\delta$ is a mixed conducting solid solution. Cobalt-doping increases the electronic conductivity in $La_{0.8}Sr_{0.2}Mn_{1-x}Co_xO_{3+}\delta$ (x = 0–1) from 10,000 to 12,000 S m^{-1} at oxygen diffusion limits of 10^{-12}… 10^{-7} cm^2s^{-1}.

Thermal expansion ($\alpha = 13 \ldots 20 \cdot 10^{-6}$ K^{-1}) causes poor adhesion of other materials such as electrodes on these solid electrolyte materials (YSZ, GCO).

When Co is substituted with Fe, Mn, or Cr in $(La,Sr)Co(Cr,Mn,Fe)O_3$, several properties such as oxygen diffusion, conductivity, catalytic activity, and thermal expansion change. The magnitude of the increase in these properties changes in the order Co >> Fe > Mn > Cr. However, chemical and thermal stability also decrease in the same order. At high temperatures, chromium and manganese are superior, while at low temperatures iron or pure LSC are better choices.

Sodium Ion Conductors

β-Alumina, a sodium aluminum oxide (1.57 $Na_2O \cdot 11\ Al_2O_3$; max. 8% Na), has the sodium ion conductivity comparable to an aqueous salt solution. The conduction mechanism involves sodium ions filling the lattice plane of interstitial oxide ions (Equation 258).

$$\text{V Al}^{3+} \cdots \text{O}^{2-} \text{(2 Na}^{+}) \cdots \text{Al}^{3+} \text{ V (V = void at aluminum)} \qquad (258)$$

Polycrystalline β-alumina, which is used in high-temperature sodium batteries, has conductivity in the range of 0.1–0.2 S cm^{-1} at 350°C. Metastable β″-alumina has the approximate composition of $Na_6Al_{32}[V_{Al}]O_{51}$ and must be stabilized by mono or bivalent cations of Li^+, Co^{2+}, Mg^{2+}, or Zn^{2+}, resulting in the composition such as $(Na_2O)_{1.67}(MgO)_{1.33} \cdot 10.33\ Al_2O_3$ (2% MgO).

The mechanism of conduction involves sodium ions hopping through aluminum vacancies. The conductivity is accomplished by the grain-boundaries in contrast to better conducting through the bulk phase (0.16 eV; 0.065 S cm^{-1} at 300°C).

The diffusion coefficient *D* obeys qualitatively the Arrhenius and Nernst-Einstein equation (Equation 259).

$$D = D_0 e^{-\frac{E_A}{kT}} \quad \text{and} \quad \sigma T = \sigma_0 e^{-\frac{E_A}{kT}} \qquad (259)$$

Doping with aluminum nitride improves sodium content and conductivity. Sodium can easily be replaced in melts by ammonium ions, Rb^+, Ga^+, In^+, Tl^+, Sn^{2+}, Pb^{2+}, Mn^{2+}, Fe^{2+}, Cu^+, Cu^{2+}, Ag^+, Zn^{2+}. In β-gallates, Al^{3+} is replaced by gallium (III).

A disadvantage of β″-alumina is its water sensitivity.

Inorganic Proton Conductors

High-temperature proton conductors have been explored for steam electrolysis. The protons are located in either disordered wet materials or ordered-crystalline structures for intrinsic conductors or originate from external sources (extrinsic conductors). Wet materials tend to dehydrate above 100°C and stop conducting protons, while water-free materials can be used between 150 and 300°C. Extrinsic proton conductors conduct above 600°C.

Super-Ion Conductors Heteropoly acids $H_3PM_{12}O_{40} \cdot nH_2O$ (M = W, Mo, Si) achieve up to 0.2 S cm^{-1} at room temperature through protons located in connecting water layers.

Hydrogen uranyl phosphate and arsenate, $HUO_2MO_4 \cdot 4H_2O$ (HUP: M = P; HUA: M = As) have conductivity of 0.001 S cm^{-1} (Table 33).

Ammonium-doped β-Alumina, $(NH_4)_{1,67-y}(H_3O)_yMg_{0,67}Al_{0,33}O_{17}$, stores water between the crystalline layers, which works as a vehicle for proton conductivity (0.01 S cm^{-1} at 200°C).

Solid Acids In water-free cesium hydrogen sulfate $CsHSO_4$ (≥141°C) and phosphate CsH_2PO_4 (≥230°C), proton-hopping along H bridges occurs (>0.01 S cm^{-1}) (Equation 260).

$$-\cdots \text{H[SO}_4]^{\ominus} \cdots \text{H[SO}_4]^{\ominus} \cdots - \rightarrow - \cdots \text{H}^{\oplus} \cdots [\text{SO}_4]^{\ominus} \cdots \text{H}^{\oplus} \cdots [\text{SO}_4]^{\ominus} \cdots - \qquad (260)$$

Extrinsic Proton Conductors Doped alkaline earth metal niobates and cerates accept protons from water vapor in their oxygen sites and vacancies (Equation 261).

$$H_2O + O_O^{\times} + V_O^{\bullet\bullet} \rightarrow 2OH_O^{\bullet} \qquad (261)$$

$H(UO_2)PO_4 \cdot 4H_2O$	$LaNbO_4(Ca)$ 800°C	$BaCeO_3$ 800°C	$CsHSO_4$	β-Alumina	Heteropoly Acids
≈ 0.001	0.001	0.01	0.001 to 0.01	≈ 0.01	< 0.2

TABLE 33 Conductivity σ (S cm^{-1}) of Solid State Proton Conductors

The protons hop between adjacent oxygen atoms.

At high temperatures the competing oxide ion conduction becomes a dominant conduction process. Examples of these systems are $BaCa_xNb_{1-x}O_3$ (BCN), $BaZrO_3$, and $M^{II}CeO_3$ (M = Sr, Ba). Calcium-doped $LaNbO_4$ is stable in CO_2 and hydrated atmosphere.

Organic Solid Electrolytes

Solid polymer electrolytes (SPE) have been the most important ion-conducting polymers since 1975. Proton exchange *membranes* (PEM) function like solid acids, but can be used as catalyst supports and separators for gases. In PEM fuel cells, the membrane is humidified by water vapor in the gas feed, whereas in the electrolysis process the membrane is exposed to liquid water.

Perfluorosulfonate Membranes (PFSA) Nafion® (DuPont) and its successor products of various companies consist of a fluoro-hydrocarbon chain with sulfonic acid groups ($-SO_3H$). Nafion is synthesized by copolymerization of tetrafluoroethene with sulfonated tetrafluorethene and unsaturated fluoro ether via addition of tetrafluoroethene to epoxidized perfluoropropene (Equations 262 and 263).

$$F_2C{=}CF_2 + SO_3 \rightarrow FO_2S\text{-}CF_2\text{-}COF \tag{262}$$

$$F_3C\text{-}CF{=}CF_2 \rightarrow F_3C\text{-}CF(O)CF_2 \xrightarrow{F_2C=CF_2} \rightarrow F_3C\text{-}CH_2\text{-}CH_2\text{-}O\text{-}CF{=}CH_2 \tag{263}$$

Finally, the hydrolysis of SO_2F-groups produces SO_3H. The polymer uptakes about 30 vol% water, which is essential for conductivity (Table 34). Swelling and structural changes at higher levels of humidity are detrimental.

Solvated protons (in acids) or Na^+ ions (in sodium hydroxide solution) migrate between adjacent SO_3 sites in water-filled, spherical cavities of about 4 nm in diameter. These cavities are connected by 1 nm long and 0.1 nm wide channels (Figure 105).

Nafion-117 is 7/1000 inch (178 μm) thick and has an equivalent mass of 1100 Da, corresponding to the ion exchange capacity of 1/1100 (0.91)·mmol monovalent cations (0.021 g Na^+) of the dry polymer. Anions such as chloride can hardly penetrate the channels.

Nafion conducts comparably with 1-molar sulfuric acid ($\sigma \approx 0.1$ S cm^{-1}, $\sigma/A \approx 2$ S cm^{-2}). The transference number is $t = 1$, because proton carries 100% of the applied current.

mol H_2O/mol SO_3H	2	2.7	5	8.4	10	14	22
σ (S/cm)	0.005	0.01	0.023	0.04	0.05	0.06	0.09

TABLE 34 Conductivity of Nafion Depending on the Water Content

Proton conductivity increases with rising temperature at given water content, according to the Arrhenius equation (Equation 264).

$$\ln \sigma = -\frac{E_A}{R} \cdot \frac{1}{T} + \ln \sigma_0 \tag{264}$$

Dissolved gases cannot penetrate through Nafion membranes at low differential pressures. Gas permeability μ (in mol m $m^{-2}s^{-1}Pa^{-1}$) differs between dry (0) and wet (w) transport zones, depending on the water content, ϕ, in m^3 H_2O/m^3 wet polymer (Equation 265).

$$\mu = k_0 e^{-\left(\frac{E_A}{RT}\right)} + \varphi\, k_w\, e^{-\left(\frac{E_{Aw}}{RT}\right)} \tag{265}$$

Nafion is mechanically stable, inert, and can be used up to 125°C, even in aggressive media. The polymer ages by deprotonation of COOH end groups, followed by loss of CO_2 and HF (Equations 266 and 267). Metal ion contaminations (Cu, Ni, Pb) gradually occupy the SO_3H sites and increase resistivity. Hydroperoxide radicals can damage the membrane if iron (II) and copper (II) ions are present (Equation 268).

$$RCF_3 + H_2O_2 \rightarrow RCF_2COOH + HF \tag{266}$$

$$RCF_2COOH + OH \rightarrow RCF_2\cdot + CO_2 + H_2O \tag{267}$$

$$H_2O_2 + Fe^{2+} \rightarrow Fe^{3+} + OH + OH^- \tag{268}$$

Advanced Composites Solid electrolytes for applications at temperatures above 100°C use PTFE-reinforced perfluorosulfonated polymers. Flemion® (Asahi Glass) is a fiber-reinfoced perfluorocarboxylate.

Anion Exchangers Hydroxide ion conductors are based on quaternary ammonium groups in polymer matrix. Sulfonated polystyrene, which is crosslinked with divinyl-benzene, can be chloromethylated and substituted by amino groups (Equation 269).

$$[\text{Poly–NR}_3^+]OH^- \rightarrow \text{Poly–NR}_3^+ + OH^- \tag{269}$$

Hydrocarbonates and carbonates migrate across the solid electrolyte as well. Anion exchange membranes are less conductive, and less chemically and thermally stable than PEM. In strongly alkaline solutions, benzyltrimethylammonium $[(C_6H_5)N(CH_3)_3]^+$ is a preferred formulation because ammonium groups detach in those solutions.

Matrix Materials for High Temperatures

Matrix or **diaphragm** is the absorbent material between the electrodes, which absorbs the electrolyte solution without any need to be ionically conducting. Until recently the best matrix materials have not reached the high current density and mechanical durability of Nafion.

Fluorinated Polyolefins Thermoplastic foils form polyolefins, and polytetrafluoroethylene (PTFE) are fabricated by solution casting or extrusion casting. Fluorinated ethylene-propylene (FEP) copolymers such as TFE-HFP, highly crosslinked with styrene/divinylbenzene, are inert and thermally stable. Perfluoroalkoxy alkanes (PFA) ($-[CF_2]_m-CF[OR]-)_n$, the copolymers of tetrafluorethylene (TFE) and perfluorovinylpropylether, withstand most chemicals and temperatures up to 310°C.

Polyethers Polyoxyphenylene oxide (PPO), $-([C_6H_3][CH_3]_2\text{-}O-)_n$, is stable in diluted acids and strong bases. They are mainly used in blends with polystyrene, butadiene, or polyamide.

Sulfonated Polymers Sulfonated phenol resins and polystyrene resins are not stable in aggressive media. Polysulfons (PSU) are used as matrix for superacids, but prone to stress cracking corrosion.

Polyether sulfones (PES) lose water and ionic conductivity above 100°C.

Sixty mol% disulfonated polyether ether ketones (SPEEK) have conductivity of 0.16 S cm^{-1}. They are resistant to sulfuric acid, but may swell and form hydrogels.

Polyphosphazenes $(-N{=}P[R_1R_2]-)_n$, with aryloxy side groups and sulfonic acid or phosphoric acid groups are used for proton-conducting membranes.

Polyimides Polybenzimidazole (PBI), $(-[C_6H_4]\text{-}[C_7H_4N_2]_2-)_n$, is able to absorb some water. They are not very stable in polar-aprotic solvents and strong acids.

PBI-phosphoric acid composites are heat-resistant to 200°C and conduct moderately (0.0046 S cm^{-1}). Proton conductivity is strongly dependent on the acid doping level.

Nafion can be doped with phosphoric acid, resulting in good proton conductivity of around 0.01 S cm^{-1}.

Ceramic Composites Glass fiber-reinforced polymers, barium titanate, calcium titanate, nickel oxide, and zirconium dioxide, as well as composite or reinforced membranes (with SiO_2, TiO_2, or WO_3) promise higher operating temperature and pressure and less gas crossover during electrolysis.

Alkaline Electrolysis

Based on the electrolyte, there are three main types of electrolyzers: alkaline, solid polymer electrolyte, and high-temperature electrolyzers.

Cell Configuration The main advantages of alkaline water electrolyzers are fast kinetics of the anodic oxygen evolution reaction and inexpensive, porous nickel electrodes. The electrodes have to be replaced every 2 to 3 years. Water is fed at the cathode and is formed at the anode (Equations 270 and 271).

(–) Cathode (Ni, Fe): $$2\,H_2O + 2\,e^- \rightarrow H_2 + 2\,OH^- \tag{270}$$

(+) Anode (Ni, C): $$2\,OH^- \rightarrow \tfrac{1}{2}O_2 + H_2O + 2\,e^- \tag{271}$$

The largest losses occur in the diaphragm, which was previously made of asbestos and presently of synthetic polymers. Technical electrolyzers work in the kilowatt and megawatt range at temperatures of 50–80°C, pressures of 1–150 bar, cell voltages of 1.8–2.4 V, and current densities of 0.1–0.45 A cm^{-2}. The main manufacturers are Brown Boveri, Norsk Hydro, Lurgi, DeNora, Electrolyser Corp., and Krebskosmo.

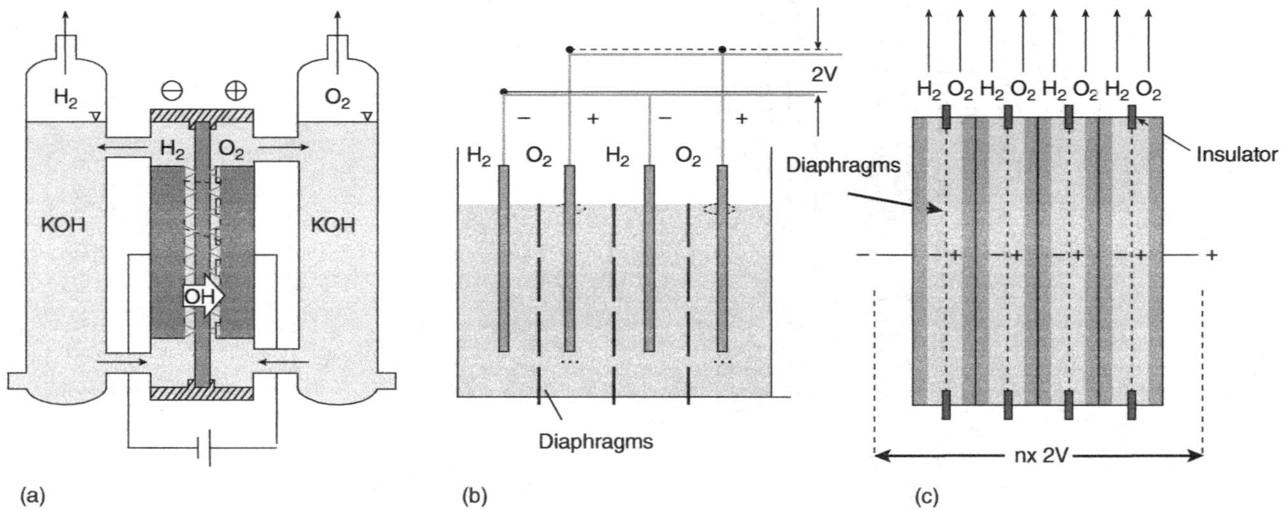

FIGURE 105 (a) Principle of alkaline electrolysis. (b) Monopolar tank cells. (c) Bipolar electrolyzer.

There are two basic designs used for alkaline water electrolyzers:

1. In the older **monopolar** configuration, the cathodes and anodes are alternatively suspended in the same electrolyte in a steel tank (Figure 105 and Table 35). The operating voltage is at about 2 V. The reduced output of a single cell is not critical because a single defective cell can be bypassed. Highly conductive electrolytes are required in these electrolyzers. Metallic frames are used as current collectors because of uneven current distribution (limit 400 cm²).
2. The **bipolar** or filter-press design uses alternating layers of electrodes and separation diaphragms that are clamped together in series to increase the stack voltage. Each electrode serves as both the cathode (on one side) and the anode for the adjacent cell (on the other side). Each cell needs its own sealed electrolyte space in order to prevent a short circuit between the individual cells. The electrolyte flows in parallel through all half cells. The advantages of bipolar design are the unlimited electrode area and reduced stack footprint, small cell resistance, high current densities, and hydrogen generation at high pressure. The weakest cell determines the overall performance of the stack.

	Monopolar Tank Cells (Obsolete)	Bipolar Electrolysis (Alkaline)	Solid Polymer Electrolysis (SPE)	Steam Electrolysis (SOEL)
Current density (A cm^{-2})	0.1–0.2	0.2–0.5	0.6–2.0	0.1–1
Cell voltage (V)	1.9–2.5	1.8–2.4	1.8–2.2	0.9–1.5
H_2 production (m^3 h^{-1})				
per cell	40	—	≤ 15	< 1
per system	—	< 800	≤ 240	< 6
Energy consumption (kWh/m^3 H_2)	5	4.1–4.5 (1 bar) 4.5–6.5 (>>1 bar)	4.0–5.5 (1 bar) 6–10 (>>1 bar)	≈ 3
Pressure at H_2 side (bar)	0.1–1	≤ 30	≤ 30 (350)	≤ 10
Temperature (°C)	70–100	75 (40–140)	50–80	700–1000
Electrode area (cm²)	—	< 40000	< 2500	≤ 100
Power density (W cm^{-2})	—	≤ 1.0	≤ 4.4	≤ 3.0
Voltage efficiency (heat of combustion base)	—	62–82 %	67–82 %	—
Lower partial load range	—	20–40 %	0–10 %	—
Service life of stack (h)	—	< 90000	< 20000	7600
Degradation rate (μV h^{-1})	—	< 3	< 14	—
Advantages	—	Mature, low-cost, stable, MW range	Compact, efficient, dynamic use	Combined heat and power, high pressure
Challenges	Voltage	Corrosion, diaphragm	Expensive, durability	Life, bulky, brittle ceramics

Table 35 General Technical Data of Electrolysis Configurations

Cell Voltage Cell voltage is determined by the oxygen electrode. The water vapor partial pressure $p(H_2O)$ above the KOH solution increases with rising temperature and decreases with rising concentration. In 75% KOH and 25°C, the water vapor pressure drops to 0.5 bar, which increases cell voltage by 18 mV (Equation 272).

$$E = \Delta E^0 + \frac{RT}{2F}\ln\frac{p(H_2)}{p(H_2O)} + \frac{RT}{2F}\ln\sqrt{p(O_2)} \qquad (272)$$

Dissolved Gases and Bubbles Concentrated KOH solutions are able to absorb oxygen, which may create outgassing and may complicate pressure control during emergency shutdown. Gas bubbles cannot be removed rapidly from the electrolyte and may cover hydrophobic electrode surface and block the active area, causing additional resistance. This may dominate the cell resistance at high current densities. With rising degree of coverage θ, the active electrode area is reduced to $A \cdot (1 - \theta)$ and additional overpotential forms according to the Tafel equation (Equation 273).

The electrolyte resistance increases as well as a result of the void fraction $\varepsilon \approx 0.4$ (hydrogen) and 0.5 (oxygen) of the bubble front layers, according to the Bruggeman equation (Equation 274).

$$\eta(\theta) = a + b\log\frac{I}{A(1-\theta)} = \eta + b\log\frac{1}{1-\theta} \qquad (273)$$

$$\frac{R_e(\theta)}{R_e(0)} = (1-\varepsilon)^{-\left(\frac{3}{2}\right)} \qquad (274)$$

Nonconductive bubbles form an insulating layer on the electrode, which is thicker on the top. Above the critical diameter, gas bubbles detach from the electrode surface and move into the electrolyte. To reduce the problem of gas bubble adherence to electrode surface, perforated electrodes have flow channels with small holes (H_2, 0.1 mm; O_2, 0.7 mm) to generate small bubbles.

In alkaline solution, a strong magnetic field of 1 to 5 T perpendicular to the electrode is used to reduce bubble size from 0.2–0.4 mm to 0.1–0.2 mm. This enables detachment of bubbles from the electrodes by magneto hydrodynamic convection.

An ultrasonic field may be used as well at low electrolyte concentration to create fine bubbles through cavitation and accelerate detachment of hydrogen and oxygen bubbles.

In a strong gravity field of 200 g, gas bubbles are separated from the electrolyte by a strong buoyancy force $\Delta\varrho\, g$ and smaller bubble volume through a special construction c in wind turbine blades or rotating wheels.

Diaphragm The separator between the electrodes is based on ceramics (ZrO_2/polysulfon, K_2TiO_3/PTFE), polymers (glass fiber-reinforced PPS, sulfonated PEEK), or nickel oxide.

Electrode Materials Perforated metal sheets (nickel, nickel-plated iron) are state-of-the-art. At the cathode side, steel is allowed. Iron inhibits the embrittlement by unwanted nickel hydrides. Alloying with molybdenum improves the corrosion stability of nickel anodes.

System Components and Operating Behavior Alkaline electrolyzer technology is mature in the megawatt range. An electrolyzer at atmospheric pressure comprises:

1. Stack and MW-power supply
2. Electrolyte circuit with water feed, water purification, hydroxide tank, gas separators
3. Gas processing with scrubber, gas container, compressor, dryer, hydrogen tank. Highly pure gases are obtained: 99.8–99.9% H_2, and 99.3–99.8% O_2.

Alkaline electrolyzers respond almost immediately to load changes and tolerate interruptions and brief overloads. Partial load operation is problematic because of gas permeation through the separator, outgassing of dissolved gases, and corrosion of the nickel anodes (Ni → NiOOH → $Ni[OH]_2$). Hence, the system is purged with nitrogen from time to time and a standby cell voltage of 1.3 V is applied.

The challenges for alkaline electrolyzer technology include low partial load range, limited current density, and low operating pressure. The diaphragms do not completely prevent cross-diffusion of oxygen into the cathode chamber and vice versa, which may cause safety problems due to the lower explosion limit > 4 mol% H_2. At low load, less hydrogen and oxygen are produced, while hydrogen permeability through the diaphragm remains constant.

Calculation Example A water electrolyzer at 120 V and 1.8 V per cell produces 180 L/h oxygen at 150 mA/cm² and a current efficiency of 99.5%. Electrochemical equivalent: $k(O_2) = 0.2089$ L/Ah.

a. What is the required electric charge and current?

$$Q = \frac{V}{\alpha k} = \frac{180 \text{ L}}{0.995 \cdot 0.2089 \text{ L Ah}^{-1}} = 866 \text{ Ah}$$

$$I = \frac{dQ}{dt} = \frac{d(V/k)}{dt} = 866 \frac{\text{Ah}}{\text{h}} = 866 \text{ A}$$

b. What is the electrode area required at a given current density?

$$A = \frac{I}{j} = \frac{866 \text{ A}}{0.15 \text{ A cm}^{-2}} = 5773 \text{ cm}^2 \approx 0.6 \text{ m}^2$$

c. What is the electric power of the electrolyzer?

$$P = UI = 120 \text{ V} \cdot 866 \text{ A} \approx 104 \text{ kW}$$

d. How many individual cells must be connected in series to obtain the desired operating voltage?

$$N = \frac{U}{U_1} = \frac{120 \text{ V}}{1.8 \text{ V}} \approx 67$$

e. What is the estimated power density of the electrolyzer, if each individual cell is 5 mm thick?

$$\frac{P}{V} = \frac{104 \text{ kW}}{0.005 \text{ m} \cdot 67 \cdot 0.6 \text{ m}^2} \approx 517 \frac{\text{kW}}{\text{m}^3}$$

f. How can waste heat of the electrolyzer be reduced?

$R = \frac{\varrho d}{A}$ is small if the specific electrolyte resistance ϱ is small (high conductivity), for small electrode distance d (narrow electrolyte gap) and large electrode surface area A.

Solid-Polymer Electrolyte (SPE) Electrolysis

The first SPE or PEM-electrolyzer was built by General Electric in 1966. The 100–400 μm thin proton exchange membrane (PEM), such as Nafion, not only serves as an ionically conductive medium ($\sigma \approx 0.1$ S cm^{-1}, pH ≈ 2), but also as the structural support for the electrodes and as a product gas separator (Figure 106). The low gas crossover rate of the membrane yields hydrogen with high purity. The proton transport across the membrane responds quickly to changes of the power input, in contrast to the permeable diaphragms in liquid electrolytes.

Cold, pure water reacts on the anode producing oxygen gas and protons. Protons are then conducted from the anode side to the cathode side, where they participate in the reduction reaction. When voltage in excess of the thermodynamic voltage of 1.23 V is applied, the electrons flow into the hydrogen generating electrode (Equations 275 and 276).

$$(-)\ 4H^+ + 4e^- \rightarrow 2H_2 \tag{275}$$

$$(+)\ 2H_2O + 4H^+ + 4e^- \rightarrow O_2 \tag{276}$$

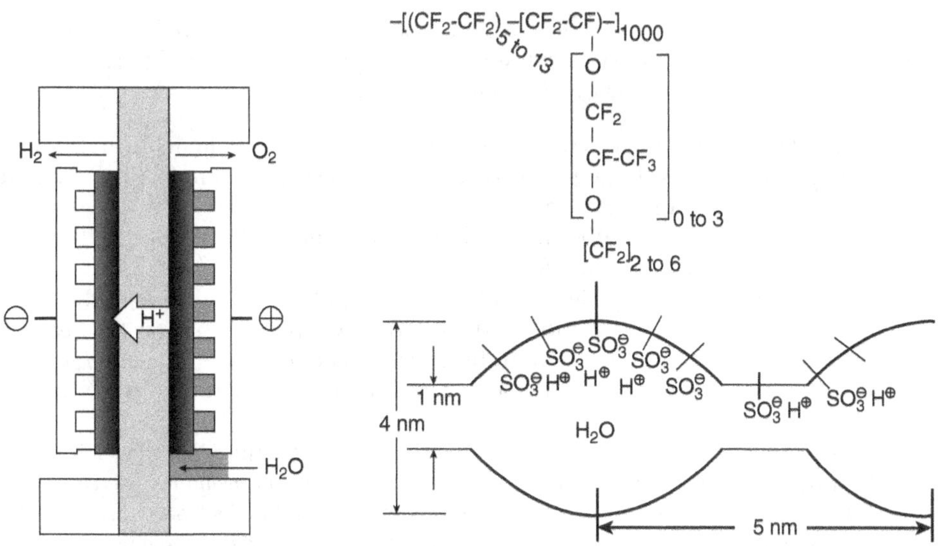

Figure 106 (a) Principle of SPE electrolysis: cathode, membrane, anode, bipolar plate with gas channels. (b) Structure of a solid polymer membrane (Nafion).

Cell Voltage Open circuit voltage can be predicted according to the Nernst equation, modified by the empirical correlation (Awasthi et al., 2011) (Equation 277).

$$E_0 = [1.23 - 9 \cdot 10^{-4} \cdot (T - 298\ \mathrm{K})] - \frac{RT}{2F} \ln \frac{p(\mathrm{H_2})\ p(\mathrm{O_2})^{1/2}}{p(\mathrm{H_2O})} \tag{277}$$

The activation potential for anode ($\alpha = 2$) and cathode ($\alpha = 0.5$) follows the Butler-Volmer equation based on experimental exchange current densities i_0 (Pt-Ir anode: 10^{-12}–10^{-3}; Pt cathode: 0.001–0.1) (Equation 278).

$$\eta_{ct} = \frac{RT}{\alpha F} \operatorname{arcsinh} \frac{i}{i_0} \tag{278}$$

The mass transport overpotential on anode and cathode is based on Fick's law. It depends on the O_2 or H_2 concentration c at the membrane electrode and a reference concentration c_{ref} (Equation 279).

$$\eta_d = \frac{RT}{zF} \ln \frac{c}{c_{ref}} \tag{279}$$

The ohmic voltage drop in the porous electrodes, current collectors, and separators obeys Ohm's law. The empirical conductivity σ of the membrane (thickness d, area A) depends on its water content w (degree of hydration) (Equations 280 and 281).

$$U_\Omega = I \cdot \left[(R_{\mathrm{anode}} + R_{\mathrm{cathode}}) + \left(\frac{d}{A\sigma} \right)_{\mathrm{membrane}} \right] \tag{280}$$

$$\sigma = (0.005139\ w - 0.00326\ ) \cdot e^{1268(303^{-1} - T^{-1})} \tag{281}$$

Practical cell voltages at 80°C are 1.5 V (at 0.2 A cm^{-2}), 1.6–1.7 V (at 1 A cm^{-1}), and 1.7–2.0 V (at 2 A cm^{-2}) using iridium anodes, platinum cathodes, Nafion membranes, and current collectors and bipolar plates of titanium.

Electrode Materials The electrodes in PEM electrolyzers contain noble metals and platinum group metal oxides, typically on carbon support, for example, Pt/C. The catalyst loadings are around 1–2 mg/cm^2.

Hydrogen evolution efficiency and lower overpotentials improve in the following order: Ni < Ru < Os < Re < Ir < Rh < Pt < Pd.

Oxygen evolution efficiency and lower overpotentials improve in the following order: Nb < Au < Pt < PtRh < Rh < Pd < Ru, Ir, RuO_2, IrO_2. Iridium is the best catalyst for the oxygen evolution reaction; however, it is one of the rarest elements in the earth's crust (0.001 mg/kg), in fact so rare that even platinum is 10 times more abundant.

Two electrodes are deposited on either side of the membrane and strictly adhere to the membrane, forming a unit or a structure known as *membrane-electrode assembly* (MEA). The interface between the catalyst particles on the electrode and the extension of the membrane material forms a large surface area available for reactions. Both anode and cathode are porous to allow gases formed at the interface of the electrode catalyst and the membrane to be removed.

Water Diffusion The electrolyzer performance is determined by the effectiveness of water management. The feed water must be purified from heavy metal ions, which come from pipe corrosion. Anodically fed water penetrates the membrane by electro-osmosis and gets to the hydrogen compartment. If dry oxygen is wanted, the cathode is flooded with water, which then diffuses to the anode side and is consumed there.

Migration of solvated protons in the electrolyte to the negative electrode (electroosmosis) is described by the Nernst-Einstein equation (D diffusion coefficient in m^2s^{-1}, u proton mobility in $m^2s^{-1}V^{-1}$, F Faraday constant, R molar gas constant) (Equation 282).

$$u = D\frac{F}{RT} \tag{282}$$

The electroosmosis in Nafion is to 2.6 g H_2O/Ah in 0.4-molar KCl solution at 30 mA cm^{-2}.

Current Collector In a PEM water electrolyzer, hot water is fed at the anode, against the oxygen flow. Water passes through channels in the porous current collector (electrode backing layer) and bipolar separator plates.

Hydrogen leaves the cell across the cathode current collector and cathode separator plate. Current collectors must be electrically conductive, and must have large porosity (20% to 50%) and pore size distribution (5…30 μm). For optimum gas diffusion, current collectors must exhibit low gas permeability (10^{-13}… 10^{-11} m^{-2}) and low electrical resistance (0.005…0.01 Ω cm).

The contact between the nano-size particles in the catalyst layer and micro porous current collector must be smooth in order to avoid hot spots. Thermally sintered titanium powders (50…75 μm) are preferred over stainless-steel grids and titanium grids.

Carbon paper, cloth, and fleece, used as gas diffusion layers (GDL) and current collectors in fuel cells, cannot be applied to PEM electrolysis due to rapid oxidation to CO_2. The cathodic current collector is prone to hydrogen embrittlement in the following order: Ti, Ta > Nb > Zr > graphite; and gold plating is necessary. The current collectors must be firmly pressed against the catalytic layers to obtain good, homogeneous electrical contacts over the surface and prevent leakage current.

Bipolar Plates The separator plates between adjacent cells must have excellent electrical conductivity, have good mass and heat transport properties, provide gas-tightness, and be low cost. When titanium is used, it gradually develops a passive oxide layer that can significantly increase the contact resistance, hence, it must be coated with graphite, platinum, or gold. Graphite comes at a low cost, but has poor mechanical strength and oxidation stability.

Cell Components and Operating Behavior PEM electrolyzers are usually configured in a series connection and bipolar design (Figure 106a), which enables current densities above 2 A cm^{-2}. The solid polymer electrolyte connects the anode of one cell with the cathode of the next. Hydrogen and oxygen gases are collected on the opposite sides of the electrolyzer.

Hydrogen storage and safety are handled by elevated pressures at both electrodes or a *differential pressure* on the hydrogen side ($p(H_2)-p(O_2) > 0$). This way, the handling of pressurized oxygen is avoided and self-ignition of titanium in oxygen prevented. In addition, the dehydration and expansion of the membrane are prevented as well.

The efficiency and cell voltage are minimally improved at high pressure (44 mV per decade), while pressures above 100 bar require thicker membranes with special fillers, used to prevent gas crossover. The catalysts for internal gas recombination are needed as well.

The complete system consists of the following:

1. Power supply
2. Atmospheric oxygen cycle with water tank, pump, heat exchanger, water purification, stack passage, gas separator, aerosol filter, oxygen outlet
3. Pressurized hydrogen cycle with filter, heat exchanger, cooling trap for electro-osmosis water, dryer, and hydrogen tank

The thin membrane enables low cell resistance and five times larger current densities than possible in alkaline electrolyzer technology. The energy consumption per hydrogen volume is approximately the same. The highly pure hydrogen (99.99 %) must contain only trace amounts of oxygen that can penetrate the membrane and enable a recombination reaction $H_2 + ½O_2 \rightarrow H_2O$.

SPE electrolysis responds quickly to load changes, works effectively at very low current densities, and allows temporary overload. Overvoltage caused by gas bubbles is less critical in PEM electrolysis than with alkaline electrolyzers.

The cost of electrolytic hydrogen is still prohibitive at present to make it a widespread, industrially important method for hydrogen production. Roughly 70% of overall cost is due to the cost of electricity, except in cases where there is excess electricity available from hydroelectric power plants or solar- or wind-generated electricity. Therefore, more efficient electrolysis steps and electrolyzer components need to be developed to make hydrogen production from water electrolysis more economically attractive.

Oxygen-Depolarized Cathode Small electrolyzers for medical oxygen production utilize the byproduct hydrogen at the oxygen electrode. Because an internal fuel cell is formed by recombination of oxygen and hydrogen at the cathode catalyst, overall cell voltage drops to 0.7 V (Equations 283 to 286).

(+) Anode: $$2H_2O \rightarrow O_2 + 4H^+ + 4e^- \quad \text{(voltage: > 1.5 V)} \tag{283}$$

(–) Cathode: $$4H^+ + 4e^- \rightarrow 2H_2 \tag{284}$$

Oxygen reduction: $$O_2 + 4H^+ + 4e^- \rightarrow 2H_2O \quad \text{(voltage: 0.8 V)} \tag{285}$$

Recombination: $$2H_2 + O_2 \rightarrow 2H_2O \tag{286}$$

Steam Electrolysis at Solid Oxide Electrolytes

It has been previously explained that water electrolysis works better at high temperature. Solid oxide electrolysis (SOEL) works at ~800°C using oxide-ion conducting ceramic materials such as YSZ (see the section "Electrolytes" above) (Equations 287 to 289).

(–) Cathode: $$H_2O + 2e^- \rightarrow H_2 + O^{2-} \tag{287}$$

(+) Anode: $$O^{2-} \rightarrow ½O_2 + 2e^- \tag{288}$$

Cell reaction: $$H_2O \rightarrow H_2 + ½O_2 \tag{289}$$

The decomposition of steam is less endothermic than the decomposition of liquid water. The production of hydrogen at 100°C by steam electrolysis requires theoretically 243 kJ mol^{-1} = 120 MJ kg^{-1} = 33.3 kWh kg^{-1} or, per volume, 10.8 MJ m^{-3} = 3 kWh m^{-3}. For comparison: aqueous electrolysis at 25°C consumes 286 kJ mol^{-1} = 5 kWh m^{-3} hydrogen.

From the Gibbs' free energy, the theoretical decomposition voltage of steam electrolysis is shown to be 0.92 V at 1000°C and 1.17 V at 100°C. The entropy term lowers the energy input in the amount of $T\Delta S$ = 0.89 kWh m^{-3} hydrogen.

Practical tubular cells operate at 1.05–1.2 V (at 0.2–0.4 A cm^{-2} and 1000°C), which is below the thermoneutral voltage, so that stack works endothermically.

Syngas and Methanation If the electrolyzer is fed with a CO_2-steam mixture, synthesis gas can be produced. CO_2 is split in a side reaction at a nickel catalyst, and methane is formed in the heat exchanger behind the cathode (Equations 290 to 294).

(–) Cathode: $CO_2 + 2e^- \rightarrow CO + 2e^-$ (290)

(+) Anode: $O^{2-} \rightarrow \frac{1}{2}O_2 + 2e^-$ (291)

Cell reaction: $CO_2 \rightarrow CO + \frac{1}{2}O_2$ (850°C, >1.48 V) (292)

Methanation: $3H_2 + CO \rightarrow CH_4 + H_2O$ (400°C) (293)

Sabatier process: $4H_2 + CO_2 \rightarrow CH_4 + 2H_2O$ (–165 kJ mol^{-1}) (294)

Cell Design The solid oxide electrolyte is applied in tubular cells or at bipolar plates (Figure 107). The different thermal expansion of electrode and electrolyte materials poses technical problems, which seriously limit the operational life of an electrolyzer. The fabrication of planar cells is less complicated than that of tubular cells, but sealing is more difficult.

The electrolyzer is fed with overheated steam from purified water (850°C, 50 bar). Residual water in the cathode gas is removed by a membrane. A small amount of 10% hydrogen in the water feed is useful to protect the cathodes against oxidation.

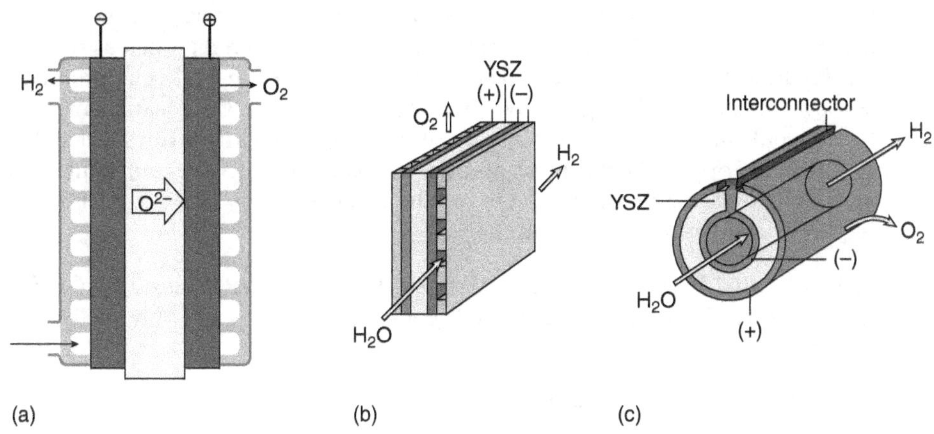

Figure 107 (a) Principle of solid oxide electrolyte electrolysis; (b) planar design; (c) tubular design.

Electrode Materials The hydrogen electrode is a porous nickel-based composite material such as the "ceramic metal" CERMET $Ce_{0.9}Gd_{0.1}O_{1.95}$ (coated on nickel-YSZ), which is fed with a steam-hydrogen mixture. Nickel-free cathode catalysts use lanthanum and strontium perovskites.

The porous oxygen electrode consists of a thin film of mixed electron-oxide conductors, such as strontium-doped lanthanum manganite LSM ($La_{0.6}Sr_{0.4}MnO_3$ on YSZ). Gradual aging is caused by the insufficient mechanical stability and adhesion of the coating.

Two fabrication methods are possible. In the first, the electrolyte (100–200 μm) supports the thin electrodes (10–40 μm). In this case the oxygen overvoltage determines the cell voltage. The second method involves a thin layer of electrolyte (10–15 μm) coated on a thick cathode (<1 mm), which then controls the cell voltage.

CERMET on heat-resistant steel is used as an interconnector between the bipolar cells.

Operating Characteristics Efficiency is usually given in relation to the heating value of hydrogen. For steam generation, SOE electrolysis can take advantage of the waste heat of solar thermal or geothermal systems. Thermoneutral mode of operation works at the same temperature of steam inlet and gas outlet, which is provided by a high-temperature heating. Autothermal operation needs only a water evaporator.

There are only a few commercial providers worldwide for this technology that appears to be still not ready for large-scale deployments.

Sustainability of Electrolysis

At present, 1 kg hydrogen produced by electrolysis costs roughly 10 times more than conventional hydrogen from steam reforming of natural gas. However, from the perspective of raw material shortage, greenhouse gas emissions, ocean acidification (caused by the uptake of CO_2 and SO_2 from the atmosphere), and finally the social costs of unsustainable processes, pricing assessment speaks in favor of electrolysis. Electrolysis makes sense if nonfossil fuel energy sources are used for electricity generation. In this scenario, oxygen should be considered a useful product and used as well.

Power to Gas Electrolytic hydrogen utilizes surplus wind and solar energy. Five percent hydrogen can be fed into the existing town gas networks as gas-heating systems tolerate up to 20% hydrogen. However, the intermediate storage of energy from the general electricity grid in the form of hydrogen has not been regarded as sustainable so far. Hydrogen storage and reconversion into electricity in fuel cells promise at best 30–40% efficiency. Utilizing 20 GW wind power (20% of the European plants) in electrolyzers would also require large amounts of iridium, which might run short in the future.

Another option is the methanation of electrolytic hydrogen, which proceeds according to the exothermic Sabatier reaction: $2\,H_2 + CO_2 \rightarrow CH_4 + 2\,H_2O$. This process requires considerable investment costs.

Carbon-Assisted Electrolysis In the 1980s, electrolysis at coal electrodes in sulfuric acid was proposed. This process theoretically consumes the energy of only 40 kJ/mol hydrogen (Equations 295 to 297).

(−) Cathode: $$2\,H^+ + 2\,e^- \rightarrow H_2 \qquad (E^0 = 0\ V) \qquad (295)$$

(+) Anode: $$C + 2H_2O \rightarrow CO_2 + 4\,H^+ + 4\,e^- \qquad (E^0 = 0.21\ V) \qquad (296)$$

Total reaction: $$C + 2\,H_2O \rightarrow CO_2 + 2\,H_2 \qquad (\Delta E^0 = 0.21\ V) \qquad (297)$$

Practical operating voltages of 0.8–1.0 V were reported. However, the composition of the carbon material is complicated, because the anodic oxidation of long carbon chains and benzene rings is difficult.

Pt–Ir (80:20) and titanium-supported IrO_2 (with Fe(III) liquid catalyst) were found to be better electrode materials for coal oxidation. The selective generation of hydrocarbons instead of CO_2 is still a problem. Solid oxide electrolysis above 600°C can use the reaction heat of coal steam reforming ($C + 2\,H_2O \rightarrow CO_2 + 2\,H_2$) for the anode reaction: $C + O^{2-} \rightarrow CO_2 + 4\,e^-$.

Further Reading

- A. Awasthi, K. Scott, S. Basu, Dynamic modeling and simulation of a proton exchange membrane electrolyzer for hydrogen production, *International Journal of Hydrogen Energy* 36 (2011) 14779–14786.
- M. Carmo, D. L. Fritz, J. Mergel, D. Stolten, A comprehensive review on PEM water electrolysis, *International Journal of Hydrogen Energy* 38 (2013) 4901–4934.
- T. Smolinka, J. Garche (eds.), *Hydrogen Production by Water Electrolysis*, Elsevier, 2021.
- M. Wang, Z. Wang, X. Gong, Z. Guo, The intensification technologies to water electrolysis for hydrogen production—A review, *Renewable and Sustainable Energy Reviews* 29 (2014) 573–588.

CHAPTER 10

Fuel Cells

A fuel cell is an electrochemical device that receives a continuous supply of reactants—fuel and oxidant—to produce power. Important fuel cells utilize the cold combustion of hydrogen and oxygen by electrochemical reactions and can be considered to be oxyhydrogen batteries. Fuel cells are not considered "rechargeable."

The first fuel cell was developed in 1839 by William Robert Grove, 20 years before Gaston Planté built the first lead-acid battery.

Fuel cells utilize polymer, liquid, solid, or molten electrolytes, which determines the abbreviation of the respective technology (Figure 108). Different ionic charge carriers transfer electricity between the electrodes depending on the type of electrolyte. Because the various electrolytes become conductive at different temperatures, the corresponding fuel cells must operate above those temperatures.

Batteries deliver a limited and predetermined amount of electricity based on the finite quantity of reactants in their enclosed casing, while the fuel cells operate as long as fuel and oxidant are supplied from the external sources. Usually, fuel cells utilize oxygen from air and have only the fuel for the anode supplied.

Fuel cells have been employed in trucks, trains/locomotives, motorcycles, boats, submarines, racing cars, and airplanes.

Polymer Electrolyte Fuel Cell (PEFC)

The most important fuel cell technology is based on a polymer electrolyte membrane.

A thin perfluorinated ionomer polymer membrane (proton exchange membrane, PEM) conducts protons from the anode to the cathode. The fuel, usually pure hydrogen, is oxidized at the anode (negative electrode) while oxygen from air is reduced at the cathode (positive electrode). The electrode reactions are exactly the reverse of the electrolysis of water (Equations 298 and 299). The carbon electrodes are coated with finely dispersed, carbon-supported platinum and are fed with humidified gases. Theoretical cell voltage for a fuel cell reaction is 1.23 V, but in practice the OCV is about 0.8 V at 70–85°C.

$$(-) \qquad 2\,H_2 \rightarrow 4\,H^+ + 4\,e^- \tag{298}$$

$$(+) \qquad O_2 + 4\,H^+ + 4\,e^- \rightarrow 2\,H_2O \tag{299}$$

Advantages Because hydrogen has the highest specific energy of all substances (33 kWh/kg), three times higher than gasoline, the fuel cell coupled with hydrogen storage is considered an ideal energy storage and conversion system.

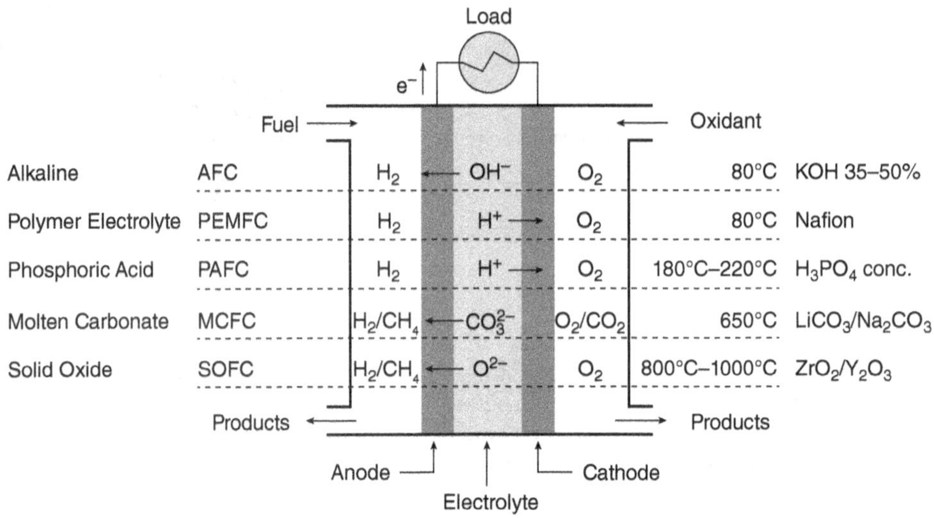

FIGURE 108 Cross-section of a fuel cell with ionic charge carriers, temperatures of operation, and electrolyte.

The compact membrane-electrode assembly, which can be stacked using bipolar plates, enables high specific power of up to 3 kW kg^{-1}. Electric efficiency of 50–68% per cell is obtained when hydrogen is used as a fuel and 43–50% when the fuel is natural gas.

Porous membranes of polybenzimidazole containing phosphoric acid within the pores allow operating temperatures up to 220°C, although electrochemical activity is lower than based on today's state-of-the-art proton conducting membranes. These fuel cells have high efficiency, simple cooling, and reduced sensitivity to dry operating conditions and carbon monoxide in the fuel.

Challenges Hydrogen can be effectively oxidized at any temperature, but it must be completely pure and free from traces of carbon monoxide, which comes from the steam-methane reforming process. Traces of CO (>5 ppm) in the hydrogen feed poison the platinum catalyst.

The membrane is sensitive to drying and freezing.

The PEM technology is still expensive. As hydrogen is not freely available in nature, it must be produced from another source in an endothermic process. Therefore, hydrogen is not a source of energy; it is an energy storage medium. The performance limitations regarding the presence of CO can be avoided if hydrogen is produced by water electrolysis.

System Components and Performance Data

Fuel cells deliver voltages of 0.6–0.7 V and current densities of 300–1000 mA cm^{-2}. Specific power varies depending on the cell geometry.

Performance degradation is several millivolts per 1000 hours. Stationary fuel cells reach efficiencies of 20–50%.

Cost target is about 35 US$/kW for automotive application and about 1000 US$/kW for stationary applications.

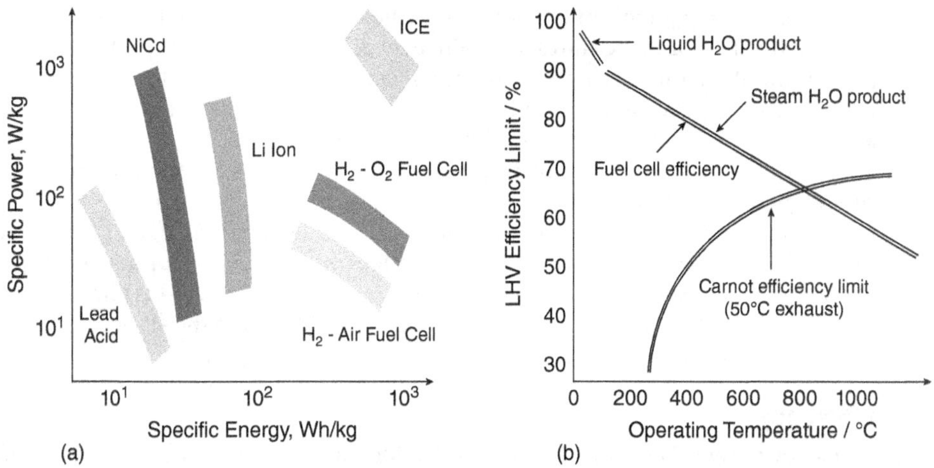

FIGURE 109 (a) Specific power against specific energy for selected energy storage and conversion technologies. (b) Efficiency of fuel cells and heat engines.

The fuel cell attributes are often represented in a graph of specific energy against specific power (Figure 109). It is obvious that fuel cells have higher specific energy than batteries as a result of the high specific energy of hydrogen and equal or better specific energy compared to gasoline. However, fuel cells are inferior in terms of specific power due to slow rates of reaction, especially on the oxygen cathode.

A practical fuel cell power plant requires four subsystems:

1. The fuel cell stack contains a series combination of individual fuel cells that generates direct current output of several hundred volts.
2. The fuel supply includes flow controls and fuel processing, if necessary. Ancillary subsystems comprise oxidant processing, thermal, and fluid management. Hydrogen is typically flown into the anodic compartment with no outlet (i.e., "dead end") at a modest pressure (10 to 50 kPa), which allows the anode to sustain the electrochemical reaction at any given rate. At selected time intervals, accumulated impurities and water are purged from the anode compartment. The unused fuel (10–40%) is combusted to generate heat needed for fuel processing. Static air supply by diffused-air is restricted to simplified low-power applications. In forced-air fuel cells, air is pumped or blown to the cathodes against the internal pressure drop (1–20 kPa), whereby 40–50% of the supplied oxygen is consumed in the cell. The exit air is generally discharged to the atmosphere.
3. The **power conditioner** converts the DC output to AC or a different type of power and quality required by the application.
4. A **control subsystem** monitors the operating parameters such as temperatures, mass flows, power conditioning, and the cooperation of the subsystems. The reactant-air flow rate and temperature must be controlled as a function of load to prevent water buildup and cell dry-out. Closed anode compartments without outlet are purged using a timer or coulometer, which opens and closes a solenoid valve in the exit line.

Water management must provide sufficient humidity in the reactant streams because the membrane requires a certain level of water content for proton conductivity. Non-humidified ambient air is used only in simple, small fuel cells. A limited flow rate of air is used to prevent drying but has to remove product water from the cathode even at high oxygen utilization rates at high current densities.

Water droplets in the air distribution channels of the cathode flow-field result in serious performance losses.

The **thermal management** requirements are associated with water management, so that individual cells are cooled to guarantee moderate and uniform temperature throughout the stack. Liquid water cooling in larger PEM fuel cells is effective while for small fuel cell applications forced air cooling is used through the stack or across finned extensions of the bipolar plates.

Efficiency

Efficiency of fuel cells is an often misunderstood and misquoted category. The simplest definition is based on the ratio of practical cell voltage U_0 and thermodynamic cell voltage ΔE (Equation 300a).

$$\eta_{\text{th}} = \left| \frac{U_0}{\Delta E^0} \right| < \frac{\Delta G}{\Delta H} \tag{300a}$$

The maximum possible, purely thermodynamic efficiency is the ratio of the Gibbs free energy change and enthalpy change for the reaction. This theoretical efficiency only shows how much energy can be obtained from a fuel cell reaction compared to the combustion of the fuel (hydrogen) in heat engines.

Fuel efficiency is the ratio of fuel that reacted in a fuel cell and total fuel volume flow that entered the cell (Equation 300b). $Q = \mathrm{d}V/\mathrm{d}t$ is the volume flow.

$$\mu_{\text{F}} = \frac{Q}{Q_{\text{in}}} \tag{300b}$$

Finally, the most complete efficiency expression includes fuel efficiency and practical voltage efficiency (Equation 301).

$$\eta_{\text{total}} = \mu_{\text{F}} \cdot \frac{U}{U_0} \tag{301}$$

Calculation Examples

1. Determine the theoretical efficiency limit for a CH_3OH/O_2 fuel cell.

$$\eta_{\text{th}} = \frac{\Delta G}{\Delta H} = \frac{(-702.0 \text{ kJ mol}^{-1})}{(-726.9 \text{ kJ mol}^{-1})} \approx 97\%$$

2. What is the theoretical cell voltage for a hydrogen fuel cell operating at 200°C if free reaction energy at that temperature is –220.4 kJ/mol?

$$E = -\frac{\Delta G}{z\,F} = \frac{220400 \text{ J mol}^{-1}}{2 \cdot 96485 \text{ C mol}^{-1}} \approx 1.14 \text{ V}$$

3. Compare fuel cell efficiency and Carnot efficiency in Figure 109.

   In heat engines, the theoretical maximum efficiency according to Carnot is $(T_1 - T_2)/T_1$, where T_1 is the maximum temperature of the heat engine and T_2 is the temperature at which the heated fluid is released. The fuel cell efficiency obviously has no such limitation.

4. Think about the entropy term $T\ \Delta S$ of the fuel cell reaction with respect to efficiency.

   To claim higher efficiency for a fuel cell compared to wind generators or heat engines is not always correct. The fuel cell reaction, $2\ H_2(g) + O_2(g) \rightarrow 2\ H_2O(l)$, has negative entropy, $\Delta S < 0$, hence Gibbs free energy change decreases with temperature, $|\Delta G| = |\Delta H| - T\ |\Delta S|$, and efficiency $\Delta G/\Delta H$ drops at high temperatures. Fuel cells should therefore be operated at room temperature.

Current-Voltage Characteristics

Operational fuel cell voltages are affected by kinetic limitations, that is, the losses due to the electrochemical processes at the three-phase boundary between electrode, electrolyte, and gas. As in case of all other electrochemical systems, fuel cells are prone to three main sources of voltage loss, which reduce the open-circuit voltage U_0 to values between 0.5 and 0.9 V. These sources of losses or overpotential are as follows:

1. Ohmic drop in the electrolyte, $I \cdot R_e$, which does not belong to the overpotential.
2. Charge transfer at both electrodes R_{ct}, also called activation polarization, which follows a Tafel line, $\eta_{ct} = a + b \ln I$ (where a, b constants); see the section "Charge Transfer Kinetics without Diffusion" in Chapter 9.
3. Mass transport at both electrodes (diffusion impedance R_d), also called concentration polarization, $\eta_d = \frac{RT}{zF} \ln\left(1 - \frac{I}{I_{\lim}}\right)$, which limits the highest possible current.

The higher the current or current density drawn from a fuel cell, the larger are the voltage losses due to the internal resistance R or overpotentials at both electrodes η (Figure 110 and Equation 302):

$$U(I) = U_0 - I\ R = U_0 - IR_e - \eta \tag{302}$$

where $R = R_e + R_{ct} + R_d$ and $\eta = \eta_{\text{anode}} + \eta_{\text{cathode}}$.

Automotive Applications

In 1959, Allis-Chalmers powered a tractor with a 15 kW-alkaline fuel cell.

The Apollo and Gemini space missions with alkaline cells and PEM cells awakened the interest of the automakers.

The first PEMFC vehicle was presented in 1993 by *Energy Partners Inc.*, which consisted of three 15 kW PEMFCs powering a pick-up truck over a 90 km range. *Daimler-Benz* started their fuel cell development in the early 1990s. Different car manufacturers released demonstration vehicles and busses. Hyundai, Toyota, and Honda have presented commercial fuel-cell cars.

In contrast to classical batteries, fuel cells can be designed individually for a given stack power and fuel tank capacity. For long-distance travel, electric vehicles require a large battery, while a fuel cell can be easily refueled. The linear relationship between

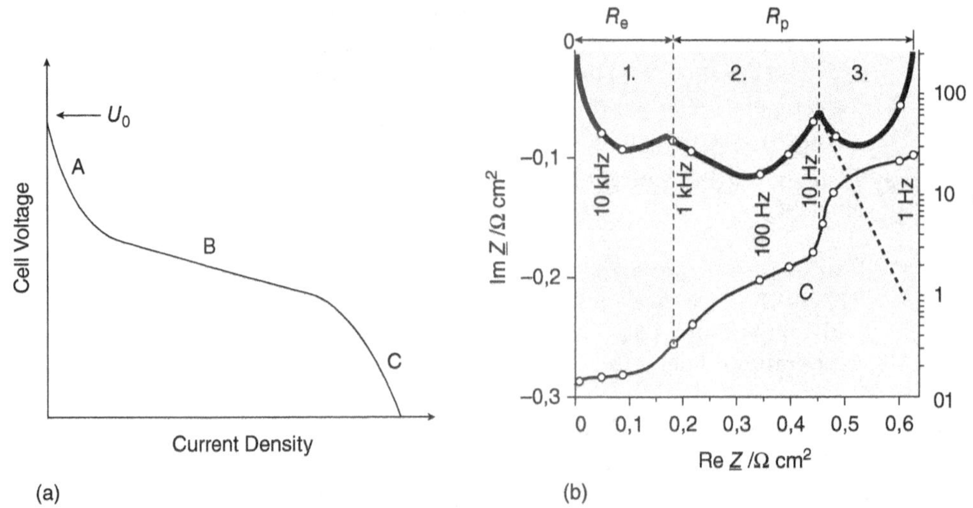

FIGURE 110 (a) General trend in fuel cell voltage due to internal losses: A, activation; B, ohmic drop; C, mass-transport-controlled region. U_0 is open-circuit voltage, theoretically 1.23 V. (b) Impedance spectrum of a PEM fuel cell in the complex plane: 1, membrane; 2, charge transfer; 3, mass transport or diffusion impedance. *C* is apparent capacitance of the three-phase boundary.

range d and mass m shows a steep slope b for the battery ($d = b\,m + m_0$), while a less steep slope for a fuel cell line starts at a higher system mass m_0. This means that batteries have an advantage at short driving ranges and fuel cells at longer driving ranges.

Direct Methanol Fuel Cell (DMFC)

Besides hydrogen as a fuel, fuel cells can also use hydrocarbons such as liquid or gaseous methanol. The fuel cell using methanol as a fuel has a theoretical cell voltage comparable to a fuel cell using hydrogen 1.186 V. In practice, cell voltage for a methanol fuel cell is about 0.5 V (Equations 303 and 304).

$$(-) \qquad CH_3OH + H_2O \rightarrow CO_2 + 6\,H^+ + 6\,e^- \qquad (303)$$

$$(+) \qquad \frac{3}{2}\,O_2 + 6\,H^+ + 6\,e^- \rightarrow 3\,H_2O \qquad (304)$$

Advantages No fuel processing is required because an aqueous solution of liquid methanol is directly oxidized at the anode.

Subkilowatt applications such as auxiliary power generators have been demonstrated.

Challenges Methanol can penetrate the membrane by diffusion and electroosmosis and is oxidized at the anode.

Present catalysts are relatively inactive.

Power and efficiency per cell (20–30%) are low.

Alkaline Fuel Cell (AFC)

In an alkaline fuel cell, a 30% potassium hydroxide solution serves as an electrolyte between two nickel electrodes, which are fed with hydrogen and oxygen.

The electrode reactions are standard for alkaline medium (Equations 305 and 306).

$$(-) \qquad H_2 + 2\,OH^- \rightarrow 2\,H_2O + 2e^- \tag{305}$$

$$(+) \qquad \frac{1}{2}O_2 + H_2O + 2\,e^- \rightarrow 2\,OH^- \tag{306}$$

Advantages

1. Low-cost catalysts
2. Fast oxygen reduction kinetics
3. Low operating temperature (70°C)
4. Up to 60% efficiency

Challenges

1. Pure oxygen is required in order to prevent formation of K_2CO_3 in the electrolyte, so CO_2 must be excluded.
2. Corrosion at the electrodes and in the gas spaces is caused by the alkali hydroxide.

Alkaline cells were used in 1966 by GM in a Chevrolet Electrovan. In 1967, Karl Kordesch built a hybrid AFC/Ni-Cd motorcycle with hydrazine as the fuel. In 1970, Union Carbide demonstrated a hybrid car with a 6 kW AFC and lead-acid battery running for a 200 km range. In Japan in 1972, Panasonic and Daihatsu manufactured a pick-up truck with a 5.2 kW hydrazine-air AFC, consuming 1 L of hydrazine for 80 km. In 1998, Zero Emission Vehicle Company (ZEVCO) launched a taxi with 5 kW AFC in London. Although the AFC is relatively cheap because it uses nonplatinum catalysts, the difficult air supply (CO_2 removal) has hindered the commercialization.

Phosphoric Acid Fuel Cell (PAFC)

PAFCs use hot phosphoric acid (190–230°C) as electrolyte. The electrodes use carbon support for a platinum catalyst. This fuel cell can use pure hydrogen or hydrogen from steam reforming of natural gas. The cathode uses oxygen from air. The hydrogen oxidation and oxygen reduction proceed in acidic medium (Equations 307 and 308).

$$(-) \qquad 2\,H_2 \rightarrow 4\,H^+ + 4e^- \tag{307}$$

$$(+) \qquad O_2 + 4\,H^+ + 4\,e^- \rightarrow 2\,H_2O \tag{308}$$

Advantages

1. Prototypes in the MW range have demonstrated the cogeneration of power and heat.
2. System efficiency is up to 85%, 40% as electricity and another 45% if the waste heat is used for building heating.

Challenges It has a moderate electrode life and relatively low electrolyte conductivity.

Molten Carbonate Fuel Cell (MCFC)

This high-temperature technology uses molten alkali carbonates in a heat-resistant matrix at about 630°C. Cell voltages of 0.75 V were demonstrated while the theoretical voltage is 1.04 V. Electrode reactions proceed in a molten salt electrolyte (Equations 309 and 310).

$$(-) \qquad H_2 + [CO_3]^{2-} \rightarrow H_2O + CO_2 + 2\,e^- \qquad (309)$$

$$(+) \qquad O_2 + 2\,CO_2 + 4\,e^- \rightarrow 2\,[CO_3]^{2-} \qquad (310)$$

Advantages

1. Low-cost electrodes are based on nickel-chromium and oxidized sinter nickel.
2. CO tolerant hydrogen generation is possible by internal reforming near the anode.
3. The system allows the cogeneration of electricity and heat.
4. Coal-based or marine diesel fuels can be used in continuously operating facilities.

Challenges A few stationary applications have been realized. CO_2 cycling is required, and short circuits caused by electrode corrosion and the sensitivity to sulfur are critical items.

Solid Oxide Fuel Cell (SOFC)

A plane or tubular solid electrolyte of yttria-stabilized zirconia, ZrO_2/Y_2O_3 (YSZ), which conducts oxide ions is used as electrolyte in solid oxide fuel cells. The solid electrolyte is coated with nickel and doped perovskites such as $LaSrMnO_3$. The electrode reactions show that oxygen ions conduct charge in the electrolyte (Equations 311 and 312).

$$(-) \qquad 2\,H_2 + O^{2-} \rightarrow 2\,H_2O + 4\,e^- \qquad (311)$$

$$(+) \qquad O_2 + 4\,e^- \rightarrow 2\,O^{2-} \qquad (312)$$

Advantages

1. Cell voltage 0.88–0.93 V (in air and pure oxygen).
2. CO tolerant fuel cell as internal reforming of natural gas takes place directly at the fuel electrode.
3. High electric efficiency of 60–65% (cell) and 55–65% (natural gas system).
4. Despite the slow start-up, once running, SOFCs provide high-grade waste heat, which can be used to heat buildings.

Challenges

1. Temperature stability of the materials.
2. Electric interconnection between adjacent cells by ceramic conductors.
3. Only a few stationary plants have been realized.

SOFC has been considered for auxiliary power generators (APU), which run on diesel or gasoline. The efficiency of belt-driven generators ranges from 50% to 30% (at maximum RPM) and 80% with an integrated starter-generator (ISG).

To generate electricity, a fuel consumption is 1.5–3 L per 100 km. Truck engines consume about 3 L of fuel per hour, or 30 kW thermal power input, when idling. A fuel cell might provide at least 40% efficiency.

Regenerative Fuel Cell Systems (RFCSs)

A regenerative or reversible fuel cell (RFC) is a combination of an electrolyzer and a fuel cell. In some RFCSs the same stack works both as a fuel cell and as an electrolyzer—discrete RFC (DRFC); however, the performance is lower compared to other fuel cell systems.

The bulkier and more complex RFCS uses two separate stacks, each designed for one function which enables better performance—unitized RFC (URFC) systems. In the charging mode, a water electrolyzer generates hydrogen. In the discharge mode, a fuel cell converts hydrogen and oxygen from air to water and electricity.

To achieve quick start, rapid response to load changes, and low cost, PEM and alkaline systems have been preferred.

RFCSs promise high energy density exceeding conventional battery systems.

Oxygen Electrode Water management in a single RFCS causes a special problem at the oxygen electrode because the hydrophilic electrolyzer electrode (anode) is flooded with water, whereas the hydrophobic PEM fuel cell electrode (cathode) must remove liquid product water.

Hydrogen Electrode The oxygen catalyst in the electrolyzer anode must function as well as an oxygen reduction catalyst in the fuel cell. The electrolyzer cathode must be able to manage hydrogen oxidation.

The polarity changes are a serious problem especially for the oxygen electrode because of the large potential differences between electrolyzer and fuel cell.

Bifunctional oxygen electrodes are designed to manage both oxygen evolution and oxygen reduction at high current densities.

1. A single-layer electrode might be easy to manufacture, but stable bifunctional catalysts and support materials are required.
2. Two separate thin-film electrodes, separated by the electrolyte, are seldom used because of the increased gas diffusion path and reduced performance.
3. Single electrodes may carry multiple catalytic layers, which combine one hydrophilic layer for oxygen evolution at electrolyte interface and a hydrophobic layer for the oxygen reduction at phase boundary toward the air.
4. Single electrodes with segmented areas (like on a chessboard) with different catalysts have been proposed, but they would likely have low catalytic activity.

Catalysts Usually, carbon-supported platinum is used at the negative electrode (cathode) for hydrogen evolution and oxygen reduction. The hydrophobic consumption layer near the gas compartment has a higher porosity than the hydrophilic gas evolution layer adjacent to the electrolyte. A slight pressure to the gas side, which breaks the electrolytic contact, prevents the oxidative attack on the oxygen reduction layer.

Pt-Ir mixtures have been used as an anodic catalyst for oxygen evolution and hydrogen oxidation. In acid or PEM systems, noble metals are required to provide fast oxygen reaction kinetics: Pt black is better than alloys and composites.

The order of effectiveness is the following: PtIr > Pt/RuO_2 > PtRu ≈ PtRuIr > Pt/IrO_2, for oxygen reduction. For oxygen evolution the order is as follows: PtIr (99:1) and Pt/IrO_2 are better than PtRu > PtRuIr > Pt/RuO_2 and Pt black. Iridium and its oxide improve the oxygen evolution but slow down oxygen reduction.

Bifunctional catalysts in alkaline solutions may consist of transition metal oxides such as perovskites, pyrochlores, spinels, and mixtures thereof.

Support The highly active oxygen produced during electrolysis corrodes the catalyst and the support material, which is typically carbon black and graphitized carbon cloth or paper as the gas diffusion layer (Equation 313).

$$\text{Carbon corrosion:} \quad C + 2\,H_2O \rightarrow CO_2 + 4e^- + 4H^+ \tag{313}$$

Sintering and dissolution of the active layer increase the electrical contact resistance between particles and current collector and destroy the hydrophobicity for water removal. Metallic substrates, borides, carbides, and nitrides (TiC, TiCN, TiN) have been investigated. Porous titanium is stable as an anodic gas diffusion electrode in fuel cell mode. URFC catalysts with reduced platinum content have been coated on Nafion membranes modified by polypyrrole, which offers electronic conduction pathways for the Pt(II)/Pt redox couple and enhances the grain-boundary and contact resistance.

Cracking and dissolution of the binder (Nafion, PVDF, PTFE), which fixes the catalyst on the membrane, is an additional problem.

Cycling Each transition between the electrolyzer and the fuel cell mode involves a large potential change of the anodic and cathodic reactions at the same electrode. To avoid this problem, a special operation mode uses one electrode always as the anode (oxygen evolution at charge and hydrogen reduction at discharge), and the other electrode is always the cathode (hydrogen evolution and oxygen reduction). The catalysts are optimized separately for anodic and cathodic reactions. When changing from charge to discharge mode, the electrode compartments are flushed with nitrogen.

Membrane Fuel cells working at normal pressure require a PEM membrane less than 50 μm thick while electrolyzers working at high pressures require a membrane thickness of more than 150 μm to avoid gas crossover.

Performance Data The problems with the single PEM-RFC systems have so far restricted the use of this technology to space applications, where high specific energy is required.

PEM reversible cells comprise (i) a solid polymer electrolyte membrane, (ii) anodic and cathodic electrocatalytic layers, (iii) microporous and gas diffusion layers, and (iv) separating bipolar plates. Theoretical energy densities reach 400 to 1000 Wh/kg and 3660 Wh/L.

Using bifunctional Pt/C and Pt-Ir catalysts, cell voltages of 1.6 (electrolysis) and 0.7 V (fuel cell) at 0.5 A cm^{-2} have been obtained. In Japan, unitized, i.e., single, RFCSs for space applications have been realized in the 100 W range (12–24 V, max. 0.5–1 A cm^{-1}, <20 cells) based on PEM technology.

Stability and durability of URFC cells are severely affected by various degradation factors related to the electrocatalysts, Nafion binder, polymer membrane, severe carbon corrosion, gas diffusion packing, and bipolar plates (graphite, titanium).

Further Reading

- K. Kordesch, G. Simader, *Fuel Cells and Their Applications*, Wiley-VCH, Weinheim, 2001.
- T. Sadhasivam, K. Dhanabalan, S.-H. Roh, T.-H. Kim, K.-W. Park, S. Jung, M. D. Kurkuri, H.-Y. Jung, A comprehensive review on unitized regenerative fuel cells: Crucial challenges and developments, *International Journal of Hydrogen Energy* 42 (2017) 4415–4433.
- W. Vielstich, A. Lamm, H. A. Gasteiger, *Handbook of Fuel Cells: Fundamentals, Technology, Applications*, Wiley, Chichester, 2003.

PART 3

Beyond Batteries and Fuel Cells

CHAPTER 11

Supercapacitors

As a link between classical capacitors and batteries, so-called double-layer capacitors or supercapacitors can be charged and discharged very quickly. Potential application areas for supercapacitors include power boosters and regenerative braking recovery in hybrid electric vehicles. The activated carbon technology is commercially available from a number of companies worldwide, whereas metal oxides and conducting polymers play no role at present.

Carbon Technology

Double-layer capacitors utilize the electric capacitance at the phase boundary between two carbon electrodes in a liquid organic electrolyte (Figure 111).

Double-layer Capacitance and Pseudocapacitance The electrochemical double layer behaves like a plate capacitor. However, adsorbed water molecules and desolvated counterions do not form an ideal capacitor. Commercial double-layer capacitors are based on the high surface area of carbon materials, where ion adsorption and battery-like redox reactions (Faradaic capacitance) take place simultaneously. The equations for the capacitance C (in Farads) and the energy W of a plate-capacitor can be applied to supercapacitors as well (Equation 314).

$$C = \frac{\varepsilon_0 \varepsilon_r A}{d}, \text{ and } W = \frac{CU^2}{2} \tag{314}$$

In Equation 314, d is the thickness of the double-layer (electrode) or the distance between the electrodes (cell), A is the real electrode cross-sectional area, and ε is the apparent permittivity of the electrode-electrolyte interface.

Capacitance and internal resistance depend on frequency, voltage, and temperature. In practice, a pseudocapacitance $C(\omega,U,T)$ is measured instead of a constant capacitance (Equation 315):

$$C(U) = \frac{\mathrm{d}Q}{\mathrm{d}U}$$

and

$$C(\omega) = \frac{\mathrm{Im}\ \underline{Y}(\omega)}{\omega} = \frac{-\mathrm{Im}\ \underline{Z}(\omega)}{\omega \cdot \left|(\mathrm{Im}\ \underline{Z}(\omega))^2 + (\mathrm{Im}\ \underline{Z}(\omega))^2\right|} \tag{315}$$

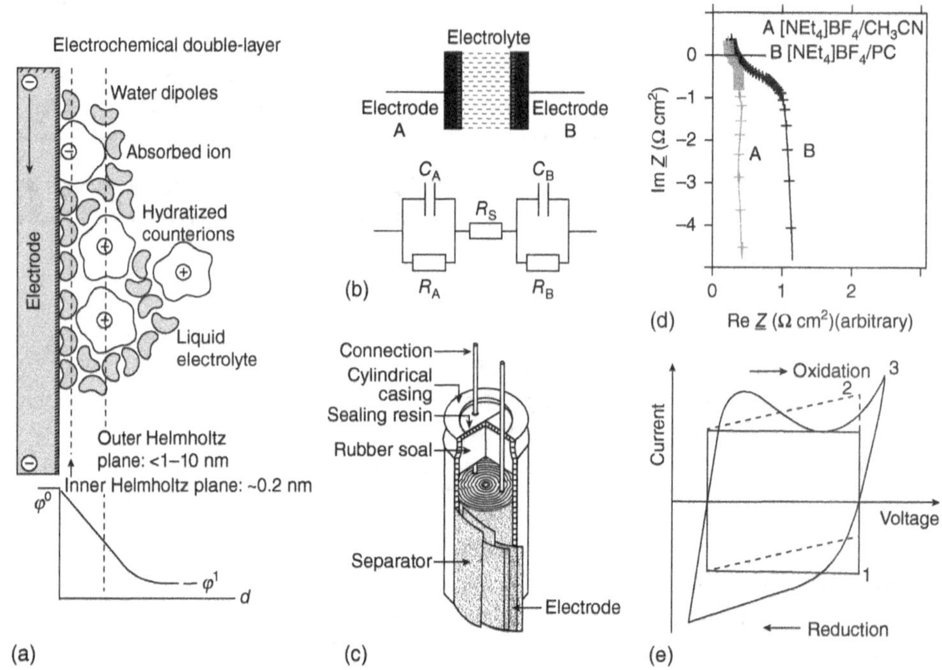

Figure 111 Double-layer capacitor. (a) Electrode-electrolyte interface. (b) Simple equivalent circuit. (c) Spiral-wound design. (d) Impedance spectrum of activated carbon electrodes in different organic solutions. (e) Cyclic voltammogram: 1, ideal capacitor; 2, capacitor with internal resistance; 3, carbon electrodes with pseudocapacitance.

where Q is the electric charge, U is the applied voltage, Z is the interfacial impedance (Re is the real part or resistance, Im is the imaginary part or reactance), Y is the admittance, and ω is the circular frequency.

State-of-the-Art Double-layer capacitors have high power capability during pulse discharge of up to 10 kW kg^{-1} and moderate energy density of typically 5 Wh kg^{-1}. Their cell voltage is 2.3–3 V for organic electrolyte cells, capacitance per unit cell is from 10 to 5000 F, internal resistance (ESR) achieves of a few milliohms, and peak power is typically 8 to 80 kW kg^{-1}.

Full discharge and deep discharge are possible and do not affect the capacitors negatively. Coulombic efficiency is above 99%, while cycle life is more than one million charge-discharge events. Calendar life is estimated at more than 10 years.

Materials Supercapacitors mainly use stable complex salts in organic solvents as electrolyte. For example, tetraethylammonium tetrafluoroborate in acetonitrile allows a rated voltage up to 2.7 V. Aqueous electrolytes have not found technical widespread use because of the small voltage window. Electrodes are made of activated carbon bonded with fluoropolymers. Additives such as carbon black on aluminum support are used. Commercial electrodes are made by spiral-wounding.

Carbonaceous fibers, metal oxides, and conducting polymers, as well as aqueous solutions and solid electrolytes, have limited practical use at present.

The separator is made of nonwoven fabrics of microporous polyolefin or stretched films of polytetrafluoroethylene (PTFE) modified by a hydrophilic treatment. Polyethylene-polypropylene (PE-PP) three-layer separators and expanded PTFE have been produced for several years.

Advantages Nanoporous carbon provides a high surface area (up to 2000 $m^2 g^{-1}$), and it has tailored pore geometry and pore size distribution, good wettability, and good conductivity. Assuming an average capacitance value of 15–25 $\mu F\ cm^{-2}$ and a specific surface area of 1000 $m^2\ g^{-1}$ for activated carbons, the achievable capacitance is 150–250 $F\ g^{-1}$. Typical capacitance values obtained are around 0.15 $F\ cm^{-2}$ (100–180 $F\ g^{-1}$, 70–130 $F\ cm^{-3}$) in solutions based on acetonitrile or propylene carbonate. Novel graphene materials provide far better value. The main advantage of supercapacitors is their power capability and rapid response time in kW range.

Double-layer capacitors are very safe because the electrolyte volume is small and any failures get resolved without an electrical short circuit.

Challenges Compared with batteries, the energy density of double-layer capacitors is small. The cell voltage is limited by the decomposition voltage of the electrolyte (<3 V) or even lower for aqueous systems.

Supercapacitors age by the gradual destruction of the polymer-bonded electrode and accidental electrolysis which may be a result of the application of incorrect operating voltage and high temperature, whereby the resistance increases and capacitance drops.

Hybrid Systems

Supercapacitor electrodes can be combined with battery electrodes to make a hybrid system with improved power capability.

Lead-Acid Capacitor Hybrid Electric propulsion applications require high currents from the battery. The battery/supercapacitor hybrid technology uses a combination of PbO_2 for the positive electrode and spongy lead, and carbon for the negative electrode. The result of this UltraBattery is improved cycle life and dynamic performance of lead-acid battery under partial state-of-charge as required in hybrid electric vehicles.

The hybrid devices are sensitive to deep discharges, which seriously limit their lifetime.

Lithium-Ion Capacitor The hybrid cell comprises an activated carbon positive electrode and a carbonaceous lithium-intercalation negative electrode. The capacitance of the hybrid capacitor is determined by the active mass m of the positive electrode, $C = C_+ m_+/(m_+ + m_-)$, because the lithium electrode has a higher capacitance than the carbon electrode, $C_- >> C_+$.

Cell voltage is in the 3.8–4.0 V range, while the intercalation electrode has nearly constant potential. Carbon electrode reverses its charge to negative below 3 V. The battery-like electrolyte is based on lithium hexafluorophosphate in propylene carbonate.

Further Reading

- P. Kurzweil, Electrochemical Double-Layer Capacitors, in *Encyclopedia of Electrochemical Power Sources*, ed. by J. Garche, Ch. Dyer, P. Moseley, Z. Ogumi, D. Rand, B. Scrosati, Vol 1, Elsevier, Amsterdam, 2009, pp. 607–633, 634–648 (carbon materials), 658–664 (hybrid capacitors), 665–678 (metal oxide capacitors), 679–684 (polymer); Vol 3, pp. 596–606 (history).

CHAPTER 12

Redox Flow Batteries

Redox flow batteries combine the principles of batteries and fuel cells. The principle of operation is based on soluble reagents that are oxidized and reduced at the electrodes and cycled between storage containers. The reagents are mixed with liquid electrolytes and pumped across the electrodes.

At present, the energy and power densities are small compared to classical batteries (Figure 112). For future electrotraction, the flow batteries might be considered based on the concept of rapid refueling at a filling station, without any delay as in the case of electric charging.

Vanadium Flow Battery This most developed system uses vanadium ions with different oxidation states in sulfuric acid. The anode and cathode electrolyte solutions are supplied to the cell from storage tanks. The anodic and cathodic compartments are separated by membranes.

The discharge reactions at carbon-felt electrodes involve oxidation state numbers of vanadium changing from +7 to +4 on the cathode and +2 to +3 on the anode (Equations 316 to 318).

(+) Cathodic reduction: $$VO_2^+ + 2\,H^+ + e^- \rightleftharpoons VO^{2+} + H_2O \qquad (E^0 = +1.00\ V) \qquad (316)$$

(–) Anodic oxidation: $$V^{2+} \rightleftharpoons V^{3+} + e^- \qquad (E^0 = -0.26\ V) \qquad (317)$$

Membrane diffusion: $$H^+\,(-) \rightleftharpoons H^+\,(+) \qquad (318)$$

In the real cell, the colors of the catholyte and anolyte reveal the oxidation states of vanadium: V(II) violet, V(III) green, V(IV) blue, V(V) yellow, and turquoise shadings at medium state-of-charge. Formation of V(III) and V(IV) (1:1) at the first charge in a sulfuric-acid electrolyte in both half-cells generates V(II) and V(V). In practice, cell voltages of 1.4 V at 50% state-of-charge and 1.6 V at full charge have been achieved. Specific energy is low (25–30 Wh kg^{-1}).

The advantage of the Vanadium Flow Battery is that no electrolyte contamination occurs when cross diffusion of anolyte and catholyte occurs.

Iron-Chromium Redox Flow Battery Since the 1970s, NASA has developed redox active salt solutions, which can be oxidized and reduced reversibly at polymer-bound graphite electrodes, in half-cells separated by a membrane. The discharge reactions

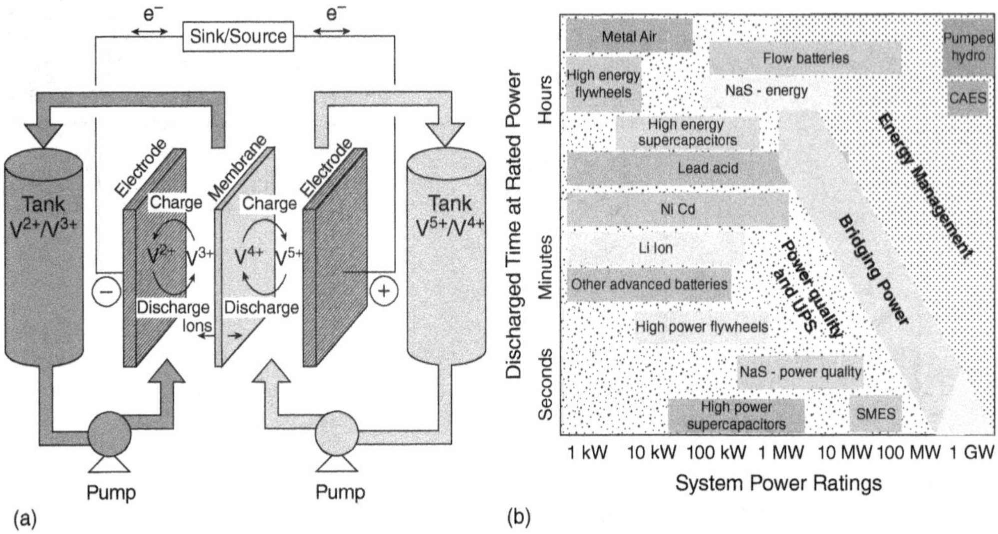

FIGURE 112 (a) Vanadium redox flow system; (b) comparison of energy storage technologies based on system power and discharge time.

show a Cr oxidation number increase on the anode and decrease on the cathode (Equations 319 to 321).

(–) Anodic oxidation:	$Cr^{2+} \rightleftharpoons Cr^{3+} + e^-$	$(E^0 = -0.41\text{ V})$	(319)
(+) Cathodic reduction:	$Fe^{3+} + e^- \rightleftharpoons Fe^{2+}$	$(E^0 = +0.77\text{ V})$	(320)
Cell reaction:	$Fe^{3+} + Cr^{2+} \rightleftharpoons Fe^{2+} + Cr^{3+}$	$(\Delta E^0 \approx 1.18\text{ V})$	(321)

Despite inexpensive chemicals, the iron-chromium redox flow battery has only limited practical importance because of its low energy density, high rate of self-discharge, and hydrogen evolution during charging.

Stationary storage systems were realized in Japan and in the United States.

Zinc-Bromine Battery In a zinc-bromine redox flow battery, a solid zinc layer is dissolved during discharge and deposited again during charge. On the other electrode, bromine forms water-soluble oligobromides $[Br_3]^-$ and $[Br_5]^-$, which are bound by organic bases.

Anode and cathode compartments are separated by a polymer membrane. The discharge reactions illustrate the processes (Equations 322 to 324)

(–) Anodic oxidation:	$Zn \rightleftharpoons Zn^{2+} + 2e^-$	$(E^0 = -0.76\text{ V})$	(322)
(+) Cathodic reduction:	$[Br_3]^- + 2\,e^- \rightleftharpoons 3Br^-$	$(E^0 = +1.09\text{ V})$	(323)
Cell reaction:	$Zn + [Br_3]^- \rightleftharpoons Zn^{2+} + 3Br^-$	$(\Delta E^0 \approx 1.85\text{ V})$	(324)

The cell voltage under the standard conditions corresponds to the theoretical specific energy of 440 Wh kg^{-1}. These flow batteries have low peak power.

The electrode and active material components are moderately expensive.

The disadvantages and challenges include formation of zinc dendrites, corrosion, high rate of self-discharge, hydrogen evolution during overcharge, and leakage currents in the electrolyte.

Premium Power demonstrated bipolar stacks (30 kW and 45 kWh). The battery was used to run an electric vehicle for over 80,000 km.

Polysulfide-Bromine Battery A regenerative fuel cell (trade name "Regenesys") was realized using redox active anions in two half-cells. The electrolyte contains sodium polysulfide and sodium tribromide. The anodic and cathodic cell compartments are separated by a cation-exchange membrane. The reaction has an open-circuit voltage of 1.36 V and a charge-discharge efficiency of 75%. The discharge reactions on the electrodes depict formation of polysulfide and bromide (Equations 325 to 327).

(–) Anodic oxidation: $2\,[S_2]^{2-} \rightleftharpoons [S_4]^{2-} + 2\,e^-$ $(E^0 = -0.27\text{ V})$ (325)

(+) Cathodic reduction: $[Br_3]^- + 2\,e^- \rightleftharpoons 3\,Br^-$ $(E^0 = +1.09\text{ V})$ (326)

Cell reaction: $[Br_3]^- + 2\,[S_2]^{2-} \rightleftharpoons 3\,Br^- + [S_4]^{2-}$ $(\Delta E^0 \approx 1.36\text{ V})$ (327)

The polysulfide-bromine battery system must be optimized to eliminate bromine and hydrogen sulfide vapors, improve low current densities, and improve monitoring.

Soluble Lead-Acid Flow Battery This simple experimental setup demonstrates the electroactivity of soluble lead salts such as methane sulfonate in methane sulfonic acid. No membrane is required, and a single electrolyte reservoir is sufficient. During discharge at carbon electrodes, lead ions pass into the solution (Equations 328 to 330).

(–) Anodic oxidation: $Pb \rightleftharpoons Pb^{2+} + 2\,e^-$ $(E^0 = -0.13\text{ V})$ (328)

(+) Cathodic reduction: $PbO_2 + 4\,H^+ + 2\,e^- \rightleftharpoons Pb^{2+} + 2\,H_2O$ $(E^0 = +1.46\text{ V})$ (329)

Cell reaction: $Pb + PbO_2 + 4\,H^+ \rightleftharpoons 2\,Pb^{2+} + 2\,H_2O$ $(\Delta E^0 \approx 1.59\text{ V})$ (330)

Technical problems include low power, optimization of bipolar electrodes, and formation of lead dendrites during charging.

Further Reading

- C. Menictas, M. Skyllas-Kazacos, T. M. Lim (eds.), *Advances in Batteries for Medium- and Large-Scale Energy Storage: Types and Applications*, Woodhead Publishing, 2014.
- Th. B. Reddy, *Linden's Handbook of Batteries*, 4th ed., McGraw-Hill, 2011.

Index

M

R

S